Lecture Notes in Applied and Computational Mechanics

Volume 101

This series aims to report new developments in applied and computational mechanics - quickly, informally and at a high level. This includes the fields of fluid, solid and structural mechanics, dynamics and control, and related disciplines. The applied methods can be of analytical, numerical and computational nature. The series scope includes monographs, professional books, selected contributions from specialized conferences or workshops, edited volumes, as well as outstanding advanced textbooks.

Indexed by EI-Compendex, SCOPUS, Zentralblatt Math, Ulrich's, Current Mathematical Publications, Mathematical Reviews and MetaPress.

Giovanni Garcea · Dieter Weichert
Editors

Direct Methods for Limit State of Materials and Structures

Advanced Computational Algorithms and Material Modelling

Springer

Editors
Giovanni Garcea
DIMES
Università della Calabria
Rende, Italy

Dieter Weichert
RWTH Aachen University
Montlignon, France

ISSN 1613-7736 ISSN 1860-0816 (electronic)
Lecture Notes in Applied and Computational Mechanics
ISBN 978-3-031-29121-0 ISBN 978-3-031-29122-7 (eBook)
https://doi.org/10.1007/978-3-031-29122-7

This Springer imprint is published by the registered company Springer Nature Switzerland AG
The registered company address is: Gewerbestrasse 11, 6330 Cham, Switzerland

Foreword

Genesis of Direct Methods: Exploration and Breakthroughs

Without going all the way back to Leonardo da Vinci, who analysed in his pioneering work published in 1493 the bending of a beam but did not provide a way of assessing its strength, the first attempt to evaluate the collapse load of a cantilever beam is due to Galileo Galilei in 1638. Although the value was overestimated due to the lack of knowledge of the material behaviour, it paved the way for the modern methods of structural analysis and may be considered as a premise of what will become the limit analysis and more broadly the Direct Methods.

An important breakthrough is Charles-Augustin Coulomb's contribution to the stability of earth retaining walls. He spent the years from 1764 to 1772 in the French colony of Martinique where he supervised the construction of new fortifications. On his return to France, studying the role of friction and cohesion, he proposed a rupture surface in the stress space. His idea is that even though the wall just shifted forward slightly, a sliding plane will form between a soil wedge at the back of the wall and the remaining soil at rest. It is the starting point of the scientific soil mechanics but also an important milestone for the limit analysis theory.

For metals, the first plasticity criterion was proposed by Christian Mohr in 1882, based on his famous graphical method of stress analysis by a circle and the concept of maximum shear stress. The next major contributions date back to the beginning of the twentieth century. Although the original idea is due to Tytus Huber and was published in Polish in 1904, the plastic yielding criterion and the flow rule was rigorously formulated by Richard von Mises in 1913 and independently by Heinrich Hencky in 1924. One of the strengths of the plasticity theory is that, thanks to a unified mathematical model, it encompasses a wide class of ductile materials such as metals, soil, rocks, concrete, granular materials, and many yielding criteria were proposed in the footsteps of pioneers.

The formulation of constitutive laws of plasticity was an essential element that allowed to develop the corresponding methods of structural mechanics.

Among them, the limit analysis is an elegant method of calculation of the strength of structures which skips the elastoplastic loading by considering directly the collapse mechanism. The exact value of the collapse load can be reached by the static and kinematic approaches thanks to the so-called lower and upper bound theorems. A key point is that it allows to assess by dual analysis the errors when only approximate values are obtained. The static method was published in Russian first by Alexeï Gvozdev in 1938. It was completed by the dual kinematic approach and formalized in a general way by Rodney Hill, William Prager and Daniel Drucker in the 1950s. It should be noted that the yield-line theory proposed by Knud Johansen in 1962 was recognized later on as an application of the kinematic approach to concrete plates.

A powerful extension of the previous method is the shakedown analysis when the structure is subjected to repeated variable loads, in particular to cyclic loads. Hans Bleich in 1932 and Ernst Melan in 1936 emphasized the crucial role played by the time-independent residual stress fields in the stabilization of the plastic strains and laid the way to the static approach. Warner Koiter gave in 1955 a rigorous proof of the static and kinematic theorems.

Developments and Maturity

All these seminal works gave rise to a wealth of research works owing to the experimental testing as well as the development of computers. The original framework was based on the elastic-perfectly plastic model in quasi-static conditions. The growth and refinement of knowledge about the material behaviour led to the need of extensions of the bound theorems to more sophisticated constitutive laws. The generalization of the shakedown theorems to the plastic materials with linear hardening was proposed by Prager in 1974, to the thermo-mechanical loading by Giulio Maier in 1969, to the dynamical loading by Corradi and Maier in 1973, to the elasto-visco-plasticity by Bernard Halphen in 1978 and Alan Ponter in 2000, to the geometrical nonlinearities by Dieter Weichert in 1986. Despite some attempts, the extension to the non-associated plasticity remains an open topic of research.

On view of the fast growth of the computer powers and the computing algorithms after the second world war, numerical approximations of the limit states of structures by Direct Methods have been extensively developed from the 1960s in parallel with the classical step-by-step method. The basic idea is to combine the finite element method to discretize the structure and the solvers of mathematical programming to find the upper and lower bounds. The advantage of the Direct Methods is especially strong for variable repeated loads. A complete step-by-step computation of the overall history up to the limit state can be very time-consuming or even exceed the capacity of the computer. Moreover, if only the extreme values of the loads are known but not the exact loading path, then shakedown analysis must be used. The main aspects characterizing the Direct Methods are the variational principle, the discretization of the fields, the analytical form of the yielding criterion, the way it is enforced and the choice of the solver, notably the simplex or more recently the interior point method and the conic programming. It is noteworthy to mention a physically-based solver algorithm, the Linear Matching Method of Ponter and Carter.

The displacement finite elements are very popular because of their simplicity and the first works using them for limit analysis are due to Hodge and Belytschko in 1967, Anderheggen in 1976, Hutula in 1976, Biron in 1976 and Nguyen Dang Hung in 1976. Although the equilibrium finite elements are not so easy to build, they are well adapted to the Direct Methods because the yielding criterion is expressed in terms of stresses. The first works based on the static approach are due to Hodge and Belytschko in 1967 and 1970, Lymer in 1970, Casciaro and Di Carlo in 1970, Zavelani-Rossi in 1974. It should be noticed that the bound properties are often lost because of the difficulty to integrate exactly the power of dissipation in the kinematic approach and to enforce anywhere the nonlinear yielding criteria in the static one. An alternative is to use two-field variational principles and mixed finite elements as in works by Christiansen in 1981 and Casciaro and Di Carlo in 1974, or hybrid finite elements as Nguyen Dang Hung in 1976.

The extensions of these powerful numerical methods to the shakedown analysis were natural and is due for displacement finite elements to Nguyen Dang Hung and Morelle in 1984 and 1986, to Karadeniz and Ponter in 1984, for equilibrium finite elements to Belytschko in 1972, to Corradi and Zavelani in 1974, to Nguyen Dang Hung and Palgen in 1980, to Morelle and Nguyen Dang Hung in 1983, to Weichert and Gross-Weege in 1987, to Franchi and Genna in 1983, and for hybrid finite elements to Morelle in 1984.

New Challenges

The Direct Methods are now a powerful tool for engineers and researchers not only to solve problems of structural engineering, but also to understand the mechanical behaviour of the materials through the homogenization techniques. This is the epistemological reversal: while the advances in the mechanical modelling of materials had led to the development of the Direct Methods for the structural analysis, these methods permit in return to make progress in the domain of constitutive laws. Thanks to the limit analysis, macroscopic plasticity criteria of composite and porous materials can be obtained in terms of the stresses and microscopic parameters such as volume fraction and material parameters of the different phases. For materials under cyclic loading, the shakedown analysis gives new insights to propose theoretically-based fatigue criteria. Recent advances in the material design, especially thanks to the additive manufacturing, open new horizons for the mechanical analysis of new architectured materials and metamaterials with high performances or surprising properties. Another challenge to face is the increasing need to take into account the interactions between the solid mechanics and other physical effects, notably dynamic, thermal, hydraulic, electric and magnetic. Even if these couplings were already considered in the literature, there is no doubt that in the future they will grow in importance.

To face these appealing but also difficult challenges, researchers need places for exchange of experiences, debates, pooling of efforts to plan joint projects. Recently, a scientific community around these research topics have progressively emerged and began to structure itself. The first Workshop on Direct Methods held at Aachen in 2007. Under the initiative of Alan Ponter and Dieter Weichert, the idea to found a scientific association is launched in Reggio Calabria during the fourth workshop in 2013. The "International Association on Direct Methods" (IADiMe) was born. This volume gives a State of the Art of the most recent research and results in this field presented in the seventh edition of the Workshop on Direct Methods held at Cosenza.

This volume gives a State of the Art of the most recent research and results in this field presented in the seventh edition of the Workshop on Direct Methods held at Cosenza.

Lille, France Géry de Saxcé

Preface

An important task for civil and mechanical engineers is the estimation of the limit states of structures and materials for the evaluation of safety and durability under any type of loading, which can cause severe inelastic deformations. Typical examples of such loads can be due to thermo-mechanical actions or loads induced by displacement, such as earthquakes, traffic, etc.

"Direct methods," spanning limits and shakedown analysis, evaluate safety factors by avoiding cumbersome and time-consuming step-by-step analyses. They are non-evolutionary and applicable to the optimal design of structures and materials. Nowadays, there is a growing interest among scientists and researchers in new mathematical formulations and high-performance numerical solution tools, which make reliable analyzes of conventional and new generation materials and structures possible.

The papers in this volume provide a state of the art on the topic and grew out of an international seminar on direct methods held on June 28, 2022 at the University of Calabria, Italy, providing insights into the latest developments in this rapidly evolving field of research. It is in line with similar books on the same topic which have been published as documentation of previous workshops, held regularly since 2008 in Aachen, Lille, Athens, Reggio Calabria, Oxford and Krakow.

Most of the contributions are related to new numerical developments that make the methods attractive for industrial design in a large panel of engineering applications. Extensions of the general methodology to new application horizons and specific technological problems are presented.

The contributions are arranged in the same format as the workshop sessions, where groups of articles cover similar topics. The chapters have all undergone a rigorous review process before being accepted for publication.

The editors warmly thank all scientists who have contributed their outstanding papers to the quality of this edition. We are also grateful to the Springer editors for their patience and guidance during the production of this volume.

Aachen, Germany — Dieter Weichert
Rende, Italy — Giovanni Garcea

Contents

Contributors

Aurora Angela Pisano Department PAU, University Mediterranea of Reggio Calabria, Reggio Calabria, Italy

Alexander Bezold Institute for Materials Applications in Mechanical Engineering, RWTH Aachen University, Aachen, Germany

Paolo Bisegna Department of Civil Engineering and Computer Science, University of Rome Tor Vergata, Rome, Italy

Jeremy Bleyer Laboratoire Navier, Ecole des Ponts, Université Gustave Eiffel, CNRS, Marne-la-vallée, France

Christoph Broeckmann Institute for Materials Applications in Mechanical Engineering, RWTH Aachen University, Aachen, Germany

Haofeng Chen School of Mechanical and Power Engineering, East China University of Science and Technology, Shanghai, China

Giuseppe Cocchetti Department of Civil and Environmental Engineering, Politecnico di Milano, Milano, Italy

Aram Cornaggia Department of Engineering and Applied Sciences, Università degli studi di Bergamo, Dalmine, Italy

Insaf Echerradi Lab Navier, Univ Gustave Eiffel, ENPC, CNRS, Marne la Vallée, France

Rosalba Ferrari Department of Engineering and Applied Sciences, Università degli studi di Bergamo, Dalmine, Italy

Paolo Fuschi Department PAU, University Mediterranea of Reggio Calabria, Reggio Calabria, Italy

Giovanni Garcea Dipartimento di Ingegneria Informatica, Modellistica, Elettronica e Sistemistica, Università della Calabria, Rende (Cosenza), Italy

Christian Gebhardt Institute for Materials Applications in Mechanical Engineering, RWTH Aachen University, Aachen, Germany

Ioannis A. Kapogiannis Department of Civil Engineering, Institute of Structural Analysis and Antiseismic Research, National Technical University of Athens, Athens, Greece

Vincent Leclère CERMICS, Ecole des Ponts, Marne-la-vallée, France

Ruiwei Liu Department of Civil and Environmental Engineering, Politecnico di Milano, Milano, Italy;
Department of Naval Architecture and Marine Engineering, Guangzhou Maritime University, Guangzhou, China

Domenico Magisano Dipartimento di Ingegneria Informatica, Modellistica, Elettronica e Sistemistica, Università della Calabria, Rende (Cosenza), Italy

Nicola A. Nodargi Department of Civil Engineering and Computer Science, University of Rome Tor Vergata, Rome, Italy

A. Oueslati Univ. Lille, CNRS, Centrale Lille, UMR 9013 - LaMcube - Laboratoire de mécanique multiphysique et multiéchelle, Lille, France

Michael Peigney Lab Navier, Univ Gustave Eiffel, ENPC, CNRS, Marne la Vallée, France

Egidio Rizzi Department of Engineering and Applied Sciences, Università degli studi di Bergamo, Dalmine, Italy

G. de Saxcé Univ. Lille, CNRS, Centrale Lille, UMR 9013 - LaMcube - Laboratoire de mécanique multiphysique et multiéchelle, Lille, France

Tobias Sedlatschek Institute for Materials Applications in Mechanical Engineering, RWTH Aachen University, Aachen, Germany

W. Q. Shen Univ. Lille, CNRS, Centrale Lille, UMR 9013 - LaMcube - Laboratoire de mécanique multiphysique et multiéchelle, Lille, France

Konstantinos V. Spiliopoulos Department of Civil Engineering, Institute of Structural Analysis and Antiseismic Research, National Technical University of Athens, Athens, Greece

Nunziante Valoroso Dipartimento di Ingegneria, Università di Napoli Parthenope, Naples, Italy

Juan Wang Ningbo Nottingham New Materials Institute, The University of Nottingham Ningbo China, Ningbo, China

Xiaoxiao Wang Department of Mechanical & Aerospace Engineering, University of Strathclyde, Glasgow, UK

Felix Weber Institute for Materials Applications in Mechanical Engineering, RWTH Aachen University, Aachen, Germany

Dieter Weichert RWTH-Aachen University, Aachen, Germany

Daniel Weisz-Patrault LMS, CNRS, École Polytechnique, Institut Polytechnique de Paris, Palaiseau, France

Hai-Sui Yu The University of Leeds, Leeds, UK

J. Zhang Key Laboratory of Ministry of Education for Geomechanics and Embankment Engineering, Hohai University, Nanjing, China

Direct Methods: History, Present and Future

Dieter Weichert

Abstract This chapter spans the arc from the first developments of Direct Methods in modern times to current tendencies in this field of research. Emphasis lies on the great lines of development and on the personal contributions of outstanding scientists. Examples of general interest illustrate the scientific and technical importance of the according methodologies.

1 Milestones and the Scientists Behind

1.1 Early Developments

The question whether a mechanical structure may fail or not due to variable or dead loads is as old as engineering activity itself and it is difficult to pinpoint the instant when mathematical models were first used to predict if a structure will resist to the acting forces or collapse. Edoardo Benvenuto's paper from 1991 [1] gives an excellent view on early developments, going back to antiquity; other information is found in e.g. Weichert and Ponter [2], or from a more general point of view, in Kurrer [3]. Here we focus on developments since the beginning of the twentieth century, when steel was increasingly used in Civil Engineering. Elastic design, commonly used at that time, was very conservative and therefore, plastic design became interesting for structural engineering. The notion of plastic hinges, of utmost importance for the plastic analysis of skeleton structures and beyond, goes back to Gábor Kazinczy (Fig. 1) [4], who is one of the founders of Limit Analysis (LA) in modern times. In

Fig. 1

D. Weichert (✉)
RWTH-Aachen University, Aachen, Germany
e-mail: dieter_weichert@hotmail.com

G. Garcea and D. Weichert (eds.), *Direct Methods for Limit State of Materials and Structures*, Lecture Notes in Applied and Computational Mechanics 101,
https://doi.org/10.1007/978-3-031-29122-7_1

particular methods to determine the load carrying capacity of beams and trusses under monotonically increasing loads, a typical LA issue, were investigated both from experimental and theoretical point of view (Maier-Leibnitz [5], Schaim [6], Fritzsche [7]). Other important contributions, in particular on limit loads for frames are due to Gvozdev [205] and Girkmann [3].

Fig. 3

Fig. 2

But as early as 1926, Martin Grüning (Fig. 2) discussed the influence of plasticity in hyperstatic truss systems under repeated loading, a kind of starting point for Shakedown-Analysis (SA), where in contrast to LA the elastic properties of the material play a role. Hans Bleich (Fig. 3) picked up and generalised in 1932 Grünings findings [8, 9], whose studies concerned only locally stationary repeated loads. This restriction was removed by Bleich [10], who also introduced the notion of "Selbstspannung", nowadays commonly used in its English equivalent "eigenstress". But it was Ernst Melan (Fig. 4), inspired to his path-breaking works in 1936 and 1938 by Grüning and Bleich, who formulated the general lower bound shakedown theorem in his key contribution "Der Spannungszustand eines Mises-Hencky'schen Kontinuums bei veränderlicher Belastung" [11].

Fig. 4

This was, theoretically, the kick-off for modern Direct Methods. However, not practically: It may surprise that Melan's powerful general theorem from 1938 remained with little resonance in the scientific community for many years. However, one must not forget the atrocious situation in Germany and, after WWII had started, in the entire world, at that time. Many brilliant German-Jewish scientists in mechanics had to struggle for their lives and to find new places to live for themselves and their families, if they had the chance to survive. In addition, the relevant papers had been published in German language, which was evidently not very popular in these and the coming years. It was Prager (Fig. 6), at that time at Brown University, who came back to the problem in his contribution "Problem Types in the Theory of Perfectly Plastic Materials" [12], presented at the Symposium on Plasticity held at Brown University in 1948, where he refers to Melan's less far reaching formulation of the lower bound theorem [13]. It was also Prager, at that time funded by the US Office of Naval Research, who introduced the denomination "Shakedown", which is known in shipyards to describe the process of accommodation of parts in new ships due to dynamic loading by engine vibration after putting them into operation for the first time (O. Mahrenholtz, private communication).

Fig. 5

Fig. 6

Fig. 7

Brown University can be considered as a kind of cradle for the further development of Direct Methods and shakedown theory in particular at that time. Besides Prager, also should be mentioned Paul S. Symonds (Fig. 7) and B.G. Neal as particularly active in the field of determination of limit loads for frame structures [14–19]. Special importance has their paper "Recent Progress in the Plastic Methods of Structural Analysis" [17] as it anticipates the upper bound theorem of shakedown. In his contribution [20], Paul Symonds gives an enlightening personal insight into the developments of shakedown theory at that time at Brown University and Cambridge University. Generally, in the U.K. to our knowledge, it was basically the group by John Baker working in structural plasticity in the same period and their interest was mainly limit analysis (see also the multi-volume work "The Steel Skeleton", by Baker et al. [21]).

W. T. Koiter (Fig. 5), who had been visitor to Brown University at that time, has the merit of not only formulating the upper bound theorem of shakedown in a general form in 1956 in his paper "A new general theorem on shakedown of elastic–plastic structures" [22], but also to revalue shakedown theory as a whole through his fundamental work "General Theorems for Elastic–Plastic Solids" [23] that summarises the state of the art at that time in a very accessible manner. We stress however that the contribution by Paul Symonds concerning the upper bound theorem should not be undervalued [24]. He formulated in fact the first time an upper bound approach in shakedown theory, not in general form, as Koiter did, but for frames. Symonds also greatly simplified the proof of Melan's lower bound theorem [15] in the general case in a form adopted by Koiter [23].

One can say that with Koiter's formulation of the lower and the upper bound theorems of shakedown theory in the context of continuum mechanics the first chapter of development was closed.

Although well known, for completeness, the two theorems are briefly reminded:

Lower bound theorem:

An elastic-perfectly plastic body shakes down if there exist a real number $\alpha > 1$ and a time independent field of self-equilibrated stresses ρ_{ij} with

$$\rho_{ij,j} = 0 \quad \text{in } v$$
$$n_j \rho_{ij} = 0 \text{ on } S_p$$

such that the superposition of the elastic stresses σ_{ij}^{E} with ρ_{ij} constitutes a safe state of stresses:

$$\alpha \sigma_{ij}^{\mathrm{E}} + \rho_{ij} \subset \mathcal{C}$$

If this relation is valid at all points of the body $\mathcal{B}$ and at any instant of time, then the total dissipation is bounded. Here, C is the convex elastic domain and σ_{ij}^{E} is the solution of a reference problem, differing from the original one only by the fact that the material behaves purely elastically. This way, σ_{ij}^{E} represents the external loads.

Upper bound theorem:

The body will not shake down over a load path contained within the load domain $\mathcal{D}$ if there exists a plastic strain rate cycle $\dot{\varepsilon}_{ij}^{\mathrm{p}}(\mathbf{x},t)$ resulting in compatible increments of plastic strains over this time interval

$$\Delta\varepsilon_{ij}^{\mathrm{p}} = \int_0^T \dot{\varepsilon}_{ij}^{\mathrm{p}}\mathrm{d}t = \frac{1}{2}(\Delta u_{i,j} + \Delta u_{j,i}) \text{ in } v$$

with

$$\Delta u_i = 0 \text{ on } S_u$$

and such that the work of the external forces represented by the purely elastic reference solution σ_{ij}^{E} exceeds the plastic dissipation in this motion over one cycle

$$\int_0^T \int_V \sigma_{ij}^{\mathrm{E}}\dot{\varepsilon}_{ij}^{\mathrm{p}}\mathrm{d}V\mathrm{d}t > \int_0^T \int_V D^{\mathrm{p}}(\dot{\varepsilon}_{ij}^{\mathrm{p}})\mathrm{d}V\mathrm{d}t$$

On the other hand, the body will shake down if there exists a real number $\alpha > 1$ such that for all loads and for all kinematically admissible strain rate cycles $\dot{\varepsilon}_{ij}^{\mathrm{p}}$ (i.e. which result in compatible strain increments) the following inequality holds

$$\alpha\int_0^T \int_V \sigma_{ij}^{\mathrm{E}}\dot{\varepsilon}_{ij}^{\mathrm{p}}\mathrm{d}V\mathrm{d}t \leq \int_0^T \int_V D^{\mathrm{p}}(\dot{\varepsilon}_{ij}^{\mathrm{p}})\mathrm{d}V\mathrm{d}t.$$

1.2 The Next Generation; Theoretical Advances and Development of Numerical Methods

The questions to be answered in the follow-up of these foundations can be roughly summarized as follows

(i) *How these powerful theorems can be applied in practical engineering?*
(ii) *How to get rid of the coercive assumptions on which the classical proofs, in particular in shakedown theory (linear elastic, perfectly plastic or linear, unlimited plastic hardening) are based on?*

To answer these questions, several lines of research had been followed: General theoretical/mathematical developments; application to specific structural elements; extensions to complex material laws and nonlinear geometrical effects; development of numerical methods. –Naturally, these lines of development are interwoven and not strictly separated. It is impossible to honor here all important contributions in this field and the scientists behind. Only a short and by no means complete overview can be given. An extended review can be found e.g. in Ponter and Weichert [2] or, from a more general point of view. Kurrer [3].

As stated before, the group around Prager at Brown University played an important role and many young scientists on visit at Brown University in the 50 and 60ties had there been inspired and carried on their research after returning home. Among them, Giulio Maier (Fig. 8) continued his work in Milano, focusing together with his co-workers on the problems of non-associated flow rules [25], geometrical non-linearities [26] and dynamic effects [27]. Non-associated flow rules are particularly important for frictional materials such as soils, masonry, but also, more recently investigated, porous and heterogeneous materials. Also, accounting for the progress in numerical methods and in view of industrial applications, Maier adapted shakedown theory to the so-called linear programming methods. His rich scientific oeuvre in the field of DM was initially motivated by problems in civil engineering but went far beyond and triggered research activities with worldwide impact [28–40]. To mention, from Italian side, are contributions e.g. by L. Corradi, R. Contro, F. Genna, A. Corigliano, U. Perego, and in recent years, V. Carvelli and G. Cocchetti.

Fig. 8

Also back from Brown University, Alan Ponter (Fig. 9) was interested in mechanical engineering design including high temperature problems and extended the lower bound theorem to creep behaviour of materials [41–43].

In cooperation with Fred Leckie, John Martin and others he derived displacement bounds in shakedown conditions and general bounding theorems in plasticity [44–49]. More recently, Ponter focused on rolling contact problems, composites and the development of methods of how to solve most efficiently shakedown and related problems by numerical methods. He introduced the so-called “Linear Matching Method” and implemented it successfully into commercial software for design purposes. As well as shakedown limits, ratchet limits in excess of shakedown are obtained [50–58]. This work continues at the University of Strathclyde by a research group led by H. F. Chen.

Fig. 9

“Variable thermal loads” was also the field of D. A. Gokhfeld (Fig. 10) from the South Ural University, USSR. He studied in the beginning of the 60ties of the last century ratchetting of mechanical parts in furnaces and observed that under moving thermal loads even without any mechanical loading large deformations may occur that render the considered part unusable [59] These studies, joining theoretical and experimental work, were extended to other types of elements such as turbine blades, parts in nuclear reactors and pressure vessels in chemical processing. Aware of the work of Ponter, Williams and Leckie on creep, he and his co-workers developed an alternative way of taking creep into account by adjusting appropriately the yield limit of the material [59–62]. Numerous papers Gokhfeld, O. F. Cherniavsky and co-workers were almost all published in Russian language and remained largely unknown outside of the Russian speaking scientific community.

Special mention deserves the book "Theory of shakedown and strain accumulation under thermal cycling" from 1980 in English which is still today an important source of information.– The work from this group is today successfully continued in a modern computational environment by A. Cherniavsky and co-workers, with new fields of application like hydrogenated metals, crack forming and use of traveling heat sources for controlled metal forming.

Fig. 10

Another question, not addressed by classical theory is the problem of dynamic (inertia) effects. Here also, in a different way than in case of viscoplasticity, time enters the formulation of the problem explicitly. This subject was addressed first by Giulio Ceradini from Rome in his theorem on dynamic shakedown in 1969 [63], still of great importance in particular in earthquake engineering. Equally active in Rome and for some time under his guidance, Raffaele Casciaro (Fig. 11) has contributed significantly to the development of DM. Particularly important was his approach to lower bound LA of plates, formulated as minimax problem and based on the discretization of the structure into triangular finite elements described in terms of stresses (moments).They are equilibrated in the interior of the element, while global equilibrium is enforced at the nodes of the elements. The works represent one of the first implementations regarding the use of mixed equilibrated finite elements in the evaluation of limit analysis multiplier. See also the extension to 2D structural and soil mechanics problems in [67] with Cascini. More recently, he proposed together with G. Garcea, L. Leonetti, G. Armentano and S. Petrolo [64–70] innovative methods for the numerical solution of shakedown problems. Among his pupils we mention here Giovanni Formica, Giovanni Garcea, Emilio Turco, Antonio Lanzo, Ginevra Salerno, Vittorio Sansalone and Antonio Madeo. Similarly active in Italy, Castrenze Polizzotto (Fig. 12) from Palermo comes from a both mathematically and engineering-driven background. He contributed and still contributes to the development of DM [71–80]. The most important topics are: structural optimization; energetic formulation of the boundary element method; elastoplastic structures under cyclic loads; variational characterization of elastic–plastic-damaging structures; non-local and gradient approaches in elasticity and plasticity. Among his pupils are Cesare Mazzarella, Teotista Panzeca, Francesco Giambanco, Santi Rizzo, Guido Borino, Paolo Fuschi, Aurora Pisano.

Fig. 11

Fig. 12

Very important for the development of DM was the Polish Academy of Sciences: The long Polish tradition in the field of Plasticity since the times of T. M. Huber was continued after WWII in particular related to limit states of plastic structures and phenomenological modeling of plastic behavior. Since 1955 W. Olszak (Fig. 13) initiated weekly scientific seminars at the Polish Academy of Sciences. The principles of limit state analysis, metal forming, yield conditions, anisotropy, inhomogeneity and flow rules were the main topics of

Fig. 13

discussion and research. In 1957 Koiter gave a set of lectures on his kinematic approach to shakedown and the upper bound theorems. Olszak, who hold his Ph.D. from Vienna Polytechnic in 1935, knew well Melan and was familiar with his theorem on shakedown. It can be supposed that this was the main inspiration to develop more intensive study of cyclic loading and shakedown as fundamental for application in structural mechanics.

In the sequel, A. Sawczuk (Fig. 14), primarily working on LA and SA of plates and shells, wrote several papers on application of kinematic theorem to specify upper bounds on load amplitudes [81–84]. It was however Jan Andrej König (Fig. 16), his doctoral student, who contributed most significantly to the advancement of shakedown theory by a large number of theoretical papers on hardening material behavior, thermal problems, bounding methods, structures and numerical methods [85–90, 95, 97–102]. His book "Shakedown of Elastic–Plastic Structures" [90] from 1987 is still of great actuality and a prime reference; it is concise and easy to read. Among the great number of Polish scientists working successfully in the field of shakedown between roughly 1960 and 2000, we only mention A. Borkowski, M. Kleiber, Z. Mróz, A. Sawczuk, M. Janas, St. Dorosz, A. Siemazsko, S. Pycko, J. Skrzypek, B. Skoczeń, J. Orkisz, J. Zwoliński, B. Bielawski, among others [81–109]. The book by Michał Życzkowski (Fig. 15) "Combined Loadings in the Theory of Plasticity" [110] gives an excellent account on research on plastic structures with more than 3000 entries of references.

Fig. 14

Fig. 15

Fig. 16

It was also König who brought shakedown theory back to Germany: In the beginning of the 80ties, Oskar Mahrenholtz (Fig. 17), then at Hannover University, was involved in studies on failure of zirconium tubes under variable thermal and mechanical loads, which is a typical problem from nuclear power engineering. König, at that time visiting scholar in Hannover, suggested solving this problem by applying the upper bound shakedown theorem and to carry out validating experiments. The experimental and numerical results they obtained were in agreement [111–113]. Some years later, König stayed again at Hannover University, this time with Erwin Stein (Fig. 18). There he initiated a series of studies focusing on the development of numerical shakedown analysis involving material hardening, cracks and structural optimization [114–121]. These works were successfully continued by Stein, Zhang, Mahnken, Wiechmann and others in the following years. (It should be noted that in the German community of civil engineers, to which Stein belonged, the application of shakedown theory was even in the 90ties far from being commonly accepted and discussed controversially [122]). Of particular interest is the

Fig. 17

Fig. 18

"reduced base"-technique, developed by Stein and co-workers, a technique that reduces significantly the numerical efforts to construct numerical solutions.

The initial boundary value problem of plasticity was the entry point for Weichert to discover indirectly shakedown theory through the influence of Pawel Rafalski [123–125], a Polish scientist from Warsaw, visiting Bochum University in the beginning of the 80ties. Weichert studied in the sequel first the problem of geometrical non-linearities in the context of shakedown theory in the framework of continuum mechanics [126], investigated then the problem of generalized material laws according to the standard material model by [127] with applications to thin-walled structures. During his stay at the American University of Beirut, he started to work with Lutfi Raad and others on problems in pavement mechanics [128–132]. Later, at Lille University in France, Weichert carried out a number of studies on numerical methods, dynamic shakedown and the problem of shakedown including material damage and cracked bodies [133–138]. He continued this work in Germany at Aachen University with applications to composites and with the aim to apply shakedown theory to large scale industrial problems [139–148]. He and his co-workers concentrated on the lower bound theorem. They used in the beginning stress-based numerical approaches, which delivered good results, but then moved to displacement-based methods for better compatibility with commercial finite element codes.

Force and stress-based methods have been intensively developed at Liège University, starting in the 60ties. There was a strong group around Baudouin Fraeijs de Veubeke (Fig. 19), Ch. Massonnet, G. Sanders, C. Fleury, mainly involved in general mechanics, limit analysis of plates and shells, and optimiza tion. M. Save and G. Guerlement continued the work at the Polytechnic School of Mons. Later, in particular Nguyen Dang Hung and Patrick Morelle applied the tools that had been developed in the innovative scientific environment of Liège University to shakedown analysis [149–154]. Their efforts were aimed at the numerical exploitation of duality principles. Manfred Staat from Jülich Research Center together with Duc Khôi Vu, coming from Liège, and others continued successfully this work in recent years [155–159]. – The path of force methods has been followed in an original manner by Kostas Spiliopoulos, with application to frame structures, however based on graph theory and linear programming [160–162]. More recently, he introduced the so called "Residual Stress Decomposition Method", which is theoretically new and numerically efficient Direct Method [163–165].

Fig. 19

Géry de Saxcé (Fig. 20), who had also started his career at Liège University before moving first to the Polytechnic School of Mons and then to Lille University, introduced the so-called bi-potential theory [166–168] as generalization of Fenchel's inequality, opening new doors to take into account more complex, friction-type material laws in limit analysis and shakedown theory. This novel approach has a high potential and is far from being fully exploited at the time being. The research by de Saxcé is therefore linked to the scientific tradition of the Belgian group, but also to the French, mathematically inclined community of mechanics, strongly involved in the 70 and 80s of the last century in the development and application of Convex Analysis based on the Fenchel inequality and so carries further the idea of classical potential-based principles. Convex Analysis had been developed in the 70s by T. Rockafellar in the context of operational research and by Moreau [169] in the context of mechanics. B. Nayroles together with O. Debordes and later D. Weichert applied this to the shakedown problem [170–172] and contributed essentially to the strengthening of the mathematical basis of the theory. Here, the important work by Quoc-Son Nguyen has to be mentioned, who contributed essentially to the understanding of the effect of hardening from mathematical point of view [173–175]. Very fruitful because well suited to extend the classical theorems to larger classes of material behavior was the introduction of the so-called Standard Material Model by Bernard Halphen and Nguyen Quoc Son [176] as had already been shown by Mandel in 1976 [177]. Radenkovic's work on non-associated flow rules from 1961 [178], although basically related to limit analysis played an important role in the sequel also in shakedown theory and should be mentioned in this place.

Fig. 20

Independently and application oriented, Joseph Zarka and his group developed the so-called Simplified Method [179, 180], particularly useful for applications involving alternating plasticity and fatigue problems in mechanical engineering as has been shown by Geneviève Inglebert and her coworkers [181]).

Coming back to typical problems of civil engineering (which, as mentioned in the beginning, in some sense has triggered limit- and shakedown analysis), there were some important but for long time spared-off areas, which are mechanics of soils, foundations and pavements. Here, just as in case of concrete and reinforced concrete structures, the complexity of the material behavior and the difficulty to develop realistic material models that fit the framework of shakedown theory are important obstacles. In particular for pavements, other effects like rutting, crack development, moisture, freezing and thawing cycles are very important aside of plasticity as to their long-time behavior; in case of foundations, mostly the fluctuation of loads is less important than gravitational "dead loads".

Apart from the groups mentioned before, there is a long tradition in the field of soil and pavement mechanics in Australia and New Zealand. John Robert Booker (Fig. 21) from Sydney University is one of the pioneers in the field and the path breaking work on shakedown analysis of pavements goes back to him and R.W. Sharp in 1984 [182], basing their approach on the particular stress pattern that develops in the rolling contact on roads. Similarly, at Newcastle University the group around Scott Sloan (Fig. 22), with K. Krabbenhøft, A.V. Lyamin and others [183–189] contributed and still contribute significantly to the development of DM. The same holds for Ian Collins from Auckland University New Zealand [132, 190, 191]. In his later work, there was a link to Ponter, Weichert and Raad through the fact that Mostapha Boulbibane, had been active in all three groups. But

Fig. 21

also shakedown of structures such as bridges has been widely investigated in Australia: It was Paul Grundy from Monash University who started as early as 1969 to study in many papers the shakedown behavior of mechanical elements, in particular linked to bridge constructions [192, 193]. His work is continued by Francis Tin Loi and his coworkers at the University of New South Wales. –It is from Australia, Newcastle University, that from Sloans group Hai-Sui Yu brought back shakedown to the U.K. at Nottingham University. Yu and his coworkers concentrate on the numerical application of Melan's theorem to compute lower bounds to the shakedown limit for rolling contact problems for Mohr–Coulomb type yield conditions for road pavement design [194–198].

Fig. 22

Fig. 23

Back to Northern Europe, Lithuania, another strand of development, combining shakedown analysis and structural optimization, has been followed by Aleksandras Čyras (Fig. 23), Juozas Atkočiūnas (Fig. 24) and co-workers from Vilnius Gediminas Technical University (Lithuania) [199–202]. It should be noted, that this group has achieved outstanding results by combining classical structural mechanics with advanced numerical methods. They can be considered as forerunners in this field.

Fig. 24

We can however have another way to look at the historical evolution of shakedown analysis, detached from the individual researcher and research groups and their connections and relations: The onset was the observation that residual stresses due to plasticity in redundant elements of hyperstatic bar-structures are beneficial for their survival under variable loads, what differentiates shakedown theory essentially from limit load theory, where the elastic properties of the material do not interfere. Applications were rare for long time due to the lack of means how to translate the theory to calculation methods. First applications appear for special types of structures like beams, plates and shells, where by appropriate assumptions and semi-analytical methods the complexity of the problem can be reduced. As the theory of plasticity is genuinely linked to metals, the fields of application were on the side of mechanical engineering pressure vessels and pipes, on the side of civil engineering steel frame structures. This first "bifurcation" was not methodological, but naturally imposed by engineering practice. In the sequel, on both sides, application-driven theoretical extensions were carried out: More complex material models for metals, for concrete and for soil-like materials, material damage and cracks, temperature influence, geometrical non-linearity are the major strands, accompanied by the development of appropriate and more rigorous mathematical foundations, such as the proper formulation of the theorems as optimization problems. The break-through to modern engineering however is due to the tremendous development of numerical methods and computer technology: Discretization of structures of almost arbitrary shape connected with fast linear solvers and highly performing optimization algorithms render the theorems of shakedown today easily applicable in practical engineering.

And it is the modern formulation of the theory which makes that the differences between shakedown theory and theory of limit analysis in practice almost vanishes: Limit analysis became a particular case of shakedown analysis and today both are subsumed under the notion of "Direct Methods".

2 Illustrative Examples of Applications

Among the numerous examples of problems solved by Direct Methods, here only few are presented to illustrate the broadness of the field of application and the pertinence of the methods from practical point of view. All examples presented are citations from published material.

Problem: A drumhead with nozzle. Determine the safe loading domain for 3 independently varying loads (internal pressure, temperature difference and axial force on the nozzle) allowing for limited local plastic deformations. This is a very common problem in process- and power engineering. *Example from* [148].

Solution: The problem has been solved using the lower bound theorem by the interior point method in connection with conventional FE methods. The drumhead was numerically discretized with 2.09 10^6 variables, 1.8 10^6 equality constraints, 0.4 10^6 inequality constraints (Figs. 25 and 26).

The graph shows that taking limited plastic deformation into account, the admissible space of independently varying load parameters is considerably larger than the load space if only purely elastic deformations are admitted.

Fig. 25 Drumhead with nozzle loaded by 3 independent variable loads (Simon et al. [148])

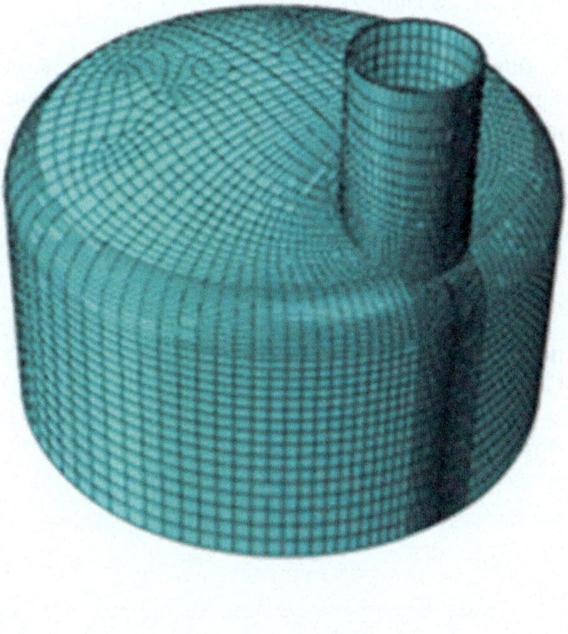

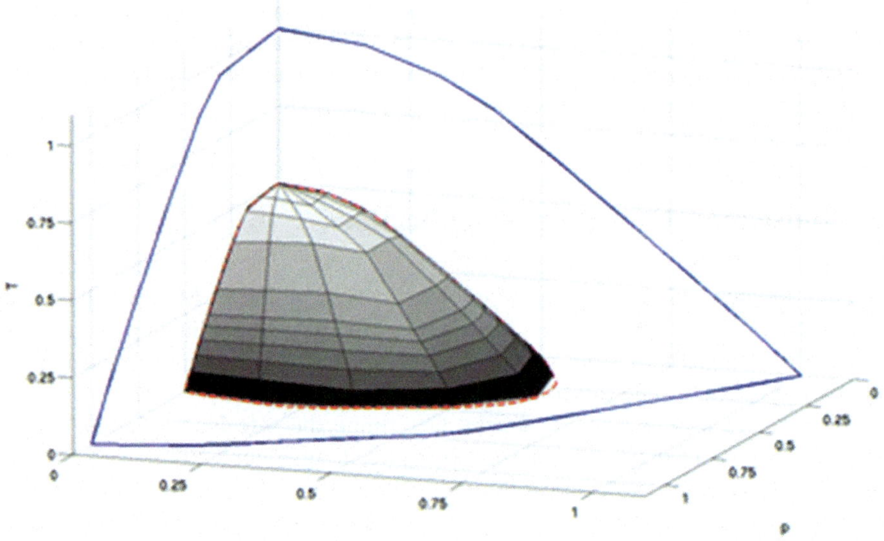

Fig. 26 3-D shakedown domain (SD-domain (blue), compared to elastic domain (grey))

Problem: Shakedown- and Limit Analysis (SDA, LA) to determine strength limits for heterogeneous composites. *Example from* [203].

Solution: Based on scanning electron microscopy (SEM) images, representative volume elements (RVE) are randomly created. These RVEs undergo SDA and LA, which are statistically evaluated. (Lower bound theorem, interior point method) (Figs. 27, 28 and 29).

Interpretation: The influence of particle size on the ultimate strength (LA) is significant as fine particles lead to a higher ultimate load. This influence is smaller in case of variable loading (SDA).

Problem: Design of a reusable spacecraft capsule allowing for limited plastic deformation; determination of admissible internal pressure variation. Example from [204, 207].

Solution: Shakedown analysis, lower bound theorem, combination of conventional FEM for the elastic reference solution and commercial optimisation solver Gurobi.

The calculated SD-limit was successfully used for tests on reduced-size model (Figs. 30, 31 and 32).

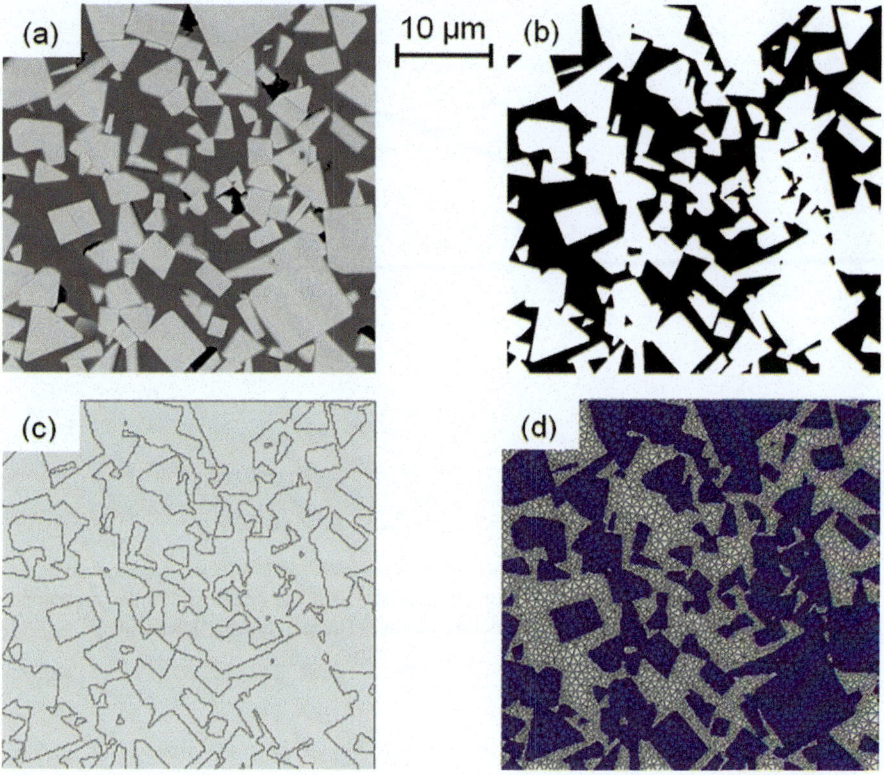

Fig. 27 Transformation of a scanned image (**a**) into a FE model (**d**). From 50 SEM images the geometric characteristics are extracted in order to create randomly 500 "virtual" RVEs

3 Perspectives

Founded independently, Limit- and Shakedown Analysis are distinct through the fundamental importance of the concept of residual stresses in SA in contrast to LA, where they do not interfere. According to the great variety of applications, different methods for calculus have been developed in both areas. Nevertheless, LA and SA have converged today from theoretical and numerical point of view, and it has been shown by Kamenjarzh [206], that LA can be regarded as particular case of SA and, vice verse, SA as an extension of LA. This is the reason why they both are subsumed as "Direct Methods".

Both, LA and SA, have theoretically bounding properties and in particular when it comes to modern numerical methods, the great majority of numerical schemes for DM rely on the combination of finite element analysis and linear or non-linear optimisation procedures.

The tremendous evolution of computing power allows now to solve problems by DM with millions of variables. This enables DM to handle industrial-scale

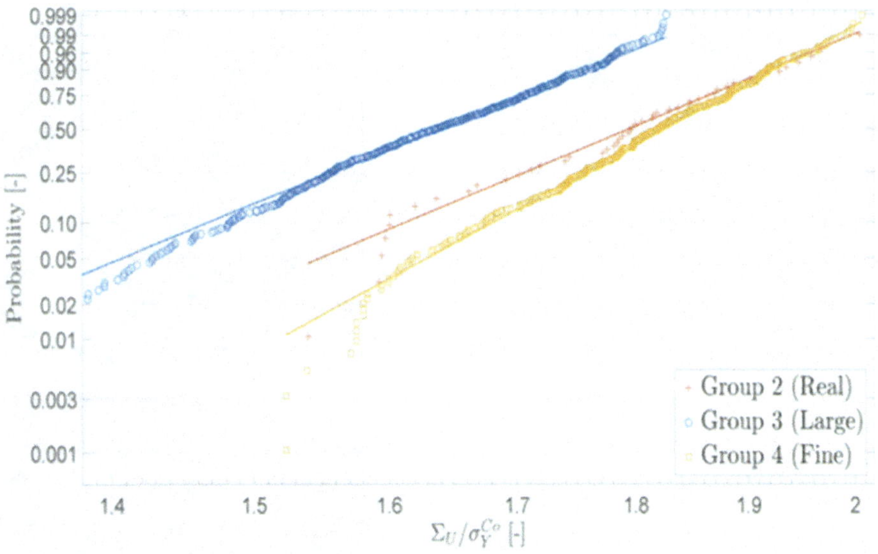

Fig. 28 Failure probability as function of ultimate stress of the composite under monotonous load (LA). Three groups: Scanned (real), coarse particles (Large), fine particles (Fine)

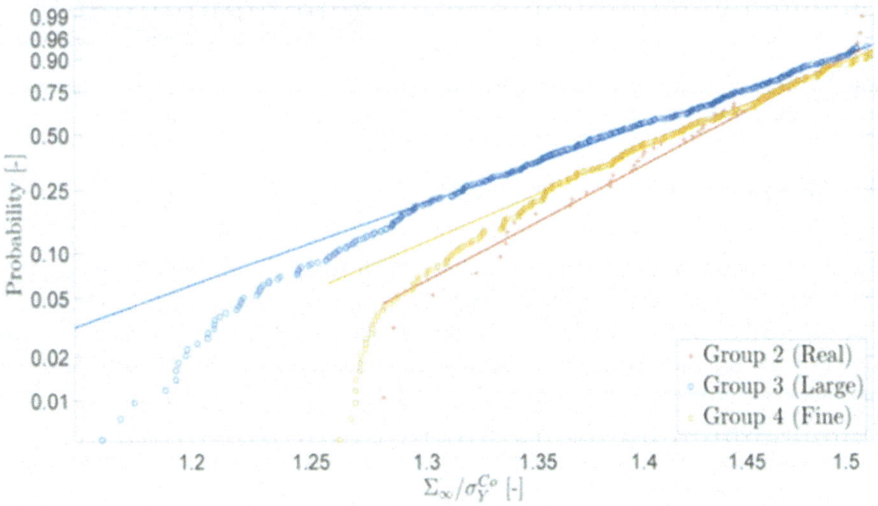

Fig. 29 Failure probability as function of relative ultimate stress of the composite under variable loading (SDA). Three groups: Scanned (real), coarse particles (Large), fine particles (Fine)

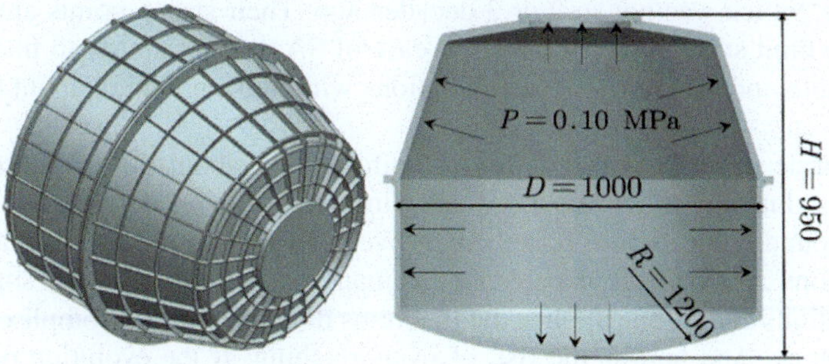

Fig. 30 The space capsule with key dimensions (unit: mm)

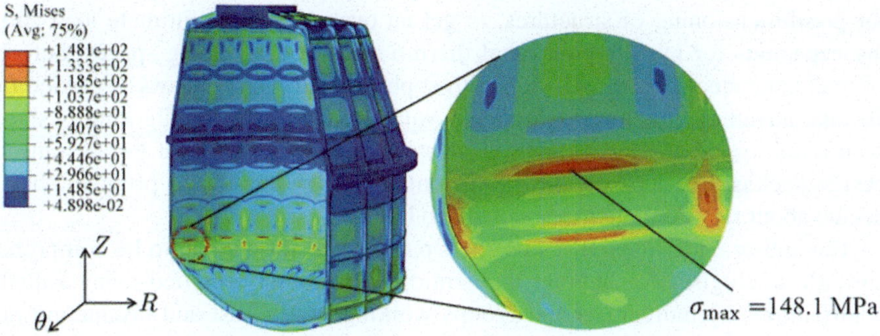

Fig. 31 Elastic stress field in the cabin model

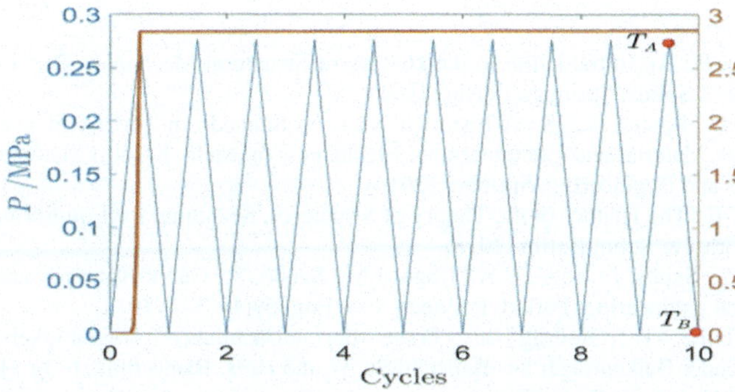

Fig. 32 Evolution of the plastic dissipation in the cabin composed of elastic perfectly plastic material (blue: load cycles, red: accumulated dissipation

problems, which was not so 2 or 3 decades ago. Their strong points are the direct access to limit states in general and, in case of SA, the possibility to find solutions for multiple, independent load combinations without information about the path of loading.

The price to pay for this is loss of information about the evolution of local quantities during the process and, most important, about deformation, although certain bounding properties have been proven. This comes along with rather harsh assumptions about material behavior, in many cases too rough for sophisticated investigations and, if one applies the theorems directly, a rather complex resolution methodology. And here is the risk of overstretching: If the evolution of a system inherently depends on the evolution of local quantities, it becomes very tricky, if not impossible, to find adequate theoretical extensions of DM. So, to find the border line of usefulness is an important issue.

DM are therefore complementary to step-by-step methods. They are well suited for pre-dimensioning of structures, to get an overview about ultimate load carrying capacities for multi-dimensional thermo-mechanical loading spaces, before a combersome step-by-step calculus with sophisticated material laws is carried out. To take advantage of these unique and valuable features, DM algorithms should be offered in a user-friendly form as easy handable complement to conventional design-packages. Then, the designer confronted to a concrete problem, has to decide about the adequate methodology to be used.

The author underlines, that this short paper does not pretend to be comprehensive. He apologises for having, unintentionally, certainly omitted many valuable contributions by individuals and groups working in this vast and fascinating field of scientific research.

References

1. Benvenuto, E.: An Introduction to the History of Structural Mechanics. Part I, Statics and Resistance of Solids, Springer, Berlin (1991)
2. Weichert, D., Ponter, A.R.S.: A historical view on Shakedown Theory, in The History of Theoretical, Material and Computational Mechanics. In: Stein, E. (ed.) Mathematics Meets Mechanics and Engineering, Springer (2014)
3. Kurrer, K.-E.: The History of the Theory of Structures, Searching for Equilibrium, 2nd edn. Wilhelm Ernst & Sohn, Berlin (2018)
4. Kaliszky, S., Sajtos, I., Lógó, B.A., Lógó, J.M., Szabó, Z.: Gábor Kazinczy and his legacy in structural engineering. Period. Polytech. Civ. Eng. **59**(1), 3–7 (2015)
5. Maier-Leibnitz, H.: Beitrag zur Frage der tatsächlichen Tragfähigkeit einfacher, durchlaufender Balkenträger aus Baustahl St. 37 und Holz, Bautechnik, 6. Jg. Heft 1, 1928, 11–14 und Heft 2, 27–31 (1928)
6. Schaim, J.H.: Der durchlaufende Träger unter Berücksichtigung der Plastizität, Stahlbau (1930)
7. Fritzsche, J.: Die Tragfähigkeit von Balken aus Baustahl bei beliebig oft wiederholter Belastung, Bauingenieur 12, 827, Berlin (1931)
8. Grüning, M.: Die Statik des ebenen Tragwerks. Springer, Berlin (1925)

9. Grüning, M.: Die Tragfähigkeit statisch unbestimmter Tragwerke aus Stahl bei beliebig häufig wiederholter Belastung. Springer, Berlin (1926)
10. Bleich, H.: Über die Bemessung statisch unbestimmter Stahltragwerke unter Berücksichtigung des elastisch-plastischen Verhaltens des Baustoffes. Der Bauingenieur, Heft **19**(20), 261–267 (1932)
11. Melan, E.: Der Spannungszustand eines Mises-Hencky'schen Kontinuums bei veränderlicher Belastung. Sitzungsberichte, Akademie der Wissenschaften, Wien **147**, 73–87 (1938)
12. Prager, W.: Problem types in the theory of perfectly plastic materials. J. Aeronaut. Sci. 337–341 (1948)
13. Melan, E.: Theorie unbestimmter Systeme aus idealplastischem Baustoff. Sitzungsberichte, Akademie der Wissenschaften, Wien **145**, 195–218 (1936)
14. Symonds, P.S., Prager, W.: Elastic-plastic analysis of structures subjected to loads varying arbitrarily between prescribed limits. J. Appl. Mech. **17**, 315–324 (1950)
15. Symonds, P.S.: Shakedown in continuous media. J. Appl. Mech. **17**, 85–89 (1951)
16. Neal, B.G.: Plastic-collapse and shake-down theorems for structures of strain hardening material. J. Aero. Sci. **17**, 297–307 (1950)
17. Symonds, P.S., Neal, B.G : Recent progress in the plastic methods of structural analysis. J. Franklin Inst. **252**(6), 469–492 (1951)
18. Symonds, P.S., Neal, B.G.: The calculation of failure loads on plane frames under arbitrary loading programs. J. Inst. Civil Engrs. **35**, 41–61 (1951)
19. Neal, B.G., Symonds, P.S.: A method for calculating the failure load for a framed structure subjected to fluctuating loads. J. Inst. Civ. Eng. **35**, 186 (1951)
20. Symonds, P.S.: An early upper bound method for shakedown. In: Weichert, D., Maier, G. (eds.) Inelastic Analysis of Structures Under Variable Loads—Theory and Engineering Applications. Solid Mechanics and its Applications, vol. 83, pp. 1–9. Kluwer Academic Publishers, Dordrecht (2000)
21. Baker, J., Horne, M., Heyman, J.: The steel skeleton, Chap. 9, vol. II. Cambridge University Press, Cambridge (1956)
22. Koiter, W.T.: A new general theorem on shakedown of elastic-plastic structures. Proc. Kon. Ne. Ak. Wet. **B59**, 24–34 (1956)
23. Koiter, W.T.: General theorems for elastic-plastic structures. In: Sneddon, I.N., Hill, R. (eds.) Progress in Solid Mechanics, pp. 165–221. North-Holland Publ. Co., Amsterdam, Chap. IV (1960)
24. Symonds, P.S.: Basic theorems in the plastic theory of structures. J. Aero. Sci. **17**, 669–670 (1950)
25. Maier, G.: Shakedown theory in perfect elastoplasticity with associated and nonassociated flow-laws, a finite element, linear programming approach. Meccanica **4**(3), 250–260 (1969)
26. Maier, G.: A shakedown matrix theory allowing for workhardening and second-order geometric effects. In: Sawczuk, A. (ed.) International Symposium, Warsaw, August 30–September 2, 1972; Foundations of Plasticity, Noordhoff, Leyden, vol. 1, pp. 417–433 (1973)
27. Corradi, L., Maier, G.: Dynamic non-shakedown theorem for elastic perfectly-plastic continua. J. Mech. Phys. Solids **22**, 401–413 (1974)
28. König, A., Maier, G.: Shakedown analysis of elastoplastic structures, a review of recent developments. Nucl. Eng. Des. **66**, 81–95 (1981)
29. Maier, G.: A generalization to nonlinear hardening of the first shakedown theorem for discrete elastic-plastic structures. Rendic. Acc. Naz. dei Lincei, Serie Ottava, 161–174 (1988)
30. Maier, G., Novati, G.: Dynamic shakedown and bounding theory for a class of nonlinear hardening discrete structural models. Int. J. of Plasticity **6**(5), 551–572 (1990)
31. Maier, G., Pan, L., Perego, U.: Geometric effects on shakedown and ratchetting of axisymmetric cylindrical shells subjected to variable thermal loading. Eng. Struct. **15**(6), 453–466 (1993)

32. Corigliano, A., Maier, G., Pycko, S.: Dynamic shakedown analysis and bounds for elastoplastic structures with nonassociative, internal variable constitutive laws. Int. J. Solids Struct. **32**(21), 3145–3166 (1995)
33. Cocchetti, G., Maier, G.: Static shakedown theorems in piecewise linearized poroplasticity. Arch. Appl. Mech. **68**, 651–661 (1998)
34. Carvelli, V., Maier, G., Taliercio, A.: Shakedown analysis of periodic heterogeneous materials by a kinematic approach. J. Mech. Eng. **50**(4), 229–240 (1999)
35. Carvelli, V., Cen, Z.Z., Liu, Y., Maier, G.: Shakedown analysis of defective pressure vessels by a kinematic approach. Arch. Appl. Mech. **69**, 751–764 (1999)
36. Cocchetti, G., Maier, G.: A shakedown theorem in poroplastic dynamics. Rend. Mat., Accademia Nazionale dei Lincei, s. 9, **13**, 43–53 (2002)
37. Maier, G., Cocchetti, G.: Fundamentals of direct methods in poroplasticity. In: Weichert, D., Maier, G. (eds.) Inelastic Behaviour of Structures Under Variable Repeated Loads, Direct Analysis Methods, pp. 91–113. CISM, Springer, Wien (2002)
38. Maier, G., Carvelli, V.: A kinematic method for shakedown and limit analysis of periodic composites. In: Weichert, D., Maier, G. (eds.) Inelastic Behaviour of Structures Under Variable Repeated Loads, Direct Analysis Methods, pp. 115–132. CISM, Springer, Wien (2002)
39. Maier, G., Pastor, J., Ponter, A.R.S., Weichert, D.: Direct Methods in Limit and Shakedown Analysis. In: De Borst, R., Mang, H.A. (eds.) Numerical and Computational Methods, vol. 3. In: Milne, I., Ritchie, R.O., Karihaloo, B. (eds.) Comprehensive Structural Integrity, Elsevier-Pergamon, Amsterdam (2003)
40. Cocchetti, G., Maier, G.: Elastic-plastic and limit-state analyses of frames with softening plastic-hinge models by mathematical programming. Int. J. Solids Struct. **40**, 7219–7244 (2003)
41. Ponter, A.R.S.: On the relationship between plastic shakedown and the repeated loading of creeping structures. Trans. ASME, J. Appl. Mech. **38**, Series E, 437–440 (1971)
42. Ponter, A.R.S.: Deformation, displacement and work bounds for structures in a state of creep and subject to variable loading. Trans. ASME, J. Appl. Mech. **39**, Series E, 953–959 (1972)
43. Ponter, A.R.S.: On the creep modified shakedown limit. In: Ponter, A.R.S., Hayhurst, D.R. (eds.) 3rd IUTAM Symposium on "Creep in Structures", Leicester University, Sept 1980, pp. 264–278. Springer (1981)
44. Ponter, A.R.S.: An upper bound on the small displacement of elastic-plastic structures. Trans. ASME, J. Appl. Mech. **39**, Series E, 959–964 (1972)
45. Ponter, A.R.S.: General displacement and work bounds for dynamically loaded bodies. J. Mech. Phys. Solids **23**, 151–163 (1975)
46. Ponter, A.R.S.: A general shakedown theorem for inelastic materials. In: Ponter, A.R.S., Karadeniz, S. (eds.) Proceedings of the SMiRT-3, London, paper L5/2, 1975, An extended shakedown theory for structures that suffer cyclic thermal loading, Part I: Theory Trans. ASME, J. Appl. Mech. **52**, 877–882 (1985)
47. Ponter, A.R.S., Karadeniz, S.: An extended shakedown theory for structures that suffer cyclic thermal loading, Part II: applications. Trans. ASME J. Appl. Mech. **52**, 883–889 (1985)
48. Ponter, A.R.S., Carter, K.F.: The ratchetting of shells subjected to severe thermal loading. In: Tooth, A.S., Spence, J. (eds.) Applied Solid Mechanics—2. Elsevier Applied Science, pp. 303–320 (1987)
49. Ponter, A.R.S., Cocks, A.C.F.: The incremental strain growth of an elastic-plastic body loaded in excess of the shakedown limit. Trans. ASME J. Appl. Mech. **51**(3), 465–469 (1984)
50. Ponter, A.R.S., Cocks, A.C.F.: The incremental strain growth of elastic-plastic bodies subjected to high levels of cyclic thermal loading. Trans ASME J. Appl. Mech. **51**(3), 470–474 (1984)

51. Ponter, A.R.S., Karadeniz, S., Carter, K.F.: The computation of shakedown limits for structural components subjected to variable thermal loading—Brussels Diagrams, Directorate General for Science, Research and Development, Office for Official Publications of the E.C., Report EUR12686EN, Brussels, 170 p (1990)
52. Ponter, A.R.S., Chen, H.F., Chiavarella, M., Specchia, G.: Shakedown analysis for rolling and sliding contact problems. Int. J. Sol. Struct. **43**, 4201–4219 (2001)
53. Ponter, A.R.S., Leckie, F.A.: Bounding properties of metal matrix composites subjected to cyclic loading. J. Mech. Phys. Solids **46**, 697–717 (1998)
54. Ponter, A.R.S., Leckie, F.A.: On the behaviour of metal matrix composites subjected to cyclic thermal loading. J. Mech. Phys. Solids **46**, 2183–2199 (1998)
55. Ponter, A.R.S., Engelhardt, M.: Shakedown limits for a general yield condition. Eur. J. Mech., A/Solids **19**, 423–445 (2000)
56. Ponter, A.R.S., Chen, H.F.: A Minimum theorem for cyclic load in excess of shakedown, with applications to the evaluation of a ratchet limit. Euro. J. Mech., A/Solids **20**, 539–553 (2001)
57. Chen, H.F., Ponter, A.R.S., Ainsworth, R.A.: The Linear Matching Method applied to the high temperature life assessment of structures, Part 1. Assessments involving constant residual stress fields. Int. J. Press. Vessel. Pip. **83**, 123–135 (2006)
58. Chen, H.F., Ponter, A.R.S., Ainsworth, R.A.: The Linear Matching Method applied to the high temperature life assessment of structures, Part 2. Assessments beyond shakedown involving changing residual stress fields. Int. J. Press. Vessel. Pip. **83**, 136–147 (2006)
59. Gokhfeld, D.A.: On the possibility of increase of plastic deformation due to thermal cycling effects. In: Calculations of strength. Issue 7. Moscow, Mashgiz publ., (in Russian) (1961)
60. Gokhfeld, D.A.: Bearing capacity of structures under thermal cycles—Moscow, Mashinostroenie publ., (in Russian), 259 (1970)
61. Gokhfeld D.A., Cherniavsky, O.F.: Theory of shakedown and strain accumulation under thermal cyclinges. In: Proceedings of the All-USSR. Symposium on low-cycle fatigue at elevated temperatures. Chelyabinsk, Issue 3. C3–31 (1974)
62. Gokhfeld, D.A., Cherniavsky, O.F.: Limit analysis of structures at thermal cycling. The Netherlands, Rockville, USA: Sijthoff and Noordhoff. Int. Publ. Alphen aan den Rijn (1980)
63. Ceradini, G.: Sull' adattamento dei corpi elasto-plastici soggetti ad azioni dinamiche. Giornale del Genio Civile **415**, 239–258 (1969)
64. Cascaro, R., Di Carlo, A.: Formulazione dell'analisi limite delle piastre come problema di minimax mediante una rappresentazione agli elementi finiti del campo delle tensioni. Nota I, Giornale del Genio Civile, vol. 108 (1970)
65. Cascaro, R., Di Carlo, A.: Formulazione dell'analisi limite delle piastre come problema di minimax mediante una rappresentazione agli elementi finiti del campo delle tensioni. Nota II, Giornale del Genio Civile (1971)
66. Cascaro, R., Di Carlo, A., Valente, G.: Un metodo statico cinematico per l'analisi limite dei continui bidimensionali piani. Giornale del Genio Civile, vol. 109 (1971)
67. Casciaro, R., Cascini, L.: A mixed formulation and mixed finite elements for limit analysis. Int. J. Numer. Meth. Engng. **18**, 211–243 (1982)
68. Casciaro, R., Garcea, G.: An iterative method for shakedown analysis. Comput. Methods Appl. Mech. Eng. **191**(49–50), 5761–5792 (2002)
69. Garcea, G., Armentano, G., Petrolo, S., Casciaro, R.: Finite element shakedown analysis of two-dimensional structures. Int. J. Numer. Meth. Eng. **63**(8), 1174–1202 (2005)
70. Leonetti, L., Casciaro, R., Garcea, G.: Effective treatment of complex statical and dynamical load combinations within shakedown analysis of 3d frames. Comput. Struct. **158**, 124–139 (2015)
71. Polizzotto, C.: Workhardening adaptation of rigid-plastic structures. Meccanica **10**, 280–288 (1975)

72. Polizzotto, C.: A unified approach to quasi-static shakedown problems for elasto-plastic solids with piecewise linear yield surfaces. Meccanica **13**, 109–120 (1978)
73. Polizzotto, C.: On workhardening adaptation of discrete structures under dynamic loadings. Arch. Mech. Stos. **32**, 81–99 (1980)
74. Polizzotto, C.: A unified treatment of shakedown theory and related bounding techniques. S.M. Arch. **7**, 19–75 (1982)
75. Polizzotto, C.: A convergent bounding principle for a class of elastoplastic strain-hardening solids. Int. J. Plasticity **2**, 357–370 (1986)
76. Polizzotto, C.: On the condition to prevent plastic shakedown of structures. ASME, J. Appl. Mech. I and II, **60**, 15–25 and 318–330 (1993)
77. Polizzotto, C., Borino, G.: Shakedown and steady state responses of elastic-plastic solids in large displacements. Int. J. Sol. Struct. **33**, 3415–3437 (1996)
78. Polizzotto, C., Borino, G., Caddemi, S., Fuschi, P.: Shakedown problems for material models with internal variables. Int. J. Mech. Sci. **35**, 787–801 (1993)
79. Polizzotto, C., Borino, G., Fuschi, P.: An extended shakedown theory for elastic-damaged models. Eur. J. Mech., A/Solids **15**, 825–858 (1996)
80. Polizzotto, C., Borino, G., Fuschi, P.: Weak forms of shakedown for elastic-plastic structures exhibiting ductile damage. Meccanica **36**, 49–66 (2001)
81. Sawczuk, A., Jäger, T.: Grenztragfähigkeitstheorie der Platten, Springer (1963)
82. Sawczuk, A.: Evaluation of upper bounds to shakedown loads of shells. J. Mech, Phys. Solids **17**, 291–301 (1969)
83. Sawczuk, A.: Shakedown analysis of elastic-plastic structures. Nucl. Eng. Design **28**, 121–136 (1974)
84. Sawczuk, A.: Mechanics and Plasticity of Structures, Ellis-Horwood/PWN, Chichester/ Warsaw (1989)
85. König, J.A.: Shakedown of strainhardening structures. 1st Canad. Cong. Appl. Mech., Quebec (1967)
86. König, J.A.: A shakedown theorem for temperature dependent elastic moduli. Bull. Acad. Polon. Sci. Sér. Sci. tech. **17**, 161–165 (1969)
87. König, J.A.: Deflection bounding at shakedown under thermal and mechanical loadings. In: Second SMiRT Conference, Berlin, paper L7/3, (1973)
88. König, J.A.: A method of shakedown analysis of frames and arches. Int. J. Sol. Struct. 327–344 (1971)
89. König, J.A.: Shakedown deflections, a finite element approach. Teoret. I Prilož. Meh. **3**, 65–69 (1972)
90. König, J.A.: Shakedown of Elastic-Plastic Structures. Elsevier, Amsterdam (1987)
91. Borkowski, A.: Analysis of Skeletal Structural Systems in the Elastic, and Plastic Range. PWN and Elsevier' Warsaw (1988)
92. Borkowski, A., Kleiber, M.: On a numerical approach to shakedown analysis of structures. Comput. Methods Appl. Mech. Eng. **22**, 101 (1980)
93. Dorosz, S.: An improved upper bound to maximum deflections of elasticplastic structures at shakedown. J. Struct. Mech. **6**, 267–287 (1978)
94. Dorosz, S, König, J.A., Sawczuk, A., Biegus, A., Kowal, Z., Seidel, W.: Deflections of elastic.plastic hyperstatic beams under cyclic loading. Arch. Mech. **33**, 611–624 (1981)
95. Dorosz, S., König, J.A.: Iterative method of evaluation of elastic-plastic deflections of hyperelastic structures. Ing. Archiv. **55**, 202–222 (1985)
96. Janas, M., Pycko, S., Zwoliński, J.: A min-max procedure for the shakedown analysis of skeletal structures. Int. J. Mech. Sci. **37**, 629–649 (1995)
97. Janas, M., König, J.A.: A cylindrical tank response as an example of shakedown of non-Clapeyronian systems. Arch. Mech. **43**, 49–56 (1991)
98. Kleiber, M., König, J.A.: Incremental shakedown analysis in the case of thermal effects. J. Num. Methods Eng. **20**, 1567 (1984)

99. König, J.A., Kleiber, M.: On a new method of shakedown analysis. Bull. Acad. Pol- Sci' Ser. Sci. Technol. **26**, 165 (1978)
100. König, J.A., Siemaszko, A.: Strainhardening effects in shakedown process. Ing. Archiv. **58**, 58–66 (1988)
101. König, J.A., Maier, G.: Shakedown of elastoplastic structures, a review of recent developments. Nucl. Eng. Design **66**, 81–95 (1981)
102. König, J.A., Pycko, S.: Shakedown analysis in the case of imposed displacements. Mech. Teor. Stos. **28**, 101–108 (1990)
103. Mróz, Z.: On the theory of steady plastic cycles in structures. In: Proceedings of the 1st SMIRT, L6, 489501 (1971)
104. Orkisz, J., Orringer, O., Holowinski, M., Pazdanowski, M., Cecot, W.: Discrete analysis of actual residual stress resulting from cyclic loadings. Comput. Struct. **35**, 397 (1990)
105. Pycko, S., Mróz, Z.: Alternative approach to shakedown as a solution of min-max problem. Acta Mech. **93**, 205 (1992)
106. Pycko, S., König, J.A.: Elastic-plastic structures subjected to variable repeated imposed displacements and mechanical loads. Int. J. Plast. **8**, 603–618 (1992)
107. Zwoliński, J., Bielawski, G.: An optimal selection of residual stress for determination limit and shakedown multiplier, (in Polish). In: Proceedings of Conference on Comp Meth. Struct. Mecł., Jadwisin p. a59 (1987)
108. Skoczeń, B., Skrzypek, J., Bielski, J.: Shakedown and inadaptation mechanisms of bellows subject to constant pressure and cyclic axial forces. Mech. Struct. Mach. **20**, 119 (1992)
109. Skoczeń, B., Skrzypek, J.: Inadaptation mechanisms in bellows subject to sustained pressure and cyclic axial loadings in terms of finite deformations. In: Mróz, Z., Weichert, D., Dorosz, S. (eds.) Inelastic Behaviour of Structures under Variable Loads, pp. 341–361. Kluwer Academic Publishers (1995)
110. Życzkowski, M.: Combined Loadings in the Theory of Plasticity. Polish-Scientific Publ. (1981)
111. Mahrenholtz, O., Leers, K., König, J.A.: Shakedown of tubes: a theoretical analysis and experimental investigations. In: Reid, S.R. (ed.) Metal Forming and Impact Mechanics, W. Johnson Commemorative Volume, pp. 155–172. Pergamon Press (1984)
112. Leers, K.: Experimentelle und theoretische Shakedownuntersuchung an Rohren. VDI-Verlag, Düsseldorf (1985)
113. Leers, K., Klie, W., König, J.A., Mahrenholtz, O.: Experimental investigations of shakedown of tubes. In: Sawczuk, A., Bianchi, G. (eds.) Plasticity Today, Elsevier Applied Science Publ., London, pp. 259–275 (1985)
114. Stein, E., Zhang, G., König, J.A.: Shakedown with nonlinear hardening including structural computation using finite element method. Int. J. Plasticity **8**, 1–31 (1992)
115. Stein, E., Zhang, G., Mahnken, R.: Shakedown analysis for perfectly plastic and kinematic hardening materials. In: Stein, E. (ed.) Progress in Computational Analysis of Inelastic Structures, Springer, pp. 175–244 (1993)
116. Stein, E., Zhang, G., Huang, Y.: Modeling and computation of shakedown problems for nonlinear hardening materials. Comput. Methods Appl. Mech. Eng. **103**, 247–272 (1993)
117. Huang, Y., Stein, E.: Shakedown of a cracked body consisting of kinematic hardening material. Eng. Fract. Mech. **54**, 107–112 (1996)
118. Huang, Y., Stein, E.: Prediction of the fatigue threshold for a cracked body by using shakedown theory. Fatigue Fract. Eng. Mater. Struct. **18**(3), 363–370 (1995)
119. Huang, Y., Stein, E.: Shakedown of a CT-specimen with St52-steel: experimental, analytical and numerical investigations. J. Strain Anal. **30**(4), 283–289 (1995)
120. Wiechmann, K., Barthold, J.-F., Stein, E.: Optimization of elasto-plastic structures using the finite element method. In: 2nd World Congress of Structural and Multidisciplinary Optimisation, pp. 1013–1018 (1997)

121. Wiechmann, K., Barthold, J.-F., Stein, E.: Shape optimization under shakedown constraints. In: Weichert, D., Maier, G. (eds.) Inelastic Analysis of Structures Under Variable Loads, pp. 49–68. Kluwer Academic Publishers (2000)
122. Stein, E.: Private communication
123. Rafalski, P.: Minimum principles for the stress field in an elastic-plastic body. Int. J. Engng. Sci. **14** (1976)
124. Rafalski, P.: Minimun Principles in Plasticity, Mitteilungen aus dem Institut für Mechanik Bochum, vol. 13 (1978)
125. Rafalski, P., Weichert, D.: On minimum principles for elastic-plastic plates. Bull. Acad. Pol. des Sci., Sci. Tech. **XXVI**(11) (1978)
126. Weichert, D.: On the influence of geometrical nonlinearities on the shakedown of elastic-plastic structures. Int. J. Plast. **2**(2), 135–148 (1986)
127. Weichert, D., Groß-Weege, J.: The numerical assessment of elastic-plastic sheets under variable mechanical and thermal loads using a simplified two-surface yield-condition. Int. J. Mech. Sci. **30**(10), 757–767 (1989)
128. Raad, L., Weichert, D., Haidar, A.: Analysis of full- depth asphalt concrete pavements using shakedown-theory, transactions of the transportation research board (NRC). Transp. Res. Rec. **1227**, 53–65 (1989)
129. Raad, L., Weichert, D., Najim, W.: Stability of multilayer systems under repeated loads, transactions of transportation research board (NRC). Transp. Res. Rec. **1207**, 181–186 (1988). Pavement Design
130. Weichert, D., Raad, L.: Extension of the static shakedown-theorem to a certain class of materials with variable elastic coefficients. Mech. Res. Comm. **19**(6), 511–517 (1992)
131. Boulbibane, M., Weichert, D., Raad, L.: Numerical application of shakedown theory to pavements with anisotropic layer properties, Paper No. 99–0342. J. Transp. Res. Board (NRC), Transp. Res. Rec. **1687**, 75–81 (1999)
132. Boulbibane, M., Collins, I.F., Weichert, D., Raad, L.: Shakedown analysis of anisotropic asphalt concrete pavements with clay subgrade. Geotech. J. **37**, 882–889 (2000)
133. Hachemi, A., Weichert, D.: An extension of the static shakedown-theorem to a certain class of damaging inelastic material. Arch. Mech. **44**(5–6), 491–498 (1992)
134. Hachemi, A., Weichert, D.: Application of shakedown theorems to damaging inelastic material under mechanical and thermal loads. Int. J. Mech. Sci. **39**(9), 1067–1076 (1997)
135. Belouchrani, M., Weichert, D.: An extension of the static shakedown theorem to inelastic cracked structures. Int. J. Mech. Sci. **41**, 163–177 (1999)
136. Belouchrani, M.A., Weichert, D., Hachemi, A.: Fatigue threshold computation by shakedown theory. Mech. Res. Comm. **27**(3), 287–293 (2000)
137. Hachemi, A., Weichert, D.: Numerical shakedown analysis of damaged structures. Comp. Meth. Appl. Mech. Engng. **160**, 57–70 (1998)
138. Hamadouche, M.A., Weichert, D.: Application of shakedown theory to soil dynamics. Mech. Res. Comm. **26**(5), 565–574 (1999)
139. Weichert, D., Schwabe, F., Hachemi, A.: Composite design by shakedown analysis for low cycle fatigue service conditions. In: Khan, A.S., Zhang, H., Yuan, Y. (eds.) Proceedings of the 8th International Symposium on Plasticity and Its Current Applications, held at Whistler, Canada, July 16–20, 2000, pp. 505–507. Neat Press (2000)
140. Weichert, D., Hachemi, A.: Shakedown- and limit analysis of periodic composites. J. Theor. Appl. Mech. **1**(40), 273–289 (2002)
141. Hachemi, A., Weichert, D.: On the problem of interfacial damage in fibre-reinforced composites under variable loads. Mech. Res. Comm. **32**, 15–23 (2005)
142. Hachemi, A., Mouhtamid, S., Weichert, D.: Progress in shakedown analysis with applications to composites. Arch. Appl. Mech. 1–11 (2005)
143. Weichert, D., Hachemi, A.: A shakedown approach to the problem of damage of fiber-reinforced composites. In: Sadowski, T. (ed.) Proceedings of the IUTAM-Symposium on "Multiscale Modelling of Damage and Fracture Processes in Composite Materials",

Kazimierz Dolny, Poland, 23–27 May 2005, vol. 135, pp. 41–48. Solid Mechanics and its Applications, Springer (2006)
144. Nguyen, A.D., Hachemi, A., Weichert, D.: Application of the interior—point method to shakedown analysis of pavements. Int. J. Numer. Meth. Eng. **2008**(75), 414–439 (2008). Published online in Wiley InterScience
145. Weichert, D., Hachemi, A.: Recent advances in lower bound shakedown analysis. In: Proceedings of Pressure Vessel and Piping Conference 2009, ASME, July 26–30, Prague, Czech Republic (2009)
146. Simon, J.-W., Chen M., Weichert, D.: Shakedown analysis combined with the problem of heat conduction. In: Proceedings of the ASME 2010 Pressure Vessels & Piping Division / K-PVP Conference PVP 2010 July 18–22, 2010, Bellevue, Washington, USA (2010)
147. Simon, J.-W., Weichert, D.: An improved interior-point algorithm for large-scale shakedown analysis PAMM. In: Proceedings of the Applied Mathematics Mechanics, vol. 10, pp. 223–224 (2010)
148. Simon, J., Chen, G., Weichert. D.: Shakedown analysis of nozzles in the knuckle region of torispherical heads under multiple thermo-mechanical loadings. Int. J. Press. Vessel. Pip. **116**, 47–55 (2014)
149. Morelle, P.: Numerical shakedown analysis of axisymmetric sandwich shells: an upper bound formulation. Int. J. Num. Meth. **23**(11), 2071–2088 (1986)
150. Hung, N.-D., König, J.A.: Finite element formulation for shakedown problems using a yield criterion of the mean. Comput. Appl. Mech. Eng. **8**(2), 179–192 (1976)
151. Hung, N.-D., Palgen, L.: Shakedown analysis by displacement method and equilibrium finite element. Trans. CSME **6**(1), 32–39 (1980–1981)
152. Hung, N.-D.: Shakedown analysis by finite element method and linear programming techniques. J. de mécanique appliquée, Paris, France **2**(4), 587–599 (1983)
153. Hung, N.-D., Yan, A.-M.: Direct finite element kinematical approaches in limit and shakedown analysis of shells and elbows, 233–254. In: Inelastic Analysis of Structures under Variable Loads, Theory and Engineering Applications. Kluwer Academic Publishers (2000)
154. Hung, N.-D., Vu, D.K.: Primal-dual algorithm for shakedown analysis of structures. Comput. Methods Appl. Mech. Eng. (Elsevier) **193**(42–44), 4663–4674 (2004)
155. Staat, M., Heitzer, M., Yan, A.M., Vu, D.K., Hung, N.-D., Voldoire, F., Lahousse, A.: Limit analysis of defects. In: Berichte des Forschungszentrums Jülich, Jül-3746 (2000)
156. Khôi Vu, D.K., Staat, M., Tran, I.T.: Analysis of pressure equipment by application of the primal-dual theory of shakedown. Commun. Numer. Meth. Eng. **23**(3), 213–225 (2007)
157. Vu, D.K., Staat, M.: Shakedown analysis of structures made of materials with temperature-dependent yield stress. Int. J. Solids Struct. **44**(13), 4524–4540 (2007)
158. Heitzer, M., Staat, M.: Reliability analysis of elasto-plastic structures under variable loads. In: Weichert, D., Maier, G. (eds.) Inelastic Analysis of Structures under Variable Loads: Theory and Engineering Applications, pp. 269–288. Kluwer, Academic Press, Dordrecht (2000)
159. Tran, T.N., Phạm, P.T., Vu, D.K., Staat, M.: Reliability analysis of inelastic shell structures under variable loads. In: Weichert, D., Ponter, A.R.S. (eds.) Limit States of Materials and Structures: Direct Methods, pp. 135–156. Springer Netherlands (2009)
160. Spiliopoulos, K.V.: On the automation of the force method in the optimal plastic design of frames. Comp. Meth. Appl. Mech. Eng. **141**, 141–156 (1997)
161. Spiliopoulos, K.V.: A fully automatic force method for the optimal shakedown design of frames. Comp. Mech. **23**, 299–307 (1999)
162. Spiliopoulos, K.V.: Force method-based procedures in the limit equilibrium analysis of framed structures. In: Weichert, D., Ponter, A.R.S. (eds.) Limit States of Materials and Structures: Direct Methods, pp. 233–252. Springer, Netherlands (2009)
163. Spiliopoulos, K., Panagiotou, K.: A direct method to predict cyclic steady states of elastoplastic structures. Comput. Methods Appl. Mech. Eng. **223–224**, 186–198 (2012)

164. Spiliopoulos, K., Panagiotou, K.: An enhanced numerical procedure for the shakedown analysis in multidimensional loading domains. Comput. Struct. **193**, 155–177 (2017)
165. Kapogiannis, I., Spiliopoulos, K.: Recent updates of the residual stress decomposition method for shakedown analysis. In: Pisano, A., Spiliopoulos, K., Weichert, D. (eds.) Direct Methods, Methodological Progress and Engineering Applications. Springer (2021)
166. De Saxcé, G.: Une generalization de l'inégalité de Fenchel et ses applications aux lois constitutives, C.R. Acad. Sci. Paris **314**(II), 125–129 (1992)
167. Bodoville, G.D., Saxcé, G.: Plasticity with non-linear kinematic hardening: modelling and shakedown analysis by the bipotential approach. Int. J. Plast. **17**(1), 21–46 (2001)
168. Bousshine, L., Chaaba, A., De Saxcé, G.: A new approach to shakedown analysis for non-standard elastoplastic material by the bipotential. Int. J. Plast. **19**(5), 583–598 (2003)
169. Moreau, J.J.: On unilateral constraints, friction and plasticity. In: New Variational Techniques in Mathematical Physics, CIME Course, pp. 173–322. Springer (1974)
170. Debordes, O., Nayroles, B.: Sur la théorie et le calcul à l'adaptation des structures élasto-plastiques. J. Mécanique **20**, 1–54 (1976)
171. Nayroles, B.: Tendences récentes et perspectives à moyen terme en élastoplasticité asymptotique des constructions, Congrès Français de Mécanique, Grenoble, France (1977)
172. Nayroles, B., Weichert, D.: La notion de sanctuaire d'élasticité et l'adaptation des structures. C.R. Acad. Sci. Paris **316**(II), 1493–1498 (1993)
173. Nguyen, Q.-S.: Extension des théorèmes d'adaptation et d'unicité en écrouissage non linéaire. C.R. Acad. Sci. **282**, 755–758 (1976)
174. Nguyen, Q.-S.: Min-Max duality and shakedown theorems in plasticity. In: Alart, P., Maisonneuve, O., Rockafellar, R.T. (eds.) Nonsmooth Mechanics and Analysis, Theoretical and Numerical Advances, Chap. 8, Springer (2006)
175. Nguyen, Q.-S., Pham, D.: On shakedown theorems in hardening plasticity. C.R. Acad. Sci. **329**, 307–314 (2001)
176. Halphen, B., Nguyen, Q.-S.: Sur les matériaux standard généralisés. J. Mécanique **14**, 1–37 (1975)
177. Mandel, J.: Adaptation d'une structure plastique écrouissable. Mech. Res. Comm. **3**, 251–256 (1976)
178. Radenkovic, D.: Théorèmes limites pour un materiau de Coulomb à dilatation nonstandardisée. Comptes Rendus de l'Académie des Sci. Paris **252**(4103–4104), 1961 (1961)
179. Zarka, J., Frelat, J., Inglebert, G., Kasmaï-Navidi, J.: A new approach to inelastic analysis of structures, CADLM edition, France (1989)
180. Inglebert, G., Zarka, J.: On a simplified inelastic analysis of structures. Nucl. Eng. **57**, 333–368 (1980)
181. Hassine, T., Inglebert, G., Pons, M.: Shakedown and damage analysis applied to rocket machines, In: Weichert, D., Maier, G. (eds.) Inelastic Analysis of Structures under Variable Loads: Theory and Engineering Applications. Kluwer, Academic Press, Dordrecht (2000)
182. Sharp, R.W., Booker, J.R.: Shakedown of pavements under moving surface loads. J. Transp. Eng. ASCE **110**(1), 1–14 (1984)
183. Krabbenhøft, K., Lyamin, A.V., Sloan, S.W.: Bounds to shakedown loads for a class of deviatoric plasticity models. Comp. Mech. **39**(6), 879–888 (2007)
184. Krabbenhøft, K., Lyamin, A.V., Sloan, S.W.: Shakedown of a cohesive-frictional half-space subjected to rolling and sliding contact. Int. J. Solids Struct. **44**(11–12), 3998–4008 (2007)
185. Zhao, J.D., Sloan, S.W., Lyamin, A.V., Krabbenhøft, K.: Bounds for shakedown of cohesive-frictional materials under moving surface loads. Int. J. Solids Struct. **45**(11–12), 3290–3312 (2008)
186. Lyamin, A.V., Sloan, S.W.: Lower bound limit analysis using non-linear programming.
187. Int. J. Num. Meth. Eng. **55**(5), 573–611 (2002)
188. Krabbenhoft, K., Lyamin, A.V., Hjiaj, M., Sloan, S.W.: A new discontinuous upper bound limit analysis formulation, 180. Int. J. Num. Meth. Eng. **63**(7), 1069–1088 (2005)

189. Podlich, N., Lyamin, A.V., Sloan, S.W.: High performance interior point methods for three-dimensional finite element limit analysis. In: COMPLAS XV: Proceedings of the XV International Conference on Computational Plasticity: Fundamentals and Applications, CIMNE, pp. 121–132 (2019)
190. Collins, I.F., Wang, A., Saunders, L.: Shakedown theory and the design of unbound pavements. Road Transp. Res. **2**, 28–39 (1993)
191. Collins, I.F., Wang, A., Saunders, L.: Shakedown in layered pavements under moving surface loads. Int. J. Numer. Anal. Methods Geomech. **17**, 165–174 (1993)
192. Grundy, P.: Shakedown of bars in bending and tension. J Eng. Mech. Div. ASCE **95**(EM3), 519–529 (1969)
193. Alwis, W.A.M., Grundy, P.: Shakedown analysis of plates. Int. J. Mech. Sci. **27**(1/2), 71–82 (1985)
194. Shiau, S.H., Yu, H.S.: Finite element method for shakedown analysis of pavements. In: 16th Australasian Conference on the Mechanics of Structures and Materials, pp. 17–22 (1999)
195. Shiau, S.H., Yu, H.S.: Shakedown analysis of flexible pavements. In: The John Booker Memorial Symposium, pp. 643–653 (2000)
196. Yu, H.S.: Shakedown theory for pavement analysis. Int. J. Road Mater. Pavement Des. **6**(1), 7–9 (2005)
197. Li, H.X., Yu, H.S.: A non-linear programming approach to kinematic shakedown analysis of composite materials. Int. J. Num. Meth. Eng. **66**(1), 117–146 (2006)
198. Yu, H.S., Wang, J.: Three-dimensional shakedown solutions for cohesive-frictional materials under moving surface loads. Int. J. Sol. Struct. **49**(26), 3797–3807 (2013)
199. Atkočiūnas, J.: Optimal Shakedown Design for Elastic-Plastic Structures, Vilnius, Technika (2011)
200. Atkočiūnas, J., Venskus, A.: Optimal shakedown design offrames under stability conditions according to standards. Comput. Struct. **89**(3–4), 435–443 (2011)
201. Alawdina, P., Atkočiūnas, J., Liudas, L.: Optimization of the structures at shakedown and Rosen's optimality criterion. Civ. Environ. Eng. Rep. **22**(3) (2014)
202. Cyras, A., Atkočiūnas, J.: Mathematical model for the analysis of elastic-plastic structures under repeated-variable loading. Mech. Res. Comm. **11**, 353–360 (1984)
203. Chen, G., Wang, H., Bezold, A., Broeckmann, C., Weichert, D.: Statistical investigation of grain size's influence on effective strengths of particulate reinforced metal matrix composites (PRMMCs). Comp. Meth. Appl. Mech. Eng. **352**(11), 691–707 (2019)
204. Chen, G., Wang, X., Liuc, F., Zhoua, Z., Zhang, L.: A numerical and experimental study of the shakedown behavior of a reusable space capsule used in the crewed spacecraft, private communication (2022)
205. Gvozdev, A.A.: The determination of the value of the collapse load for statically indeterminate systems undergoing plastic deformation (in Russian). In: Proceedings of the Conference on Plastic Deformations, December 1936, p. 19. Moscow/Leningrad: Akademiia Nauk, SSSR 1938. In: Engl.: Int. J. Mech. Sci. **1**, 322–335 (1960)
206. Kamenjarzh, J.A.: Limit Analysis of Solids and Structures. CRC Press, London (1996)
207. Lia, W., Zenga, F., Chen, G., Deng, Y., Liua, G., Zhanga, X., Bezold, A., Broeckmann, C.: Shakedown analysis for structural design applied to a manned airtight module. Int. J. Press. Vessel. Pip. **162**, 11–18 (2018)

A Unified Shakedown Limit Equation for Pavements and Railways Under Repeated Traffic Loads

Juan Wang and Hai-Sui Yu

Abstract In 1943, Terzaghi proposed a simple, unified equation for determination of the bearing capacity of soils considering static surface loads. From then, this classical equation and its later extensions have been widely used in the design of various foundations against instantaneous failure. With the fast development of transportation industry, researchers have been interested in the evaluation of shakedown limits of pavements and railways under repeated traffic loads, which are much smaller than Terzaghi's bearing capacity. Noting that various shakedown limits for different problems share some common trends and key factors, this paper proposes a simple, unified shakedown limit equation, in a format analogous to Terzaghi's equation. The shakedown limit equation includes three terms, which represent the contributions from cohesion, self-weight of the underlying soil, and self-weight of any superficial rigid layers, respectively. Numerical results indicate that the coefficient in the cohesion term N_c^{sd} depends on the soil friction angle; while the coefficient in the self-weight term N_γ^{sd} is controlled by soil friction angle and a dimensional factor $\gamma a/c$. Values of N_c^{sd} and N_γ^{sd} for a typical rolling point contact problem also explain the different contribution ratios from the soil self-weight to the shakedown limits of pavement and railway problems.

1 Introduction

Pavement and railway investment is considered as a key driver of economic growth in many countries. For example, China has seen a continuous growth in pavement and railway mileages in the past 15 years, reaching 5280 K km for pavements and

---

J. Wang
Ningbo Nottingham New Materials Institute, The University of Nottingham Ningbo China, Ningbo, China
e-mail: juan.wang@nottingham.edu.cn

H.-S. Yu (✉)
The University of Leeds, Leeds, UK
e-mail: DVC@leeds.ac.uk

G. Garcea and D. Weichert (eds.), *Direct Methods for Limit State of Materials and Structures*, Lecture Notes in Applied and Computational Mechanics 101,
https://doi.org/10.1007/978-3-031-29122-7_2

150 K km for railways including 41 K km of high-speed railways, as shown in Fig. 1. Meanwhile, the increase of traffic volume and the shortage of natural resources (e.g. granular fill for embankments) arise more challenges in transportation infrastructure industry. The design of pavements and railways has to be performed in an efficient way, structurally and financially. Existing design methods for pavements and railways are based on elastic theory and/or empirical equations, and therefore potential benefits from plasticity of soils and granular materials were not well recognized or fully utilized [1, 2].

Recently, shakedown theory, based on elastic–plastic theory, has attracted lots of attentions from researchers in fields of pavement and railway engineering. A number of laboratory tests and field data has demonstrated shakedown and non-shakedown phenomena of pavements and railways due to different load levels [3–8]. Meanwhile, different shakedown analysis methods based on lower-bound or upper-bound shakedown theorems have been developed to obtain shakedown limits of different pavement and railway structures, some of which have been well validated [5, 6, 9–33]. Various factors that influence the shakedown solutions have also been investigated in detail. Despite those studies, many works were dedicated to very specific cases,

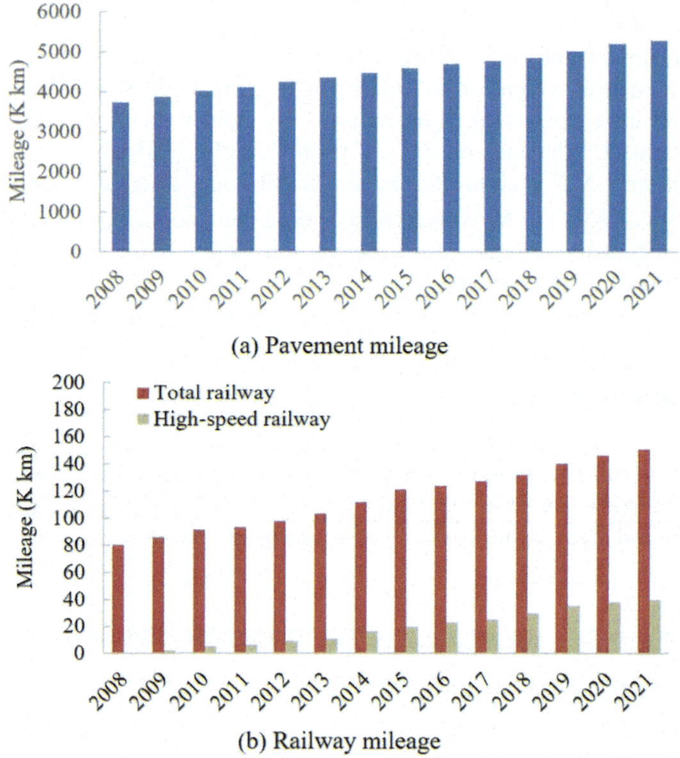

Fig. 1 Pavement and railway mileages in China

leading to distinct shakedown solutions for different problems; and some common trends and key factors shared by the problems were overlooked. Moreover, there have been no general equations that would consider various contributing factors and explain their different contributions to the final shakedown limit.

In this paper, a new, simple, unified equation for shakedown limits of soils under repeated moving surface loads will be proposed. This equation will consider the contributions from several key factors. It can be used to interpret the reason for the different shakedown limits for pavement and railway problems. It can also be employed to guide the improvement of designs.

2 Key Factors for Pavement and Railway Shakedown Solutions

Wang and Yu [25] reviewed the development of shakedown analysis methods for pavements and railway engineering problems and summarized different key factors for shakedown solutions of pavement and railway problems, respectively. Table 1 present the key factors. It has been noticed that the effects of unit-weight, depth-dependent properties, and traffic moving velocity on shakedown limit are significant in railways but not in pavements. Meanwhile, temperature plays an important role only in pavement shakedown.

Table 1 Different key factors for pavement and railway engineering problems

Category	Item	Pavement	Railway
Material properties	Elastic parameters	√	√
	Plastic parameters	√	√
	Unit weight		√
	Depth-dependent properties		√
Load	Contact area/shape	√	√
	Load spacing/pressure profile	√	√
	Frictional coefficient	√	√
	Moving velocity		√
Configuration	Layer thickness	√	√
	Other structural components	√	√
Environment	Temperature	√	
	Water	√	√

3 Terzaghi's Bearing Capacity Equation and Correction Factors

Terzaghi's bearing capacity equation gives the ultimate capacity q_{ult} of a shallow foundation under static loading. It was derived by Terzaghi in 1934 considering general shear failure of soil. As shown in Fig. 2, the foundation in the analysis was a strip footing of width $2a$ (i.e., shallow continuous footing), embedded at a depth D. The soil above the foundation level was replaced by a surcharge q_0, since the shearing resistance of the soil above the foundation level was neglected. The soil beneath the footing was a homogeneous semi-infinite mass with a unit weight γ. Its shear strength was described by the Mohr–Coulomb failure criterion, in which the cohesion and the friction angle of the soil were denoted as c and ϕ, respectively. The load Q was applied in the vertical direction through the center of the footing, and it could be converted into a uniform pressure on the contact area. Based on these assumptions, Terzaghi found the ultimate capacity of the strip foundation per unit length can be written as [34]:

$$q_{ult} = N_c c + N_q q_0 + N_\gamma \gamma a \tag{1}$$

where N_c, N_q, and N_γ are bearing capacity coefficients, giving the resistances due to the material cohesion c, the overburden stress q_0, and the self-weight of the material, respectively. Their values are dependent on friction angle of soil, as exhibited in Fig. 3.

Although this equation was initially developed for strip footings, it can be modified by multiplying each term by a shape factor, to consider the influence of the shape of the contact area for circular, square or rectangular foundations. Similarly, for cases where the load is applied at an inclined angle, inclination factors can be introduced. Other extensions of Terzaghi's bearing capacity equation can be also

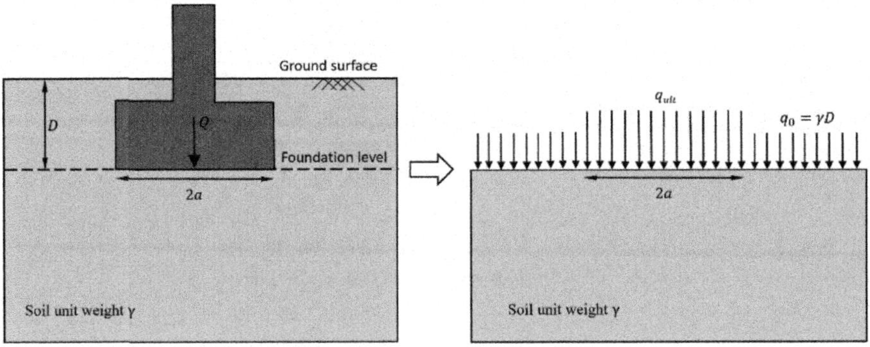

Fig. 2 Idealisation of a shallow foundation problem

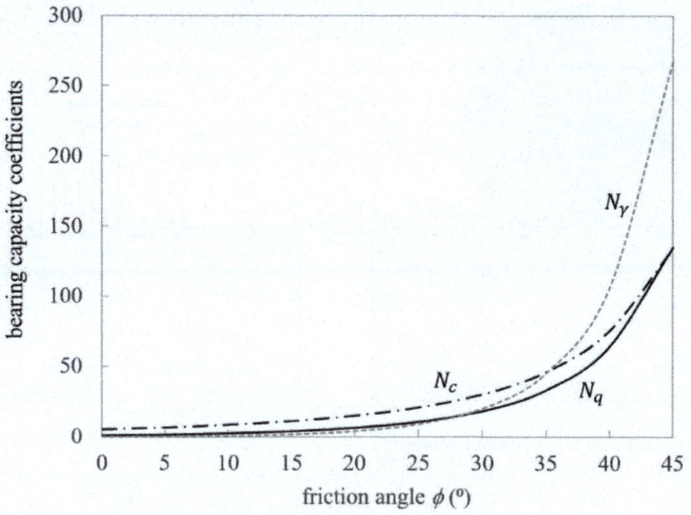

Fig. 3 Bearing capacity coefficients for shallow foundations [35]

found in many literatures. Based on those works, Terzaghi's bearing capacity equation and its extensions have been used in the design of shallow foundations in many countries. Overall, Terzaghi's bearing capacity has successfully brought together the contributions of different key factors in a concise manner. It is basic and powerful, each term of which has a clear physical meaning.

4 A Unified Shakedown Limit Equation

Fundamentally, the shakedown limit of a pavement or a railway is the bearing capacity of the foundation under the action of one or more repeated moving surface loads. For pavement problems, a moving wheel load can be converted into a moving Hertz pressure on pavement surface, as shown in Fig. 4, where q_0 is zero. For slab track problems, all superstructure components (i.e., rails, track slab and concrete base) can be simplified as a Euler–Bernoulli beam; and thus the four axle loads belonging to two adjacent bogies on two carriages can be converted into a bimodal pressure distribution on its substructure surface [22, 31], while the self-weight of the superstructure can lead to a surcharge q_0, as shown in Fig. 5. Those pressure distributions in Figs. 4 and 5, though in different profiles, are comparable to that in Fig. 2. Therefore, a natural choice of the format of the unified equation for shakedown limit q_{ult}^{sd} of the problems would be the one that is analogical to Terzaghi's bearing capacity equation. Assuming yield criterion of the underlying material can be also described by the Mohr–Coulomb model, the unified shakedown limit equation can be expressed as [26]:

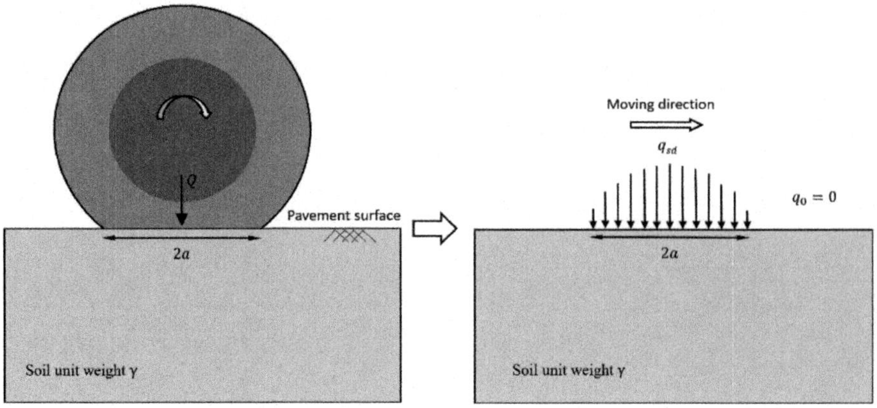

Fig. 4 Idealisation of a pavement problem

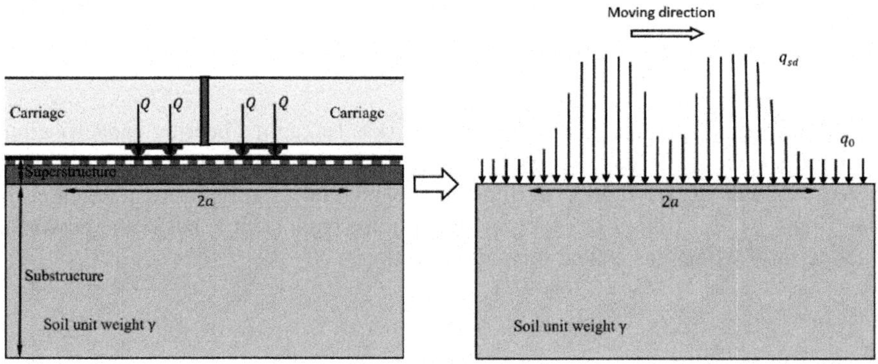

Fig. 5 Idealisation of a slab track problem

$$q_{ult}^{sd} = N_c^{sd} c + N_q^{sd} q_0 + N_\gamma^{sd} \gamma a \tag{2}$$

where N_c^{sd} and N_γ^{sd} stand for the resistances from the cohesion and the self-weight of the underlying material, respectively; N_q^{sd} represents the resistance from overburden stress due to self-weight of any structural components above the cohesive-frictional materials. The overburden term only applies if there exists one or more layers of rigid materials on the top of the cohesive-frictional materials, such as slabs and concrete base on the top of the substructure in a slab track, or a concrete layer above a granular layer in a pavement structure.

Similar to Terzaghi's bearing capacity equation, by including a set of correction factors, this equation can be extended to consider the effects of other factors, such as the shape of the contact area, the distribution of the pressure, the horizontal component of the load, anisotropic property of soil and so on. Consequently, it can be readily used to estimate the shakedown limit (or cyclic capacity) of various pavements and railways under traffic loads.

5 Coefficients for a Rolling Contact Problem

The coefficients in the unified shakedown limit equation can provide information of the contribution of each factor. In this section, one basic rolling point contact problem will be used as an example to illustrate the procedure to determine the coefficients and to reveal the contributions of key factors.

5.1 A Rolling Contact Problem

Figure 6 shows the rolling contact problem. In this problem, the soil is assumed to be homogenous and isotropic with a unit weight γ. This unit weight will lead to at-rest stresses (vertical stress σ_{zz}^{0}, and horizontal stresses σ_{xx}^{0} and σ_{yy}^{0}) in the soil prior to any load applications. Commonly, an at-rest coefficient of lateral earth pressure k is used to correlate the horizontal stresses with the self-weight induced vertical stresses:

$$k = \frac{\sigma_{xx}^{0}}{\sigma_{zz}^{0}} = \frac{\sigma_{yy}^{0}}{\sigma_{zz}^{0}} \tag{3}$$

A three-dimensional (3D) vertical contact loading is moving in the x-direction on the soil surface. The contact loading, limited within a circle of radius a, is in Hertz distribution (Eq. 4) with a maximum pressure $p_0 = 3P/2\pi a^2$ at the center of

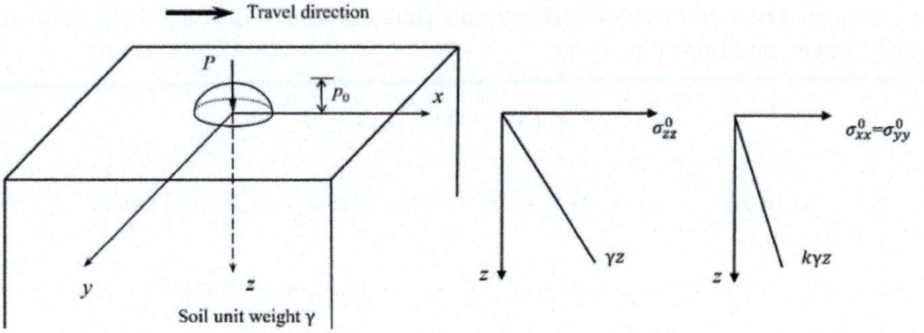

Fig. 6 A rolling contact problem

contact area. It is also assumed that the soil is an elastic-perfectly material following Mohr–Coulomb model.

$$p = \frac{3P}{2\pi a^3}\left(a^2 - x^2 - y^2\right)^{1/2} \tag{4}$$

In this problem, there is no superstructure above the semi-infinite soil mass, so the term due to surcharge in Eq. 2 will be zero. Therefore, only N_c^{sd} and N_γ^{sd} will be calculated and discussed.

5.2 Lower-Bound Shakedown Analysis

Based on Melan's lower-bound shakedown theorem [36], Yu and Wang [16] proposed a shakedown analysis method for a rolling and sliding point contact problem with zero unit weight. This method can be extended to consider the influence of the at-rest stresses. According to the lower-bound shakedown theorem, shakedown occurs if the total stress at every point satisfied yield criterion. For the problem considering soil unit weight, the at-rest stress fields should be included in the total stress, so that Melan's static shakedown theorem can be written as:

$$f(\lambda\sigma_{ij}^e + \sigma_{ij}^r + \sigma_{ij}^0) \le 0 \tag{5}$$

where $f \le 0$ is the yield criterion for the material; σ_{ij}^e is the elastic stress field induced by a unit load; λ is a dimensionless load multiplier; σ_{ij}^r is the time-independent self-equilibrated residual stress field; σ_{ij}^0 is at-rest residual stress field which is also self-equilibrated.

Following the method of Yu and Wang [16] and the numerical studies of Shiau [37] and Liu et al. [38], the symmetry and other considerations impose some constraints on the residual stresses: (1) all residual stresses must be independent of x; (2) σ_{zz}^r and σ_{xz}^r must be zero; (3) residual stresses in combination with at-rest stresses must satisfy yield condition, i.e., $f(\lambda\sigma_{ij}^r + \sigma_{ij}^0) \le 0$. Assuming that the critical planes are x–z planes, Eq. 5 and the Mohr–Coulomb yield criterion will lead to the following shakedown condition (Eqs. 6–8):

$$f = \left(\sigma_{xx}^r + \mathrm{M}\right)^2 + \mathrm{N} \le 0 \tag{6}$$

$$\mathrm{M} = \lambda\sigma_{xx}^e + \sigma_{xx}^0 - \lambda\sigma_{zz}^e - \sigma_{zz}^0 + 2\tan\phi\left(c - \left(\lambda\sigma_{zz}^e + \sigma_{zz}^0\right)\tan\phi\right) \tag{7}$$

$$\mathrm{N} = 4(1 + \tan^2\phi)\left[(\lambda\sigma_{xz}^e)^2 - (c - (\lambda\sigma_{zz}^e + \sigma_{zz}^0)\tan\phi)^2\right] \tag{8}$$

where σ_{xx}^r is a time-independent and self-equilibrated residual stress field in the travel direction; σ_{zz}^0 is at-rest stresses in the z direction, which is γz; σ_{xx}^0 is at-rest stress in the x direction, which is equal to $k\sigma_{zz}^{0}$. Then the horizontal residual stress at any point in the half-space must satisfy Eq. 9.

$$-\mathrm{M}_i - \sqrt{-\mathrm{N}_i} \le \sigma_{xx}^{r} \le -\mathrm{M}_i + \sqrt{-\mathrm{N}_i} \tag{9}$$

Since σ_{xx}^{r} must be independent of x when the applied load is no larger than the shakedown limit, all possible σ_{xx}^{r} must be bracketed by two critical stress fields σ_{xx-l}^{r} (denoted as 'maximum smaller root') and σ_{xx-u}^{r} (denoted as 'minimum larger root'):

$$\sigma_{xx-l}^{r} = \max_{z=j}^{-\infty\le x\le\infty}\left(-\mathrm{M}_i - \sqrt{-\mathrm{N}_i}\right) \tag{10}$$

$$\sigma_{xx-u}^{r} = \min_{z=j}^{-\infty\le x\le\infty}\left(-\mathrm{M}_i + \sqrt{-\mathrm{N}_i}\right) \tag{11}$$

where i represents a point in the half-space; and j represents a depth. When the applied load is the shakedown limit, the critical point of the half-space is located at the depth where the two critical residual stresses just intersect.

By substituting Eq. 10 or Eq. 11 into Eq. 6, the shakedown problem can be converted into the following optimization mathematical formulation:

$$\begin{aligned} &\lambda_{sd} = \max(\lambda) \\ &\text{s.b.}\begin{cases} f\left(\sigma_{xx}^{r}(\lambda\sigma^{e}), \lambda\sigma^{e}\right) \le 0 \text{ for all points} \\ \sigma_{xx}^{r}(\lambda\sigma^{e}) = \sigma_{xx-l}^{r} \text{ or } \sigma_{xx-u}^{r} \end{cases} \end{aligned} \tag{12}$$

In this formulation, since the elastic stress fields σ^{e} and the critical residual stress fields all depend on the load multiplier λ, Eq. 12 can be easily solved by using the procedure suggested in Wang [39] and Yu and Wang [16].

In order to obtain the two coefficients N_c^{sd} and N_γ^{sd}, shakedown limit calculation were conducted using the method considering two different situations: (1) zero unit weight; (2) non-zero unit weight. By applying a zero unit weight in the calculation, the obtained shakedown limit λ_{sd} is $N_c^{sd}c$; and therefore the coefficient N_c^{sd} can be determined. When a unit weight is considered, the calculated shakedown limit includes the resistances from both terms:

$$q_{ult}^{sd} = \lambda_{sd} = N_c^{sd}c + N_\gamma^{sd}\gamma a \tag{13}$$

The contribution of the self-weight then can be determined by deducting the contribution of cohesion from the obtained shakedown limit, as follow:

$$N_\gamma^{sd} = \left(q_{ult}^{sd} - N_c^{sd}c\right)/\gamma a \tag{14}$$

5.3 Coefficients N_c^{sd} and N_γ^{sd}

Figure 7 shows the coefficient N_c^{sd} is increased at an accelerate rate with rising friction angle, similar to the trend of the bearing capacity coefficient N_c. However, the values of N_c^{sd} is much smaller than those of N_c at the same friction angle. For example, when the friction angle is zero, $N_c = 5.14$ and $N_c^{sd} = 4.0$. Also, the increase rate of N_c is much more significant, leading to $N_c = 133.87$ when the friction angle is 45°. At this friction angle, N_c^{sd} is only 25.88. Figure 8 present the numerical results demonstrating the dependence of the coefficient N_γ^{sd} on material friction angle and a dimensionless factor $\gamma a/c$. This is distinct from the results of N_γ, which are only dependent on the friction angle.

A direction comparison of Figs. 7 and 8 reveals that N_γ^{sd} is only one fifth of N_c^{sd} at its maximum. Since typical asphalt mixtures have high values of cohesion, say 200~1000 kPa, the contribution of the self-weight term will be very small compared to the cohesion term. This explains why self-weight is barely considered in the shakedown solutions of asphalt pavements. Railways normally do not have any asphaltic layer, so the contribution of the self-weight term should have similar order of magnitude to the cohesion term. And therefore, the influence of self-weight needs to be considered in railway problems.

It can also be deduced from Fig. 8, when material cohesion c is equal to zero, no resistance can be provided by the self-weight of the material (i.e., $N_\gamma^{sd} = 0$ when $\gamma a/c =$ infinite). Meanwhile, the resistance from cohesion is also zero according to Eq. 13. As a result, the shakedown limit is zero at $c = 0$. This means, theoretically speaking, purely frictional soils will always fail if it is directly under a repeated

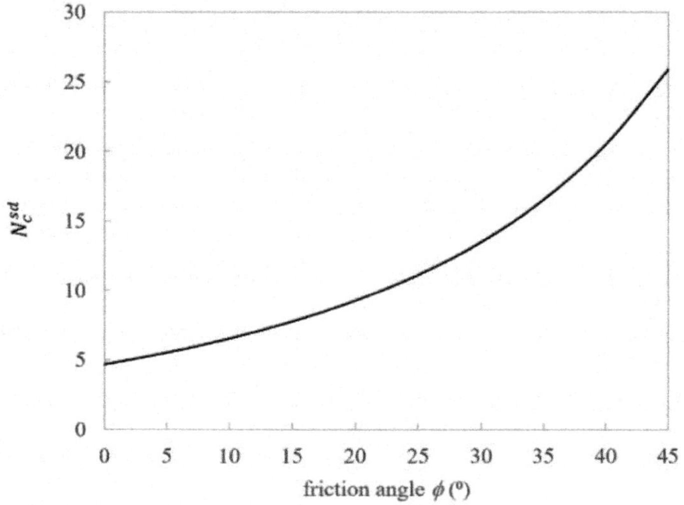

Fig. 7 Variation of N_c^{sd}

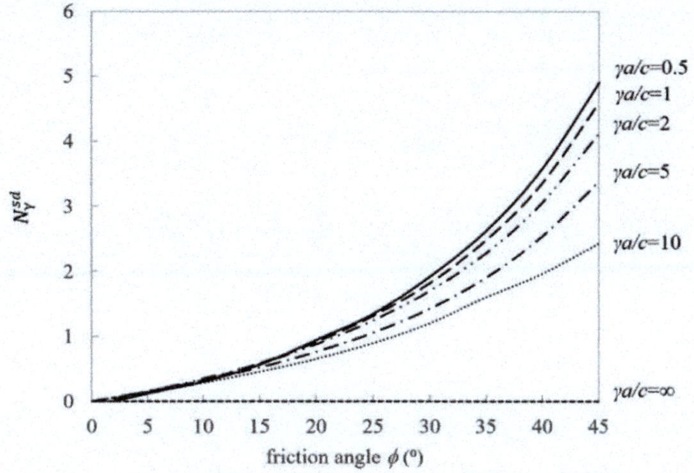

Fig. 8 Variation of N_γ^{sd}

moving load. Therefore, a layer of bound materials on the surface could be beneficial as it contributes to the surcharge term.

For a certain homogenous soil, the size of the contact area affects the self-weight term but not the cohesion term, according to Eq. 14 and Fig. 8. Shakedown limit of an enlarged contact area is competitively affected by increasing a (Eq. 14) and decreasing N_γ^{sd} (Fig. 8). When a is very large, its effect on N_γ^{sd} becomes negligible; and therefore shakedown limit tends to increase proportionally with a if the contact area is very big. This also explains that the shakedown limits of slab track problems are much more significant than those for pavement problems. Apart from that, the self-weights of superstructure components in slab tracks will also contribute to the surcharge term in Eq. 2 and thus a higher shakedown limit.

5.4 *Effect of At-Rest Lateral Earth Pressure Coefficient*

It is worth noting that, theoretical shakedown limits will not be affected by the at-rest lateral earth pressure coefficient k. Nevertheless, the critical residual stress fields (Eqs. 10 and 11) are dependent on k. Figure 9 shows the distributions of critical residual stress fields at an identical shakedown limit. A relatively small k leads to a smaller value (i.e., more compressive) of σ_{xx}^r. When the critical residual stress fields are combined with the at-rest horizontal stress fields $\sigma_{xx}^0 = k\gamma z$, the distributions are identical, as shown in Fig. 10. This explains why the shakedown limits are not changed by k. It indicates that soil always tends to deform in a way that facilitates structural shakedown.

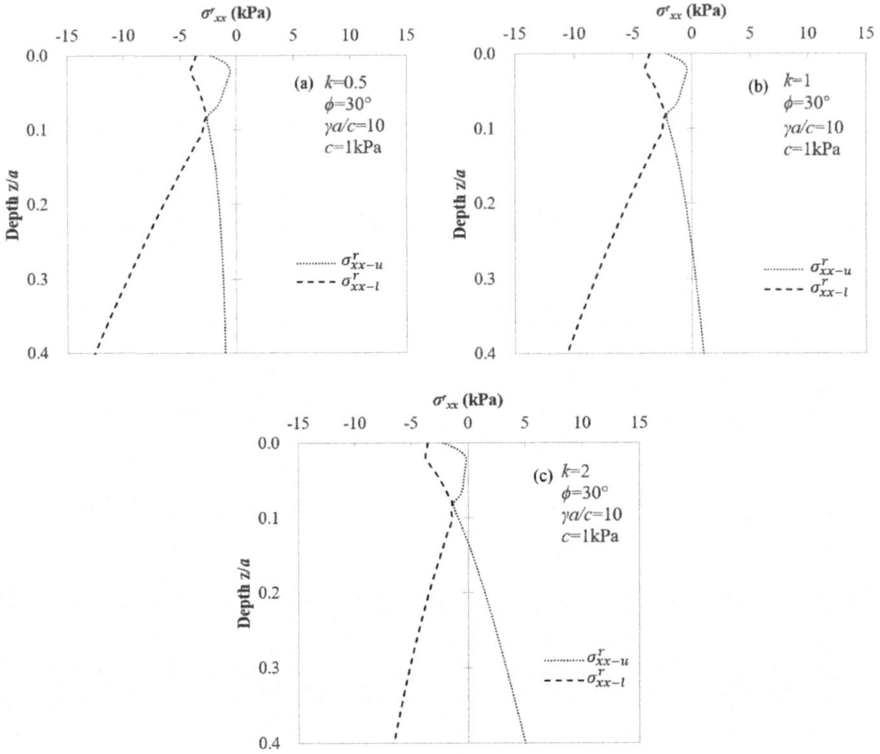

Fig. 9 Distributions of critical residual stress at shakedown limit $q_{ult}^{sd} = 25.4\,\text{kPa}$

6 Conclusion

A unified equation for shakedown limit (or cyclic capacity) of soils under repeated moving surface loads is proposed, which is in a form analogous to Terzaghi's bearing capacity equation. The contributions from cohesion, self-weight of the underlying soil, and self-weight of any superficial rigid layers are combined in the equation. It can be applied in a similar manner to Terzaghi's bearing capacity equation, to provide shakedown capacities for different pavement or railway problems.

By using the low-bound shakedown analysis method of Wang and Yu [16] while considering weight-induced at-rest stresses in soil, numerical static shakedown solutions for a rolling point contact problem are obtained. Based on the numerical results, the coefficients N_γ^{sd} in the self-weight term and N_c^{sd} in the cohesion term are calculated. It is found that both coefficients increase with rising friction angle, and they are much larger than the bearing capacity coefficients N_γ and N_c. Different from Terzaghi's equation, N_γ^{sd} also depends on the dimensionless factor $\gamma a/c$, and it is much smaller than N_c^{sd}. The lateral earth pressure coefficient k does not change the shakedown limit, because the sums of residual stresses and at-rest stresses remain identical.

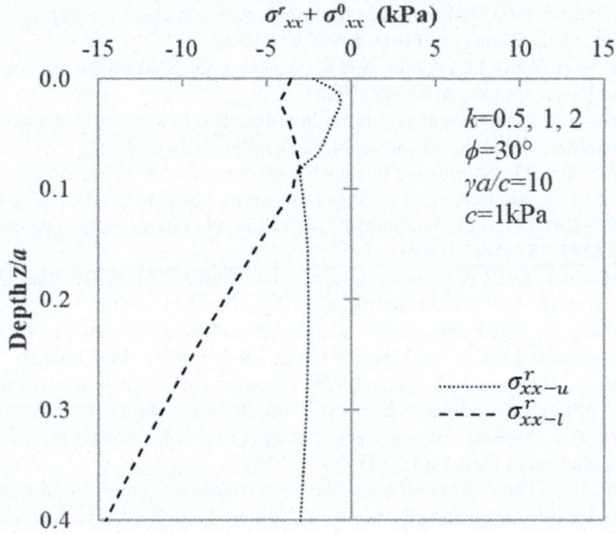

Fig. 10 Distributions of critical residual stress plus at-rest horizontal stress at shakedown limit $q_{ult}^{sd} = 25.4\,\text{kPa}$

Acknowledgements Financial support from Ningbo Natural Science Foundation (2021J184) and Zhejiang Natural Science Foundation (LY22E080018) are acknowledged. The Zhejiang Provincial Department of Science and Technology is also acknowledged for this research under its provincial Key Laboratory Program (2020E10018).

References

1. Brown, S.F.: Soil mechanics in pavement engineering. Géotechnique **46**(3), 383–426 (1996)
2. Yu, H.S.: Plasticity and Geotechnics. Springer, USA, US, New York (2006)
3. Brown, S.F., Yu, H.S., Juspi, S., Wang, J.: Validation experiments for lower-bound shakedown theory applied to layered pavement systems. Géotechnique **62**(10), 923–932 (2012)
4. Juspi, S.: Experimental validation of the shakedown concept for pavement analysis and design. University of Nottingham, UK (2007)
5. Sharp, R.W., Booker, J.R.: Shakedown of pavements under moving surface loads. J. Transp. Eng., ASCE **110**(1), 1–14 (1984)
6. Liu, S., Wang, J., Yu, H.S., Wanatowski, D., et al.: Shakedown of asphalt pavements considering temperature effect. Int. J. Pavement Eng. **23**(5), 1572–1583 (2022)
7. Xiao, J., Wang, B., Liu, C., Yu, Z.: Influences of subgrade form and ground stiffness on dynamic responses of railway subgrade under train loading: field testing case study. Procedia Eng. **143**, 1185–1192 (2016)
8. Bian, X., Duan, X., Li, W., Jiang, J.: Track settlement restoration of ballastless high-speed railway using polyurethane grouting: full-scale model testing. Transp. Geotech. **26**, 100381 (2021)
9. Boulbibane, M., Ponter, A.R.S.: The linear matching method for the shakedown analysis of geotechnical problems. Int. J. Numer. Anal. Meth. Geomech. **30**(2), 157–179 (2006)

10. Collins, I.F., Cliffe, P.F.: Shakedown in frictional materials under moving surface loads. Int. J. Numer. Anal. Meth. Geomech. **11**(4), 409–420 (1987)
11. Boulbibane, M., Collins, I.F., Ponter, A.R.S., Weichert, D.: Shakedown of unbound pavements. Road Mater. Pavement Des. **6**, 81–96 (2005)
12. Li, H.X., Yu, H.S.: A nonlinear programming approach to kinematic shakedown analysis of frictional materials. Int. J. Solids Struct. **43**(21), 6594–6614 (2006)
13. Raad, L., Weichert, D.: Stability of pavement structures under long term repeated loading. In: Mroz, Z. (ed.) Inelastic Behaviour of Structures under Variable Loads, pp. 473–496 (1995)
14. Raad, L., Weichert, D., Najm, W.: Stability of multilayer systems under repeated loads. Transp. Res. Board **1207**, 181–186 (1988)
15. Yu, H.S., Houlsby, G.T.: A large strain analytical solution for cavity contraction in dilatant soils. Int. J. Numer. Anal. Meth. Geomech. **19**(11), 793–811 (1995)
16. Yu, H.S., Wang, J.: Three-dimensional shakedown solutions for cohesive-frictional materials under moving surface loads. Int. J. Solids Struct. **49**(26), 3797–3807 (2012)
17. Krabbenhøft, K., Lyamin, A.V., Sloan, S.W.: Shakedown of a cohesive-frictional half-space subjected to rolling and sliding contact. Int. J. Solids Struct. **44**(11–12), 3998–4008 (2007)
18. Wang, J., Yu, H.S.: Shakedown analysis for design of flexible pavements under moving loads. Road Mater. Pavement Des. **14**(3), 703–722 (2013)
19. Wang, J., Yu, H.S.: Three-dimensional shakedown solutions for anisotropic cohesive-frictional materials under moving surface loads. Int. J. Numer. Anal. Meth. Geomech. **38**(4), 331–348 (2014)
20. Liu, S., Wang, J., Yu, H.-S., Wanatowski, D.: Shakedown solutions for pavements with materials following associated and non-associated plastic flow rules. Comput. Geotech. **78**, 218–226 (2016)
21. Qian, J., Wang, Y., Wang, J., Huang, M.: The influence of traffic moving speed on shakedown limits of flexible pavements. Int. J. Pavement Eng. **20**(2), 233–244 (2017)
22. Wang, J., Liu, S., Yang, W.: Dynamics shakedown analysis of slab track substructures with reference to critical speed. Soil Dyn. Earthq. Eng. **106**, 1–13 (2018)
23. Liu, S., Wang, J.: Application of shakedown theory in track substructure design. Proc. Inst. Civ. Eng.-Ground Improv. **172**(2), 116–123 (2019)
24. Tang, X., Wang, J.: A general shakedown approach for geo-structures under cyclic loading using ABAQUS/Python. Acta Geotech. **17**, 5773–5788 (2022)
25. Wang, J., Yu, H.S.: Shakedown analysis and its application in pavement and railway engineering. Comput. Geotech. **138**, 104281 (2021)
26. Nguyen, A.D., Hachemi, A., Weichert, D.: Application of the interior-point method to shakedown analysis of pavements. Int. J. Numer. Anal. Meth. Geomech. **75**(4), 414–439 (2008)
27. Zhao, J., Sloan, S.W., Lyamin, A.V., Krabbenhøft, K.: Bounds for shakedown of cohesive-frictional materials under moving surface loads. Int. J. Solids Struct. **45**(11–12), 3290–3312 (2008)
28. Costa, P.A., Lopes, P., Cardoso, A.S.: Soil shakedown analysis of slab railway tracks: Numerical approach and parametric study. Transp. Geotech. **16**, 85–96 (2018)
29. Rahmani, R., Binesh, S.M.: Mesh-free shakedown analysis of cohesive-frictional pavement under moving traffic loads: deterministic and probabilistic frameworks. Road Mater. Pavement Des. **21**(4), 1096–1134 (2020)
30. Connolly, D., Yu, H.: A shakedown limit calculation method for geogrid reinforced soils under moving loads. Geotext. Geomembr. **49**(3), 688–696 (2021)
31. Liu, S., Wang, J., Yu, H.-S., Wanatowski, D.: Shakedown for slab track substructures with stiffness variation. Geotech. Res. **5**(1), 31–38 (2018)
32. Wang, J., Yu, H.S.: Residual stresses and shakedown in cohesive-frictional half-space under moving surface loads. Geomech. Geoengin. **8**(1), 1–14 (2013)
33. Ponter, A.R.S., Hearle, A.D., Johnson, K.L.: Application of the kinematical shakedown theorem to rolling and sliding point contacts. J. Mech. Phys. Solids **33**(4), 339–362 (1985)
34. Terzaghi, K.: Theoretical Soil Mechanics. Wiley, New York (1943)
35. Craig, R.F.: Craig's soil mechanics. CRC Press (2004)

36. Melan, E.: Der spannungsgudstand eines Henky-Mises schen Kontinuums bei Verlandicher Belastung. Sitzungberichte der Ak. Wissenschaften Wien (Ser. 2A), **147**, 73 (1938)
37. Shiau, S.H.: Numerical methods for shakedown analysis of pavements under moving surface loads. University of Newcastle: Australia (2001)
38. Liu, S., Wang, J., Yu, H.S., Wanatowski, D.: Numerical shakedown and non-shakedown responses of a Tresca half-space to a three-dimensional moving load. In: Proceedings of 19th International Conference on Soil Mechanics and Geotechnical Engineering, pp. 1385–1388. Seoul, Korea (2017)
39. Wang, J.: Shakedown analysis and design of flexible road pavements under moving surface loads. University of Nottingham, UK (2011)

Elastic–Plastic Optimisation of a Cable–Rib Satellite Antenna

Giuseppe Cocchetti, Ruiwei Liu, Aram Cornaggia, Rosalba Ferrari, and Egidio Rizzi

Abstract Cable–rib satellite antennas are facing a growing challenging quest of technological operation and connected structural performance, specifically in the context of large–scale, deployable aerospace applications. In the framework of Limit Analysis, as a specific tool of structural modelling, within the Theory of Plasticity, accounting for potential material non–linearity, up to structural collapse, as an assumed general paradigm of structural resilience, the present contribution analyses a specific cable–rib satellite antenna through a novel evolutive algorithm, in order to assess possible activations of plastic joints, under perfect–plasticity conditions, and to consistently estimate static and kinematic features at incipient collapse. Modelling both beam elements (with a tubular cross–section) and cable elements, the underlying elastic–plastic analysis is then coupled to an optimisation process, toward the minimisation of displacements at incipient collapse, with respect to initial cable shortenings or, in an equivalent manner, cable pretensions. Further investigations are also developed, within the optimisation tool, re–joining dynamic modal properties with elastic–plastic outcomes, in order to minimise the total mass of the structure, at varying cross–section properties for each structural element, at given material fea-

G. Cocchetti · R. Liu
Department of Civil and Environmental Engineering, Politecnico di Milano, Milano 20133, Italy
e-mail: giuseppe.cocchetti@polimi.it

R. Liu
Department of Naval Architecture and Marine Engineering, Guangzhou Maritime University, Guangzhou 510700, China
e-mail: liuruiwei8124@163.com

A. Cornaggia · R. Ferrari · E. Rizzi (✉)
Department of Engineering and Applied Sciences, Università degli studi di Bergamo, Dalmine 24044, Italy
e-mail: egidio.rizzi@unibg.it

A. Cornaggia
e-mail: aram.cornaggia@unibg.it

R. Ferrari
e-mail: rosalba.ferrari@unibg.it

G. Garcea and D. Weichert (eds.), *Direct Methods for Limit State of Materials and Structures*, Lecture Notes in Applied and Computational Mechanics 101,
https://doi.org/10.1007/978-3-031-29122-7_3

tures, under plastic admissibility constraints. The adopted evolutive elastic–plastic algorithm, combined with optimisation tools, is proven to constitute an efficient structural modelling paradigm and computational strategy, both from numerical and design standpoints, allowing for general effective analyses of deployable satellite antennas, even beyond the scopes of the investigation on the characteristic structural sample here targeted.

Keywords Limit Analysis · Evolutive perfectly–plastic structural analysis · 3D truss–frame structures · Non–linear elastic–plastic FEM model · Structural optimisation · Satellite antenna · Deployable structure

1 Introduction

Space antennas nowadays constitute important physical platforms for specific given missions of satellites, such as Earth observation, deep space exploration and telecommunications (see, e.g., [6, 25, 35]). Typical needed characteristics, such as large diameter, reduced weight and high technological precision have made deployable antennas significantly attractive in the application of satellite antennas in the last decades [5]. Representative examples of space deployable antennas, currently operating in orbit, include the AstroMesh antenna of TRW Astro Aerospace [43], the ETS–VIII antenna of the Japan Aerospace Exploration Agency [32] and the TerreStar and SkyTerra antennas of Harris [38]. The deployable structure, as a fundamental component of deployable space antennas, has always constituted a key subject of research investigation, both from design and technological engineering standpoints.

With the rapid development in aerospace applications, space antennas with a large diameter are demanded to meet the increasing requirements for high–gain, high–resolution and long–distance space communications. Several concepts of large–scale deployable structures have recently been proposed. In [19, 37], a novel large–ring deployable structure, based on six–bar linkages, has been developed, providing a detailed design and a consistent kinematic analysis of the ring deployable structure, validated by a 3.9 m prototype, processed for test deployment repeatability. A new lightweight deployable structure for communication satellites, based on the ETS–VIII antenna, has been proposed in [33, 34], following the construction of a 30 m diameter structure, using tri–fold deployable truss modules, and providing a deployment analysis, in order to verify the practical feasibility of the innovative structure. A new double–pantograph concept, based on a ring design, has been proposed in [9], allowing for buckling and modal analysis checks, on the basis of parametric finite element models. Such double–pantograph structural design has been proven suitable to be applied for deployable structures with large antennas, covering a diameter range from 4 to 50 m. In order to further extend the applicability field, namely to construct novel deployable antenna structures with a diameter larger than 100 m, in [39], a structural topology optimisation has been proposed, based on graph theory, gathering a series of novel double–ring deployable structures, appropriate for use as structural modules of large–scale antennas. Other deployable structures, such as those of Bennett [17, 26], Myard [23, 36], Bricard linkages [4], Tensegrity [44] and Origami [2] have also been investigated, to assemble large–scale space antennas. However, a space mesh antenna consists of a self–balancing prestressed structure that frequently experiences large displacements, particularly for large–scale antennas. Given that a large number of tensioned cables and membranes are employed to achieve lightweight, the increase in stress level may exacerbate the deformation of deployable antennas and may even lead to potential collapse. Therefore, further than technological and design issues, determining how to maintain structural stability and reliability, under diverse loading configurations, appears to assume an essential role.

Some literature studies, not limited to satellite antennas, have considered the effects of deformations in deployable structures. In [41], a modelling approach has been proposed by combining cable and beam elements, for the analysis of the static behaviour of complex structures, providing a deformation analysis of the support structures, under the tension of cable networks, through a finite element model. Aiming at achieving a high–accuracy cable net, a mesh shape design method has been proposed in [24], by iteratively calculating the deformation of a deployable structure, toward the shape finding of the cable net; the comparison of the discussed results has shown that the accuracy of cable nets can substantially be improved by considering the consistent deformability of deployable structures. Moreover, the mechanical properties of deployable cable–rib structures have been studied in [1, 3], by investigating displacement and axial force fields under different loading levels and globally examining the strength of cable–rib structures. However, the main focus of the existing research on the structural analysis of satellite antennas has been centred on linear elastic analyses, with only a few studies regarding the discovery of potential (material) non–linearities in deployable structures. Therefore, a mechanical analysis of large–scale antennas, particularly with respect to accounting for an elastic–plastic behaviour and assessing the corresponding collapse configurations, appears to be of a wide interest for the present conceptual and practical development of such peculiar structures.

In the broad field of non–linear structural analysis and Limit Analysis (LA) approaches (see, e.g., [7, 8, 27, 28, 30]), in order to re–promote Limit Analysis as a modern and functional computational tool serving increasingly ambitious purposes in the current structural engineering panorama, step-by-step elastic–plastic methods have successfully been applied to the analysis of the mechanical characteristics of large–scale structures. Resting on a unifying basis provided by the Linear Complementarity Problem (LCP), an evolutive algorithm for the LA of structures able to reconstruct, during the loading path, the whole sequence of activation of localised plastic joints and to trace the corresponding piece–wise linear load–displacement curve of the structure has first been proposed [10]. It proved rather successful, in performing the elastic–plastic analysis of large 3D truss–frame structures with a considerable complexity, involving up to more than thirteen thousand degrees of freedom [11, 13, 14].

Next, a "direct" non–linear programming algorithm for the LA of 3D truss–frames has been developed, as derived from an efficient approach originally presented by

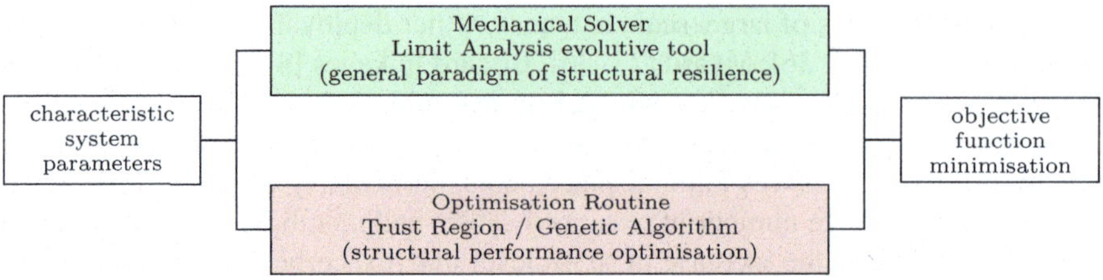

Fig. 1 Nested loop of structural optimisation, relying on effective Limit Analysis

Zhang et al. [45] for the LA of continua, then thereby tailored and enhanced for the analysis of discrete systems. The procedure deserves a particular attention for its both straightforward implementation and elevated speed of convergence, which allow for the algorithm to be proposed as a handy tool in dealing with 3D truss–frame structures endowed with several degrees of freedom [12, 13]. The evolutive LA algorithm was initially developed by considering a piece–wise linear uncoupled behaviour adopted for the internal static variables (i.e. a Rankine–type boxed–form yield domain was assumed in the space of the static variables). Then, recent improvements are extending the procedure to make it employable in general and real specific cases, by considering the yield domain of the plastic joint to be modelled through a smooth ellipsoid [15].

Such methodological formulations and implementations display efficient and robust features, which allow for expedited utilisation also in the mechanical analysis of satellite antennas. To address the requirements for large–scale lightweight deployable space antennas, several developments, such as deployable structure synthesis [17, 39], dynamic analysis [16] and optimisation design [20–22], have been investigated in previous studies. To further analyse the mechanical behaviour of space antennas and optimise their structural design, specifically with respect to elastic–plastic and collapse behaviours, the present investigation employs a devoted implementation of the above–mentioned evolutive algorithm [10] and achieves related results on the specific case study of an instance of a prototype large deployable antenna. Consistently, the developed methodology, based on elastic–plastic and Limit Analysis approaches, highlights its advantageous design features, in combination with optimisation loops (Fig. 1), being suitable to be extended and applied to other structural schemes of large–scale space antennas, as it will be illustrated by results and comments delivered in the manuscript.

The present contribution indeed states a general paradigm, of structural resilience, based on Limit Analysis. Beyond classical static elastic analysis, devoted to identify basic characteristic features, such as stiffness, in the usual range of daily operation under standard loading, a mechanical criterion may be stated, to express the margins of safety toward pushing the load bearing capacity, and monitoring the corresponding increasing deformation, up to potential collapse. LA shall become useful for that, within the assumption of local and global ductile behaviour, once it may be declined in practical forms, translating theoretical formulations into effective computational tools, specifically in the realm of large–scale structures. This may be achieved either by "evolutive methods", able to trace the whole non–linear force–

displacement response of the structure or even by more immediate "direct methods", focussing just on the determination of the static and kinematic features at incipient plastic collapse.

Once this shall properly be achieved, as a general "mechanical solver", within the assessment of assumed paradigm of structural resilience, for a given structure, it could be coupled to appropriate "optimisation tools", in looped routines, in view of seeking a structural optimisation of the resilience quest, at variable underlying parameters and features of the structure (Fig. 1). Of course, a key feature shall be that, computationally, the two steps of the loop be truly effective, specifically that of the mechanical solver (even meant for descriptions with several degrees of freedom, in particular for 3D large–scale structures), especially once representing the outcomes of a non–linear analysis. If this is truly achieved, LA may be brought up to assume a general role of structural resilience assessment, far beyond that of classical elastic analysis, and to lead down pure academic research endeavours, to real and practical design scenarios, even in the current practices of the engineering profession.

This paper aims to state that, and showing the feasibility and effectiveness of such a general paradigm of structural resilience, in the considered peculiar context of the analysis of aerospace antennas, whereby a specific sample prototype is taken under target of investigation. Thus, assuming the concept of structural resilience, via LA, optimisation is right–away concatenated, in two ways, by looking at plain elastic–plastic response up to plastic collapse and also at concomitant modal dynamic properties, to interpret achievable stiffness and basic vibration features, and desired light–mass.

The outline of the paper is as follows. Section 2 briefly describes the specific structural model and relevant features adopted for the selected case study of a deployable satellite antenna; moreover, specifically in Sect. 2.2, the formulated numerical procedure for the elastic–plastic analysis of the structure, with specific adaptations and evolutions to the present particular structural contest, is illustrated. Section 3 presents the main results out of various elastic–plastic analyses on the investigated antenna, as employed as well within optimisation algorithms, either for static elastic–plastic loading configurations and for modal dynamic analyses, respectively in Sects. 3.1 and 3.2. Section 4 gathers the salient collected results, showing their further general validity toward the potential adoption of the proposed analysis and design concept and computational strategy, in the case of other structural schemes of cable–rib space antennas.

2 Structural Modelling and Analysis of a Space Antenna

A typical satellite system displays a configuration built by the satellite body, two solar array paddles and a deployable antenna. In the present study, only the latter is considered for the structural analysis, proposing a structural configuration of a 18 m cable–rib tension deployable antenna, consistently with a previous similar design discussed in [22]. The space antenna consists of three main elements: a deployable structure, a cable network and a metal mesh surface. The latter, together with technological detailing such as deployable joints, is not considered as a structural component, in the model described in the following subsection, due to its expected lack of salient influence in the global structural behaviour of the whole antenna.

The selected cable network is composed by a series of tied cables, aiming at defining and maintaining a specific shape for the structure in the deployed configuration and for the metal mesh for its functionality. The deployable structure is configured as a setup of tubular beams (ribs) creating the supporting structure, suitably stiffened by tied cables. The combination of a supporting structure and a cable system can advantageously provide an improved structural stiffness, together with a reduced total weight (mass), also ensuring, through the pretension of cables, an optimal shape accuracy of the antenna surface. Referring, for conceptual design and geometry, to a previously designed configuration (see [20–22]), a system with six evenly distributed radial ribs is herein considered.

In the following paragraphs, a specific description of the model of the selected space antenna adopted for the present case study and a general illustration of the specific devised approach for the elastic–plastic analyses are provided.

2.1 Finite Element Modelling of the Antenna

The structural model of the antenna, composed of the supporting rib structure and the cable system, is implemented by a self–assembled FEM code within a Matlab environment, as depicted in Fig. 2. The elastic background is formulated with the adoption of classical Euler–Bernoulli beam finite elements, with tubular cross–sections, for the supporting structure, and of cable–truss elements for the cable system network, not allowing for compression in such elements, while providing possible pretensions by initial shortenings. The global size of the model is represented by 50 nodes, 79 elements and 300 degrees of freedom, although the self-made modelling tool could also run for more refined analyses with an increased size of the numerical problem (see, e.g., [13, 14]).

Consistently, the non–linear mechanical behaviour of the structure, i.e. the concentrated plasticity spreading, is localised at potential plastic joints placed at the finite element edges, as described in details in following Sect. 2.2. It is worth observing that the desired level of refinement, namely the number and location of adopted plastic joints, to be possibly activated toward developing localised plastic deformation, can be selected at the implementation stage of the model, therefore without any real limitation, on the achievable accuracy of tracing the true local and global elastic–plastic behaviour description and response results, within the elastic–plastic range, up to structural plastic collapse.

The mechanical constitutive properties of the structure of the antenna are selected for an aluminium alloy material, consistently with some previous design choices available in the literature (see, e.g., [20–22, 42]); therefore, by assuming Young's modulus E equal to 75000 N/mm^2, Poisson's ratio ν equal to 0.3, mass density ρ equal

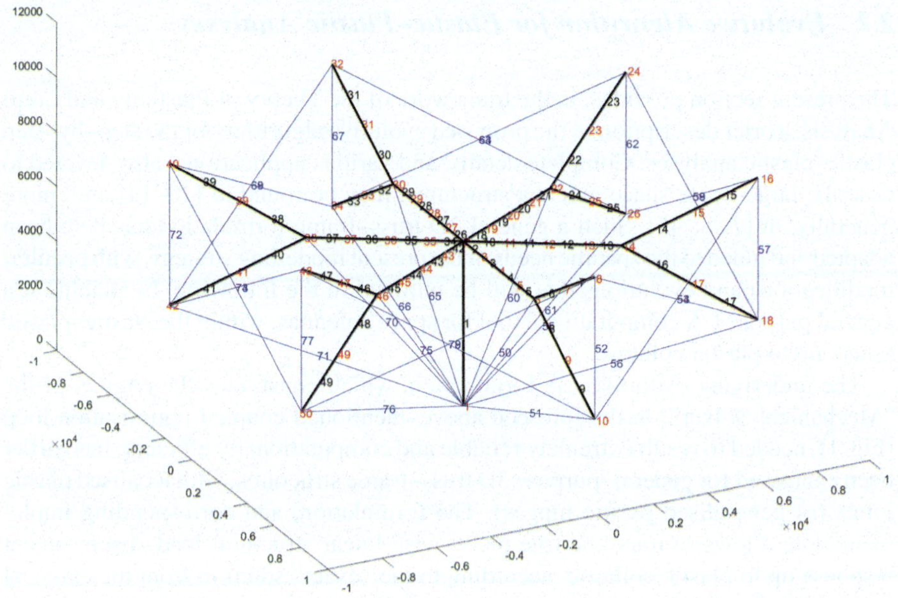

Fig. 2 Structural modelling of the space antenna (black lines—beam elements, blue lines—cable elements, red labels—node number); coordinates in mm; self–implementation within Matlab

to 2700 kg/m^3 and yield limit σ_y equal to 250 N/mm^2, with no hardening behaviour (perfect plasticity hypothesis). Although sometimes considered in the literature, an alternative design configuration based on carbon fibre and kevlar materials (see, e.g., [20–22, 40]) is not here implemented, due to the expected reduced material ductility, regarding the plastic behaviour, as a peculiar key feature of the proposed underlying design concept herein considered.

The model is completed by the definition of the boundary conditions, providing a clamping constraint at the base of the central vertical tubular beam of the antenna, namely at Node 1, in the scheme depicted in Fig. 2. The relevant loading configuration, in the absence of self–weight, as for space applications, is obtained by a system of distributed beam loads produced by the upper mesh tensioning, combined with cable shortenings. Such loading system is consistently assumed with reference to previous works, as in [22]. Namely, the distributed loading is applied as vertical upward uniformly distributed line loads to the upper branch beams of the satellite antenna (specifically, beams 2–7, 10–15, 18–23, 26–31, 34–39 and 42–47, with reference to Fig. 2), equivalently transferred to concentrated nodal actions.

2.2 Evolutive Algorithm for Elastic–Plastic Analyses

The present section provides, in the framework of the Theory of Plasticity and Limit Analysis, a brief description of the proposed evolutive algorithm for the step–by–step elastic–plastic analyses. Complete details, and various applications, also devoted to describe large–scale space antenna structures, may be found in [10–14] and, more generally, in [7, 8, 45]. Such a general 3D truss–frame formulation has here been adapted and tuned to the specific needs of the present modelling strategy, with peculiar modifications and evolutions that will be outlined in the following, by making it a general purpose LA computational tool for space antennas, within the elastic–plastic range, up to plastic collapse.

The underlying evolutive LA formulation, which constitutes the engine, of the "Mechanical Solver", in the foreseen above–mentioned coupled optimisation loop (Fig. 1), needed to result extremely reliable and computationally efficient, has earlier been expressed for general–purpose 3D truss–frame structures, with localised plastic joints (as generalised plastic hinges). The formulation, and corresponding implementation, allows to trace the true piece–wise linear structural load–displacement response up to plastic collapse, according to the "exact" solution from the classical theory of LA [10, 11, 14]. Such an evolutive (step–by–step) computational procedure, in the spirit of LA, has earlier been assessed and compared to a competent "direct method", on the side of the kinematic theorem of LA, meant to achieve, directly, just the characteristics at incipient collapse [12, 13]. It has been shown that collapse features truly match and that fastness and convergence, of the kinematic algorithm, is truly dramatic, in precipitating from above on the collapse mechanism, and associated load multiplier, showing that it may constitute an ideal choice, for the mechanical solver, in the afore–mentioned optimisation loop.

Despite, and even since more challenging, in the present effectiveness assessment endeavour, of the above–stated structural resilience paradigm, the evolutive information is more costive, but also far more richer, in as well providing the amount of deformation achievable before collapse, which is meant to constitute a characteristic quest, in the context of space antennas resilience. Thus, the present calculations have been developed just with the adoption of the evolutive algorithm, as per the underlying mechanical solver, within the assembled optimisation loop (Fig. 1). The main features of the underlying evolutive procedure are resumed below.

The constituting FEM approach is based on classical Euler–Bernoulli beam finite elements (linear straight elements, constant cross–sections, isotropic homogeneous material properties, neglected shear strain effects, cubic shape functions for transverse displacements and linear functions for axial displacements), while possible plastic deformations are conceived to arise only at pre–selected nodes, modelled as plastic joints (as generalised plastic hinges), located at the finite element edges (see, e.g., [7, 18, 29, 31]).

The LA algorithm is based on an evolutive elastic–plastic formulation developed by ruling a step–by–step analysis. The procedure exploits very simple relationships that exist between static and kinematic variables of a beam element (Fig. 3), also

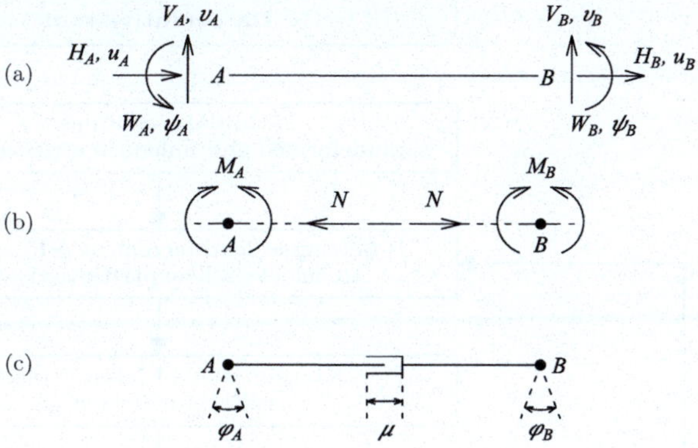

Fig. 3 Schematic description of a (2D) beam finite element with plastic joints: **a**) static (external) variables $\boldsymbol{H}$ and kinematic (external) variables $\boldsymbol{u}$; **b**) static internal variables $\boldsymbol{N}$; **c**) kinematic plastic internal variables η at localised plastic joints. Redrawn from [10]

considering the onset of possible localised plasticisations, here resumed in following Eqs. (1–5).

$$\boldsymbol{H} = \boldsymbol{k}\boldsymbol{u} + \boldsymbol{D}\eta \tag{1}$$

$$P_{ext} = \boldsymbol{H}^{\mathrm{T}}\dot{\boldsymbol{u}} = \boldsymbol{N}^{\mathrm{T}}\dot{\eta} = P_{int} \quad \forall\, \boldsymbol{H} \text{ self-equilibrated} \tag{2}$$

$$\boldsymbol{N} = \boldsymbol{Q}\boldsymbol{H} \tag{3}$$

$$\boldsymbol{D} = -\boldsymbol{k}\,\boldsymbol{Q}^{\mathrm{T}} \tag{4}$$

$$\Delta\boldsymbol{H} = \left[\boldsymbol{k} - \frac{\boldsymbol{k}\,\boldsymbol{Q}^{\mathrm{T}}\boldsymbol{n}\boldsymbol{n}^{\mathrm{T}}\boldsymbol{Q}\boldsymbol{k}}{\boldsymbol{n}^{\mathrm{T}}\boldsymbol{Q}\boldsymbol{k}\,\boldsymbol{Q}^{\mathrm{T}}\boldsymbol{n}}\right]\Delta\boldsymbol{u} = \boldsymbol{k}_{ep}\Delta\boldsymbol{u} \tag{5}$$

With reference to a beam element, in Eq. (1), $\boldsymbol{H}$ represents the self–equilibrated nodal forces associated to nodal displacements $\boldsymbol{u}$, $\boldsymbol{k}$ is the elastic stiffness matrix, and $\boldsymbol{D}$ represents the plastic stiffness matrix, obtainable as in Eq. (4), through straightforward mathematical steps related to the relationships described in above Eqs. (2–3) (see [10] for the details). In Eq. (2), nodal displacements $\boldsymbol{u}$ (associated to $\boldsymbol{H}$) and static internal variables $\boldsymbol{N}$ (associated to plastic internal variables η) are related to each other through the Virtual Power Principle; in Eq. (3), $\boldsymbol{N}$ is related to $\boldsymbol{H}$ by equilibrium through a Boolean matrix, represented by $\boldsymbol{Q}$. Finally, Eq. (5) gives an incremental form of governing Eq. (1), in particular through a direct force/displacement (linear) incremental relationship governed by a symmetric (associativity is assumed, where $\boldsymbol{n}$ gathers the gradients of the yield functions) elastic–plastic stiffness matrix $\boldsymbol{k}_{ep}$, related to a generic finite element. Given the elastic–plastic stiffness matrix of a specific beam element, the global stiffness matrix of a structure is then obtained by an appropriate assembly procedure, in the classical spirit of a finite element analysis.

The algorithm employed in this contribution develops an iterative procedure in which global elastic–plastic stiffness matrix $\boldsymbol{K}_{ep}$ is iteratively updated, based on the plastic modes that became active during the increment of the applied loads ($\boldsymbol{F}$).

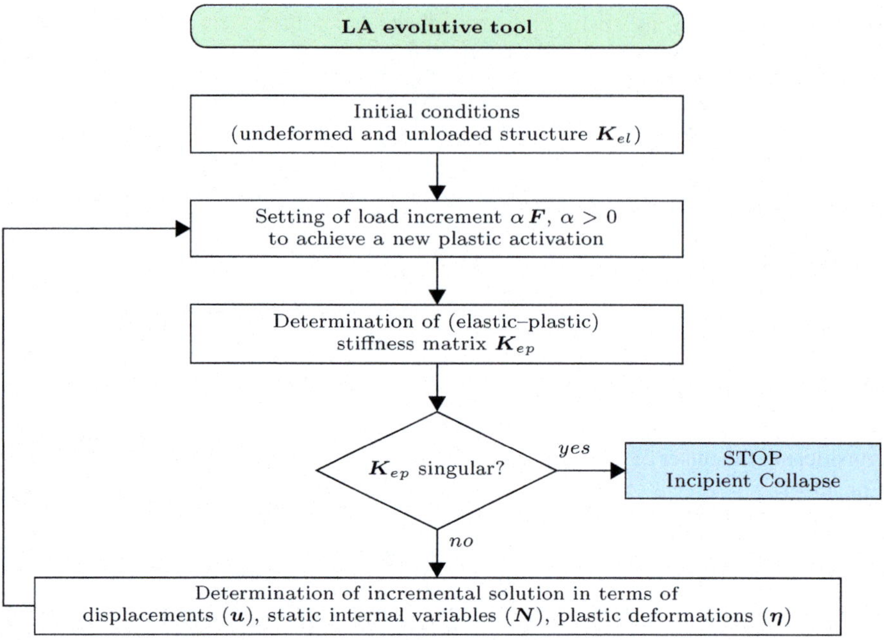

Fig. 4 Flow chart of the evolutive LA algorithm. Adapted from [10]

The macroscopic steps of the algorithm are synoptically depicted in the flow chart represented in Fig. 4. Basically, at the first step of the analysis the stiffness matrix of the structure of interest is elastic ($\boldsymbol{K}_{el}$). Then, the second step is calibrated, in order to achieve the first plastic activation within the structure; the (global) stiffness matrix is then updated, becoming elastic–plastic ($\boldsymbol{K}_{ep}$). Each of the subsequent steps is still calibrated, in order to achieve a new plastic activation within the structure, until collapse is reached, namely, from the computational point of view, when the minimum eigenvalue of global (updated) matrix $\boldsymbol{K}_{ep}$ vanishes.

As a characteristic feature, the procedure does not involve plastic multipliers as primarily variables, which generally imply the presence of a densely–populated governing matrix. Instead, a narrow–band $\boldsymbol{K}_{ep}$ governs the problem, thus improving the computational efficiency of the procedure and making it truly employable for the analysis of large–scale structures (see, e.g., [10–12]).

Moreover, toward the present context, of the Limit Analysis (and optimisation) of space antennas, the above–mentioned general formulation has been tailored, to specific needs, to formulate a novel dedicated implementation for the computational evolutive analysis of cable–rib space antennas, subsequently coupled within optimisation nested loops (Figs. 1 and 5). The following characteristic strategies and features have then been implemented.

- Cables are described by special truss elements. Such an element is kept when in the current step the axial force is of a pulling type. Otherwise, the element is eliminated (i.e. its stiffness matrix is not assembled within the global one). This procedure represents a complementarity condition, enforced in finite terms, between the axial force (N) and the fictitious "permanent" shortening (ΔL_{fict}), used to keep the axial force to zero in a "potential" compressive condition, "$-\Delta L_0$" being the initial given pre–tensioning of the cable:

$$N = \frac{EA}{L_0}(\Delta L_{tot} - \Delta L_0 - \Delta L_{fict}),\ N \geq 0,\ \Delta L_{fict} \leq 0,\ N\Delta L_{fict} = 0 \quad (6)$$

If $N > 0$, it results: $\Delta L_{tot} = \frac{N}{EA} L_0 + \Delta L_0$.
If $N = 0$, it results: $\Delta L_{fict} = \Delta L_{tot} - \Delta L_0 \leq 0$.
In this way, the truss element can transmit only a tensile axial force, as expected from a cable element.

- According to the real sequence, in which the antenna is unfolded from the initially straight configuration, the cable shortenings (producing tensile pre–stresses in the cables) or elongations (producing no pre–stresses in the cables and $\Delta L_{fict} < 0$) are applied as a first step. It is worth noting that each initial shortening ($\Delta L_0 < 0$) (or elongation, $\Delta L_0 > 0$) represents a reduced (or augmented) initial length of the cable (then computed by the optimisation procedure, discussed in the sequel, in order to minimise the maximum displacements due to permanent distributed loads). Then, from the computational viewpoint, the application of subsequent increasing distributed loads follows.
- The structural collapse mechanism is not formally activated, because of the large number of tensioned (elastic) cables, preventing rigid–body motions. However, when a first set of plastic joints is simultaneously activated (more than a single joint, as a consequence of the symmetry of the structure and of the applied loads), the correct functionality of the antenna is assumed to be lost, in the control of the algorithm, and the analysis is stopped, although in principle it may further continue, within the elastic–plastic range.
- The algorithm, previously developed by the Authors for other structural analyses in the context of 3D truss–frames (see, e.g., [10, 11, 13]), performs a check on the positive power dissipation at each plastic joint. Instead, according to the previous remark, in the current dedicated version of implementation, employed for the analyses of the cable–rib antenna, the check on the positive dissipation in the plastic joints can be avoided. This computationally speeds up the whole processing, within the mechanical solver of the stated optimisation loop.

As that assembled, for the specific use of the mechanical analysis of cable–rib space antennas, the conceived LA evolutive tool can now be coupled, as a proper "Mechanical Solver", to appropriate optimisation routines, in view of forming a nested loop toward structural optimisation, as earlier stated in the Introduction (Fig. 1).

3 Structural Optimisations of the Space Antenna

The elastic–plastic modelling concept of the space antenna structure, through the methodological and computational procedure outlined in previous Sect. 2, is here adopted toward formulating an innovative design tool to achieve an optimised solution of the antenna configuration, in terms of mechanical response. Although several optimisation criteria may be considered, for different goals within the design procedure, in the current section, two main strategies are adopted, further fulfilling the aims of the present research, with a specific concern placed on the structural characteristics and on the achievable non–linear and collapse behaviour of the conceived antenna. Therefore, an optimisation approach based on an elastic–plastic analysis and an optimisation strategy centred on combined dynamic modal analysis and elastic–plastic analysis are herein discussed, respectively in following Sects. 3.1 and 3.2.

The present optimisation steps refer to the general paradigm of structural resilience optimisation widely stated in the Introduction. Here, then, the "Mechanical Solver" is coupled to an optimisation routine, in view of deciphering global optimum solutions, leading to the best choice of the underlying structural parameters (see Fig. 1). In this step, either a classical Trust Region Algorithm (Sect. 3.1) or a Genetic Algorithm (Sect. 3.2) is employed, within the context of plain resilience assessment (Sect. 3.1) or concomitant also to modal dynamic evaluation toward light–mass and appropriate stiffness achievement (Sect. 3.2). The optimisation process (see charts sketched in Fig. 5) is shown to deliver consistent results, in terms of force–displacement response and final resilience, of the sampled space antenna, and to outline a set of liable optimum parameters that shall be useful for first–range tuning or end–use employment, in terms of practical applications and design.

3.1 Elastic–Plastic Optimisation Strategy

A first approach is proposed in the design of the space antenna as a constrained optimisation process to minimise the amount of structural displacements at incipient collapse, as a function of cable initial shortenings, namely equivalent to cable pretensions. Such a strategy involves the elastic–plastic analysis of the structure with a self–implemented code within a Matlab environment, according to the LA evolutive algorithm earlier presented in Sect. 2.2, while the minimisation problem is solved by a classical Trust Region Algorithm approach, still within an autonomous Matlab implementation (Fig. 5a).

The results of the proposed optimisation criterion, minimising the objective function expressed in the following equation, are reported in Table 1 (and later in Fig. 8, where it is also possible to observe the obtained type of collapse mechanism):

$$\omega(\Delta L_1, \Delta L_2, \Delta L_3, \Delta L_4) = \max\{|\boldsymbol{U}(\Delta L_1, \Delta L_2, \Delta L_3, \Delta L_4)|\} \tag{7}$$

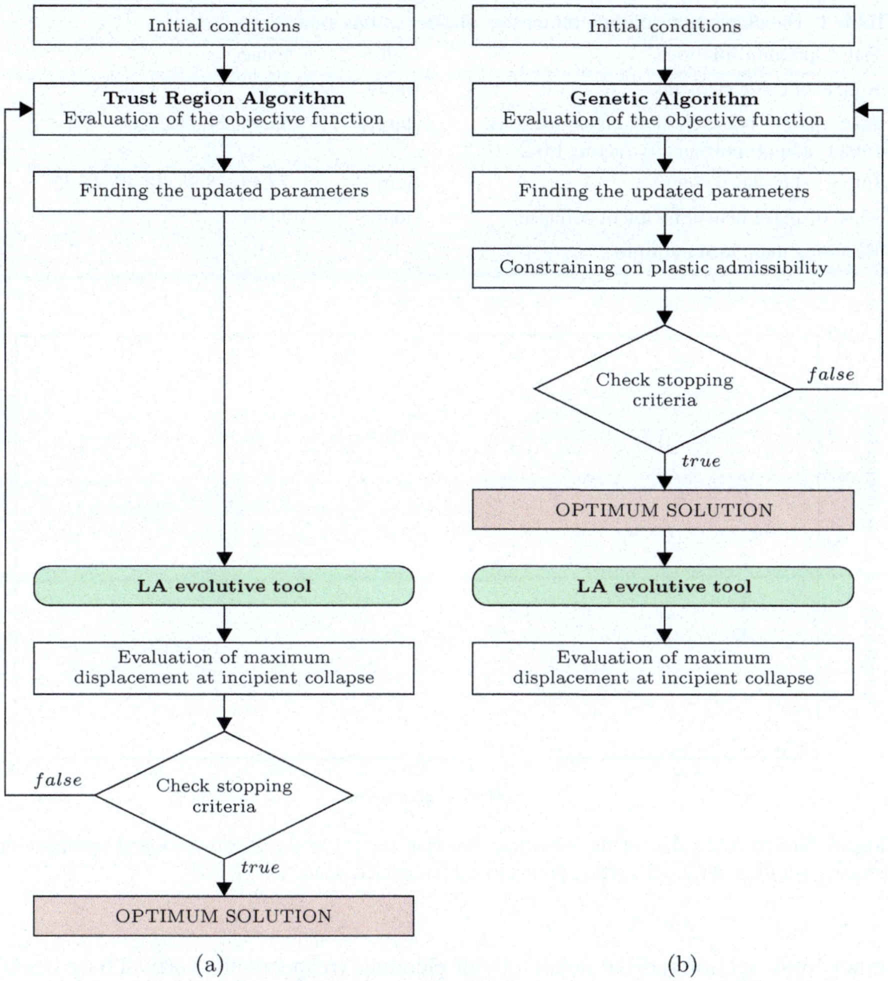

Fig. 5 Flow chart of the nested optimisation strategy: **a**) elastic–plastic optimisation; **b**) combined dynamic modal analysis and elastic–plastic optimisation

In Eq. (7), ω refers to the objective function, to be minimised, ΔL_i are the initial shortenings for the cable sets, $\boldsymbol{U}$ is the global displacement field of the structure and $|\cdot|$ means absolute value. Therefore, in the current optimisation strategy, a four–dimensional parametric space is adopted, assuming as free variables the initial cable shortenings, given, according to symmetry, four types of cables, namely the cables connecting the base node to the lower external nodes, the cables connecting the base node to the lower internal nodes, the vertical bracing cables and the inclined bracing cables. Consistently, the other geometrical features of the antenna are a priori selected, according to a previous design configuration [22], in particular regarding the

Table 1 Data/Optimum solution for elastic–plastic optimisation

Data/Optimum solution	Units	Values
Radius of cable elements	[mm]	2.50
Inner radii of beam element cross–sections (outer radii proportionality factor: 1.07)	[mm]	53.00, 24.44, 20.00
Initial cable shortenings	[mm]	*78.65, 18.98, 13.49, 13.32*
Max. displacement at incipient collapse	[mm]	2.004
Resulting max. load multiplier	[–]	23.03

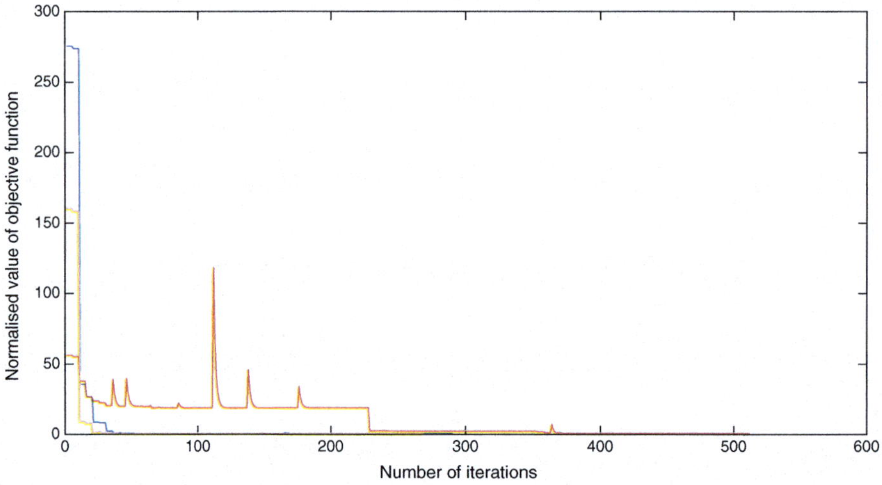

Fig. 6 Normalised value of the objective function Eq. (7) in the iterative looped optimisation process, for three different starting points in the parametric space

inner cross–section radii of beam tubular elements (respectively, vertical base beam, horizontal branch beams and inclined branch beams), the proportionality factor for beam outer radii and the radius of cable elements, as also reported in Table 1.

The results further reported in Figs. 6 and 8, in terms of optimisation iteration loops and incipient collapse configuration, highlight the efficiency of the proposed algorithm, both for understanding the structural behaviour of the satellite antenna and for coupling the approach within an optimisation loop. In particular, Fig. 6 displays that, for diverse starting points of the optimisation loop, the whole process converges to the identified optimum solution, although the large number of iterations and the presence of abrupt jumps in the evaluation of the objective function may highlight a possible lack of robustness within the optimisation strategy, for the adopted TRA routine, in the present implementation. This may point out to further inspect other optimisation routines.

The activation of the plastic joints involves the lower branch of the top antenna beams, as an effect of the combination between the geometrical features of the optimum solution for the structural design and the cable pretensions. As previously

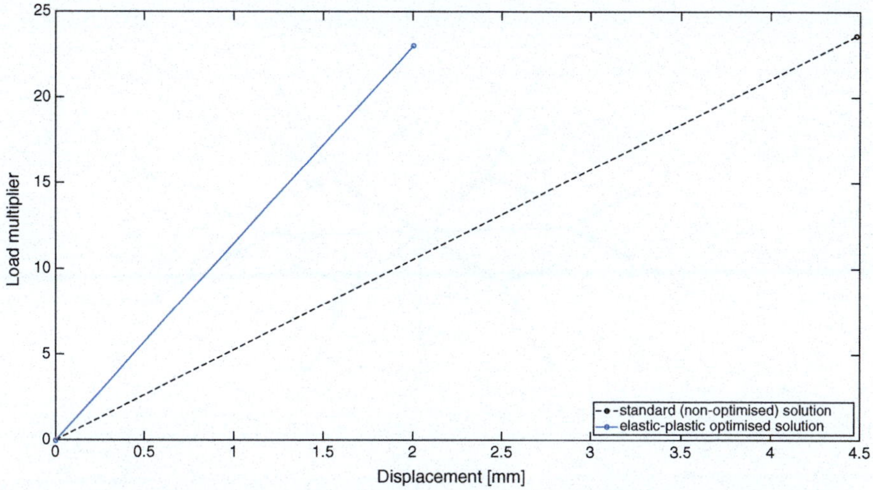

Fig. 7 Load–displacement response curve (load amplification factor versus displacement) for the optimum design configuration of the antenna, according to the elastic–plastic optimisation strategy, compared with standard (non–optimised) solution

observed in Sect. 2.2, although the structural collapse mechanism is not formally activated, the analysis is ended when the functionality of the antenna is meant to be lost. Such a stage, due to the lack of redundancy in the antenna beam substructure (ribs) and to the presence of several tensioned (elastic) cables, occurs at the activation of a first set of plastic joints. This is consistently displayed also by the resulting load–displacement curve depicted in Fig. 7. Thereby, the comparison to the standard, non–optimised, configuration shows improved characteristics, in terms of reduced maximum displacement and augmented resulting global stiffness of the structure.

The present optimisation procedure, coupled to evolutive computational LA, has demonstrated the possibility to implement the conceptual paradigm meant in the scopes stated in the Introduction and delivered consistent results, resumed as follows.

- The load–displacement curves of the space antenna are conveniently traced, and display a rather acceptable yield capacity, for the assumed and calibrated material properties, which is optimised, through the devised procedure.
- On the morphological features, the first that may be considered, in terms of radii of the rib tubular elements and callipers of the tensioning cables, they may be fixed or also be involved in the conceived optimisation strategy, fixing the values that may be assumed or possibly tuned, within a range of practical applications that may be confined by empirical rules or experience, in tiding them down, to specific values that shall maximise the structural performance, in terms of resilience.
- On the tiding of wires (initial cable shortening), a rather subtle issue, to be tuned for the specific space antenna configuration, which here has been taken as a challenging task in the optimisation process, the optimisation tool provides a clear guideline, amongst the multiple possible ways that may be attempted for the tun-

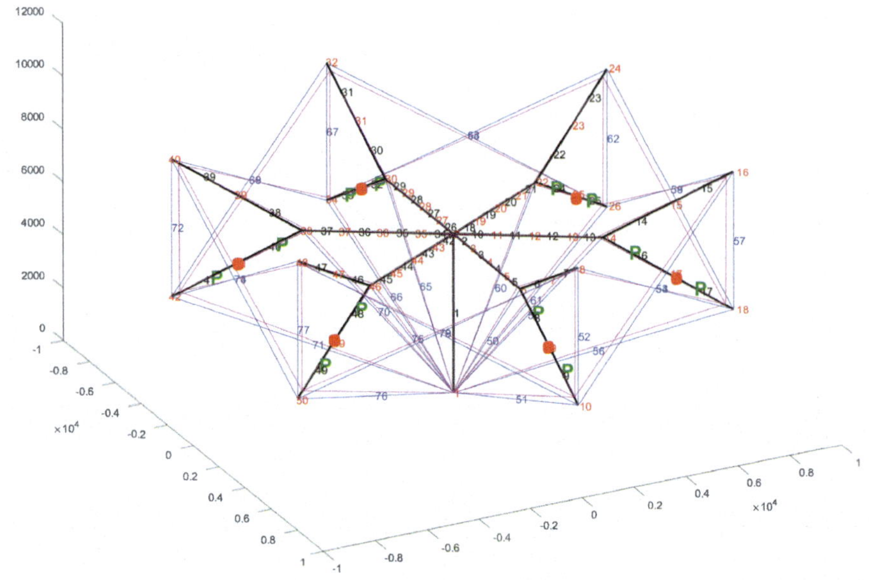

Fig. 8 Incipient collapse configuration for optimum elastic–plastic solution (magenta lines—deformed shape, green labels—plasticised elements, red dots—active plastic joints); coordinates in mm; scale factor 100

ing. Specifically in this respect, the potential implications of the present approach seem worthwhile to be pursued, and inspected.

- On the amount of maximum deformation at incipient collapse, the procedure is rather useful, to effectively arriving at marking the most possible reduced value of larger structural displacement, as targeted in the objective function, and resulting higher stiffness.
- Overall, the general process, of structural optimisation, relying on a LA tool as a mechanical solver, is able to deliver consistent optimum solutions, in the sought range of structural parameters. This looks crucial, in sharpening down various possible choices, which may appear as multiple possibilities or anyway uneasy to be set from scratch, at design and verification stages, of the targeted cable–rib space antenna.

3.2 Combined Dynamic Modal Analysis and Elastic–Plastic Optimisation Strategy

To further improve the design approach and to deepen possible alternative optimised solutions, a second optimisation strategy is herein proposed, combining modal dynamic analyses of the antenna with elastic–plastic analyses, in order to define a procedure meant to minimise as well the total mass of the structure, as a function of cross–section radii, under plastic admissibility constraints. Namely, while the evaluation of the objective function to be minimised is obtained by the solution of a dynamic modal analysis, (part of) the elastic–plastic analysis of the antenna is adopted within the optimisation constraints, in the constrained optimisation process, to guarantee that plastic admissibility is never violated within the structure, at each iteration stage, and also, in the end, at the final optimum configuration.

Such an approach assumes as objective function ω to be minimised the quantity expressed as:

$$\omega(r_1, r_2, r_3) = \frac{M_t(r_1, r_2, r_3)}{f_{min}(r_1, r_2, r_3)} \tag{8}$$

where, r_1, r_2 and r_3 are the cross–section inner radii of the ribs, M_t represents the total mass of the structure, and f_{min} is the lowest structural frequency, accounting for a compromise effect between achieving a lightweight solution and preserving a proper required stiffness of the structure. It should be noted that in the current optimisation strategy, the set of sought parameters (free optimisation variables) is composed by the three cross–section inner radii of the beam structural elements (namely, vertical base beam, horizontal branch beams and inclined branch beams), while the other geometrical features of the antenna are kept constant within the optimisation loop (Table 2). Moreover, considering design data of a previously proposed solution for an antenna structure [22] and the results of several optimisation tests, the initial cable shortenings are adjusted to a new set of values, with respect to the optimised solution from the earlier elastic–plastic strategy, then kept at such fixed values during the optimisation loop. Furthermore, it is worth to mention that, within the optimisation process, the elastic–plastic analysis of the structure is employed just for enforcing optimisation constraints, within the optimisation loop, ruled by plastic admissibility, and determining, in the end, load factor and displacement at incipient collapse stage, while the objective function is evaluated only based on the dynamic modal analyses of the space antenna (Fig. 5b).

Table 2 Data/Optimum solution for combined modal dynamic analysis and elastic–plastic optimisation

Data/Optimum solution	Units	Values
Radius of cable elements	[mm]	2.50
Inner radii of beam element cross–sections (outer radii proportionality factor: 1.07)	[mm]	*58.27, 26.82, 13.11*
Initial cable shortenings	[mm]	7.55, 6.78, 0.04, 91.57
Max. displacement at incipient collapse	[mm]	1.796
Resulting max. load multiplier	[–]	23.54
Total mass of the structure	[kg]	79.91
First structural frequency	[Hz]	0.3637

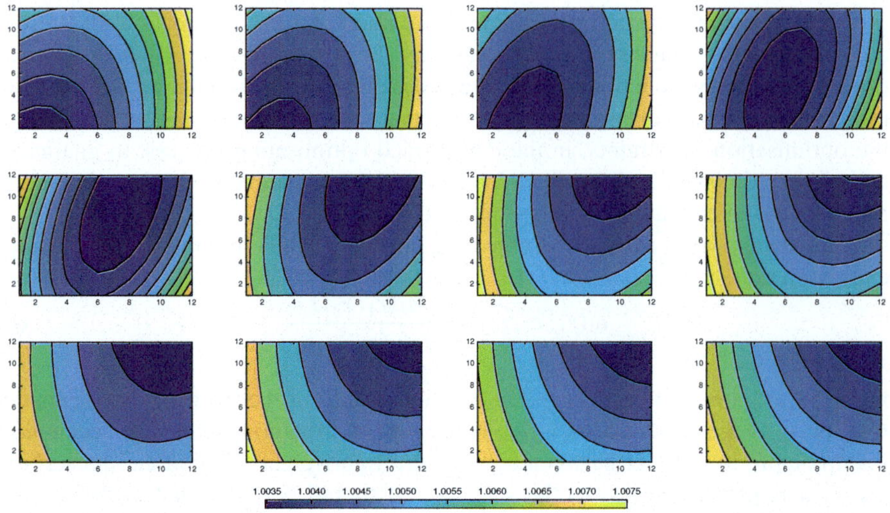

Fig. 9 Contour maps of normalised objective function ω, as in Eq. (8). Each slice is obtained as normalised contour plot, in the space of cross–section rib inner radii r_1 and r_2, for a fixed value of inner radius r_3

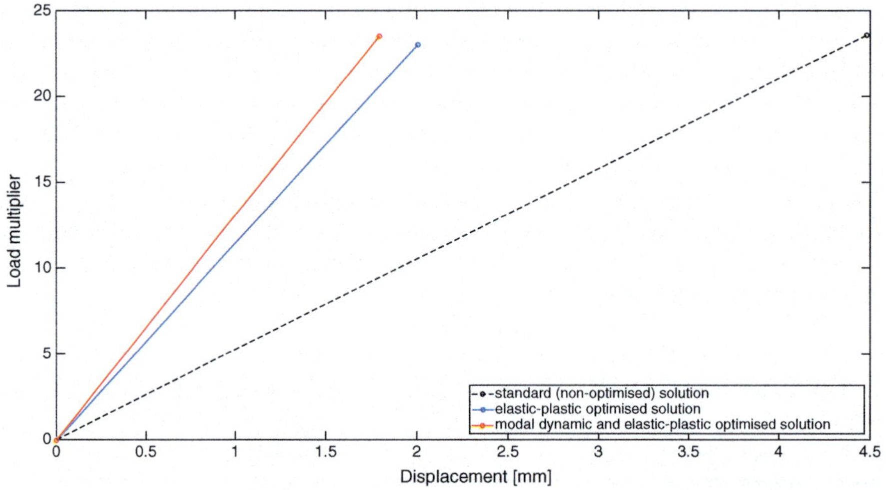

Fig. 10 Load–displacement response curve (load amplification factor versus displacement) for the optimum design configuration of the antenna, according to the modal dynamic and elastic–plastic optimisation strategy, compared with standard (non–optimised) solution and elastic–plastic strategy optimum solution

Given the more complex shape of the new objective function to be optimised, the present optimisation process is herein tackled by a Genetic Algorithm approach (always within Matlab), which has been observed to provide more robust results with respect to the specific analysed problem. The second optimisation strategy leads to a novel design solution of the space antenna, as reported in Table 2 (and later in Fig. 11).

The objective function to be optimised, despite its appreciable smoothness features, highlights several difficulties in the optimisation loop analysis, due to the competing roles of lightweight, stiffness and strength quests (Fig. 9), partially overcome by the adoption of a Genetic Algorithm approach, here preforming more effectively than a Trust Region Algorithm, and partially acceptable at a design stage, among various optimum design strategies, namely without a strict formal request to identify a true "global" minimum of the objective function. In particular, the sliced contour maps of the objective function, reported in Fig. 9, display a lack of clear identification for a "global" minimum, with respect to the ranges assumed into the (three–dimensional) parametric space, observing diverse minimum points at varying slice and/or minimum points at the domain borders.

The second here proposed optimisation strategy confirms the aforementioned (see Sects. 2.2 and 3.1) observations about collapse configuration and structural behaviour of the space antenna. Newly, the structural collapse mechanism is not directly activated as an effect of the presence of several tensioned (elastic) cables. However, at the activation of a first set of plastic joints, the structure shall exhibit, in practice, a clear loss of functionality, basically equivalent to a collapse condition. Nevertheless, it is worth to observe that the load amplification factor, for a monotonically increasing load, results quite large at the first activation of the plastic joints (Fig. 10), ensuring a robust design of the space antenna.

The activation of plastic joints, at the achieved optimum solution (Fig. 11), here involves the upper branch beams of the satellite antenna, as an effect of a diverse combination of the geometrical properties of the structure and the cable tidings, with respect to the solution discussed in Sect. 3.1. From a purely mechanical standpoint, in consideration of the elastic–plastic analysis outcomes, no specific preference can be selected between the two proposed optimisation strategies, providing in both cases a quite consistent structural design of the satellite antenna. Although the combined dynamic modal and elastic–plastic optimisation strategy exhibits larger numerical difficulties within the minimisation process, it shall appear as the preferred one, from an engineering design point of view, according to multiple analysis criteria of the structural behaviour of the satellite antenna.

Moreover, Fig. 10 globally shows and resumes the outcomes of the attempted optimisation processes, from the standard configuration, to the two optimised solutions, demonstrating a considerable improvement in the global load–displacement response and associated resulting stiffness. Further, this is here achieved together with a lighter mass and a higher first natural frequency, in this last optimum configuration, as discussed below.

Comparisons among the various solutions are indeed finally elucidated in Table 3, where it is observed that both optimisation strategies provide an improved design of

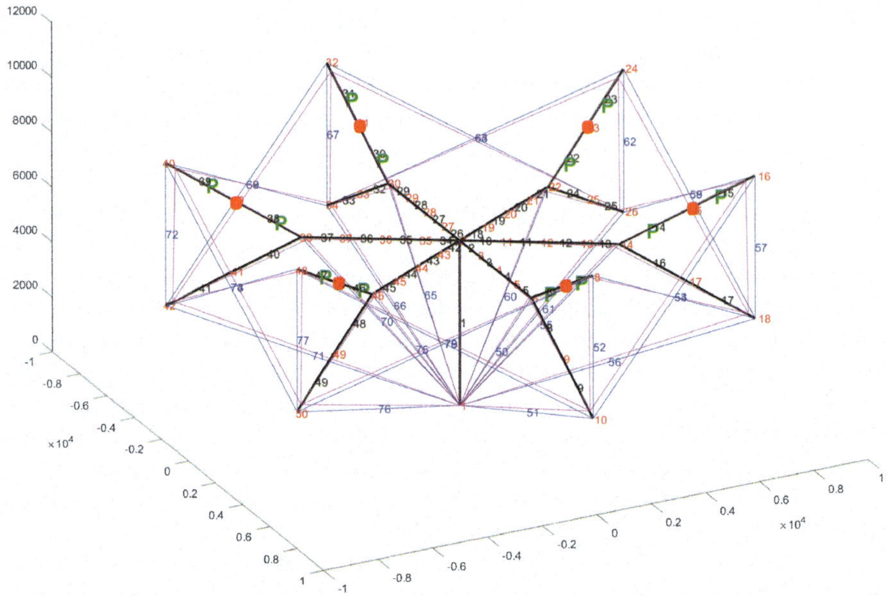

Fig. 11 Collapse configuration for modal dynamic analysis and elastic–plastic optimum solution (magenta lines—deformed shape, green labels—plasticised elements, red dots—active plastic joints); coordinates in mm; scale factor 100

the antenna, with respect to the standard configuration, obtaining a reduced maximum displacement at incipient collapse (namely a higher global stiffness) and a lowered mass (globally connected to a maximised first natural frequency of the structure). The resulting maximum load multiplier, with a limited variation among the diverse solutions, appears sufficiently large, for structural safety assessment, also in consideration of space satellite antenna scopes and functionality/robustness requirements. Moreover, the combined optimisation strategy, involving all together modal dynamic analysis and elastic–plastic analysis, globally appears to perform in a more effective way, pointing out to an improved solution, with a further gain, regarding a lower maximum displacement at incipient collapse (and increased load multiplier/displacement ratio, as global stiffness of the structure) and an augmented first natural frequency of the antenna.

Adding to the previous ending comments in Sect. 3.1, the present further attempt extends the optimisation analysis in also delivering, at the same time, information regarding basic modal dynamic properties, mainly read in terms of total structural mass and lowest natural frequency of vibration of the antenna. The following main remarks are then in order.

- The estimations of morphological parameters (radii) keep consistent, with the values in the preceding elastic–plastic optimisation analysis, showing a somehow

Table 3 Comparison of characteristic solution values between standard (non–optimised) design, elastic–plastic optimised (*ep*) design and modal dynamic and elastic–plastic optimised (*mdep*) design of the cable–rib antenna

Optimum solution	Units	Standard sol.	*ep* sol.	*mdep* sol.
Max. displacement at incipient collapse	[mm]	4.480	2.004	1.796
Resulting max. load multiplier	[–]	23.57	23.03	23.54
Load multiplier/displacement ratio	[1/mm]	5.261	11.49	13.11
Total mass of the structure	[kg]	226.2	88.40	79.91
First structural frequency	[Hz]	0.2337	0.2793	0.3637

different inner radii distribution that allows, at the same time, to address the concept of light–mass.

- The setting of cable shortenings locate a rather different best solution, for the integrated scopes of resilience and light–mass, showing the need of a guideline, in a tensioning tuning that may display a wide variability, at design stage.
- The total mass of the structure is confined to a value of 79.91 kg, for the considered material to be employed (aluminium), providing a lighter solution with respect to the one gathered by the previous optimisation strategy (88.40 kg; thus, with about a 10% saving). Different material properties, then, may be inspected, to foresee further solutions (and possibly structural schemes) that may help in lowering down this final characteristic feature (total mass), which shall always display a crucial role in space transportation and handling. This could lead to further coupling and iterations, on the practical side, with interaction among real quests and modelisation/optimisation running. In other words, the prototype antenna could then be further analysed, and optimised, even beyond the present demonstration purposes of methodological optimisation context, which has already advanced basic solution features and final structural properties.
- The lowest structural frequency, in terms of a main vibrational feature, for the linear dynamic response, shows a quite well–defined value. This may actually be targeted, for a further attention, if needed, in the design process, if felt, a posteriori, as a significant structural feature, in the process of mounting and regular operation on field, especially toward confining vibrational ranges.

4 Conclusions

The present attempt has stated, and demonstrated, a general paradigm of structural resilience, based on effective Limit Analysis computational tools, as possibly coupled to efficient optimisation loops, in the structural application field of large–scale deployable cable–rib aerospace antennas, with a specific prototype sample being analysed, as a perspective application, in terms of conceptual and practical design, seeking further high–performance and optimised solutions. Once the present principle shall be established, communicated and inherited, here by the aerospace structural engineering community, it should take off, in a flight bringing the above–mentioned computational tools in constituting an essential modelisation counterpart, for tuning design and practical solutions. In all that, LA should regain modern momentum and visibility, on the side of engineering practice, in specific fields, once released through effective and reachable algorithms. This project shall have contributed to that, in delivering various outcomes and considerations, as ruled down in the body of the paper and briefly resumed below.

In this investigation, a devoted numerical LA algorithm has been implemented, for performing the evolutive elastic–plastic analysis of deployable cable–rib satellite antennas. To significantly improve the design of the analysed structure, the computational tool has then been employed as looped to optimisation strategies, taking advantage of the underlying achieved efficiency of the LA algorithm, toward locating optimum design solutions. From the collected results, some conclusions can be derived, specifically related to the considered application case study and, besides the problem–driven implementation, regarding the general validity of the present methodology, as a design tool, in the framework of non–linear analysis and Limit Analysis of lightweight structures, such as space antennas.

In particular, it is possible to observe that the evolutive algorithm for elastic–plastic analysis truly identifies the activations of the plastic joints and consistently estimates displacements, and load bearing capacity, up to incipient collapse. Therefore, the approach displays significant advantages, being able to fully trace the evolutive non–linear (piece–wise linear) mechanical behaviour of the structure, both from evolution and collapse standpoints, computing all together internal action distribution and displacement field.

The robustness and efficiency of the elastic–plastic strategy has actively been explored within diverse optimisation criteria and objective functions, allowing for a suitable design perspective, considering non–linear structural behaviour and collapse configuration toward serviceability conditions. Such a procedure has been fundamental, in order to highlight different possible design outlines of the antenna, according to various optimisation strategies and features to be observed and expected. At the same time, this is meant to define a novel conceptual approach in the specific application field, within an advancement with respect to the classical computational tools of static elastic analysis.

From a more general point of view, further extending the obtained conclusions than for the specific problem–driven development, an efficient computational Limit Analysis, possibly combined with structural optimisation loops, shall newly lead to a convenient paradigm for structural design (see Introduction), once the algorithmic implementation is truly effective, in tracing down the non–linear mechanical response, as a structural behaviour, even for large–scale and peculiar structures, such as for cable–rib space antennas. The proposed algorithm, as a robust and practical design approach, may rapidly be extended to various structural application fields, both in research approaches and engineering design prospects.

This last observation allows to prompt out some perspectives of future developments and applications, toward further implementations of devoted specific algorithms, such as for those based on a general Linear Complementarity Problem formulation, and toward the employment of the devised methodologies for "form–finding" and design purposes, in the combination of classical structural analysis, Limit Analysis and optimisation strategies.

Acknowledgements The financial support by "*Fondi di Ricerca d'Ateneo ex 60%*" and "*PRIN Life-long optimized structural assessment and proactive maintenance with pervasive sensing techniques*" at the University of Bergamo is gratefully acknowledged. The financial support by the National Natural Science Foundation of China (Grant No. 52105011) is gratefully acknowledged by R. Liu.

References

1. Cai, J., Ma, R., Deng, X., Feng, J.: Static behavior of deployable cable-strut structures. J. Constr. Steel Res. **119**(3), 63–75 (2016). https://doi.org/10.1016/j.jcsr.2015.12.003
2. Cai, J., Ren, Z., Ding, Y., Deng, X., Xu, Y., Feng, J.: Deployment simulation of foldable origami membrane structures. Aerosp. Sci. Technol. **67**(8), 343–353 (2017). https://doi.org/10.1016/j.ast.2017.04.002
3. Cai, J., Zhang, Q., Zhang, Y., Hoon Lee, D.S., Feng, J.: Structural evaluation of a foldable cable-strut structure for kinematic roofs. Steel Compos. Struct. **29**(5), 669–680 (2018). https://doi.org/10.12989/SCS.2018.29.5.669
4. Chen, Y., You, Z., Tarnai, T.: Threefold-symmetric Bricard linkages for deployable structures. Int. J. Solids Struct. **42**(8), 2287–2301 (2005). https://doi.org/10.1016/j.ijsolstr.2004.09.014
5. Chu, Z., Deng, Z., Qi, X., Li, B.: Modeling and analysis of a large deployable antenna structure. Acta Astronaut. **95**(2), 51–60 (2014). https://doi.org/10.1016/j.actaastro.2013.10.015
6. Clasen, G., Langley, R.: Meshed patch antennas. IEEE Trans. Antennas Propag. **52**(6), 1412–1416 (2004). https://doi.org/10.1109/TAP.2004.830251
7. Cocchetti, G., Maier, G.: Elastic-plastic and limit-state analyses of frames with softening plastic-hinge models by mathematical programming. Int. J. Solids Struct. **40**(25), 7219–7244 (2003). https://doi.org/10.1016/S0020-7683(03)00363-9
8. Cocchetti, G., Maier, G., Shen, X.P.: Piecewise linear models for interfaces and mixed mode cohesive cracks. Comput. Model. Eng. Sci. **3**(3), 279–298 (2002). https://doi.org/10.3970/cmes.2002.003.279
9. Datashvili, L., Endler, S., Wei, B., Baier, H., Langer, H., Friemel, M., Tsignadze, N., Santiago-Prowald, J.: Study of mechanical architectures of large deployable space antenna apertures: from design to tests. CEAS Space J. **5**(10), 169–184 (2013). https://doi.org/10.1007/s12567-013-0050-9
10. Ferrari, R., Cocchetti, G., Rizzi, E.: Limit analysis of a historical iron arch bridge: formulation and computational implementation. Comput. Struct. **175**(10), 184–196 (2016). https://doi.org/10.1016/j.compstruc.2016.05.007
11. Ferrari, R., Cocchetti, G., Rizzi, E.: Computational elastoplastic Limit Analysis of the Paderno d'Adda bridge (Italy, 1889). Arch. Civ. Mech. Eng. **18**(1), 291–310 (2018). https://doi.org/10.1016/j.acme.2017.05.002
12. Ferrari, R., Cocchetti, G., Rizzi, E.: Effective iterative algorithm for the Limit Analysis of truss-frame structures by a kinematic approach. Comput. Struct. **197**(2), 28–41 (2018). https://doi.org/10.1016/j.compstruc.2017.11.018
13. Ferrari, R., Cocchetti, G., Rizzi, E.: Evolutive and kinematic Limit Analysis algorithms for large-scale 3D truss-frame structures: comparison application to historic iron bridge arch. Int. J. Comput. Methods **17**(5), 1940020 (2020). https://doi.org/10.1142/S0219876219400206

14. Ferrari, R., Cocchetti, G., Rizzi, E.: Reference structural investigation on a 19th-century arch iron bridge loyal to design-stage conditions. Int. J. Arch. Herit. **14**(10), 1425–1455 (2020). https://doi.org/10.1080/15583058.2019.1613453
15. Ferrari, R., Cocchetti, G., Rizzi, E.: Enriched computational limit analysis implementation for large–scale 3D truss–frame structures. In: Proceedings of the 25th International Congress of Theoretical and Applied Mechanics (25th ICTAM), ICTAM Milano 2020+1. International Union of Theoretical and Applied Mechanics (IUTAM), Milano, Italy (2021)
16. Guo, H., Shi, C., Li, M., Deng, Z., Liu, R.: Design and dynamic equivalent modeling of double-layer hoop deployable antenna. Int. J. Aerosp. Eng. **2018**(1), 2941981 (2018). https://doi.org/10.1155/2018/2941981
17. Guo, H., Song, X., Li, L., Deng, Z., Liu, R., Geng, J.: Type synthesis of deployable single-loop overconstrained linkages based on Bennett linkages. Mech. Mach. Theory **120**(2), 1–29 (2018). https://doi.org/10.1016/j.mechmachtheory.2017.09.013
18. Hodge, P.G., Jr.: Automatic piecewise linearization in ideal plasticity. Comput. Methods Appl. Mech. Eng. **10**(3), 249–272 (1977). https://doi.org/10.1016/0045-7825(77)90072-X
19. Li, B., Qi, X., Huang, H., Xu, W.: Modeling and analysis of deployment dynamics for a novel ring mechanism. Acta Astronaut. **120**(2), 59–74 (2016). https://doi.org/10.1016/j.actaastro.2015.11.023
20. Liu, R., Guo, H., Liu, R., Tang, D., Wang, H., Deng, Z.: Design and form finding of cable net for a large cable-rib tension antenna with flexible deployable structures. Eng. Struct. **199**(11), 109662 (2019). https://doi.org/10.1016/j.engstruct.2019.109662
21. Liu, R., Guo, H., Liu, R., Wang, H., Tang, D., Deng, Z.: Structural design and optimization of large cable-rib tension deployable antenna structure with dynamic constraint. Acta Astronaut. **151**(10), 160–172 (2018). https://doi.org/10.1016/j.actaastro.2018.05.055
22. Liu, R., Guo, H., Liu, R., Wang, H., Tang, D., Song, X.: Shape accuracy optimization for cable-rib tension deployable antenna structure with tensioned cables. Acta Astronaut. **140**(11), 66–77 (2017). https://doi.org/10.1016/j.actaastro.2017.07.047
23. Liu, S.Y., Chen, Y.: Myard linkage and its mobile assemblies. Mech. Mach. Theory **44**(10), 1950–1963 (2009). https://doi.org/10.1016/j.mechmachtheory.2009.05.001
24. Liu, W., Li, D.X., Yu, X.Z., Jiang, J.P.: Exact mesh shape design of large cable-network antenna reflectors with flexible ring truss supports. Acta. Mech. Sin. **30**(3), 198–205 (2014). https://doi.org/10.1007/s10409-014-0029-6
25. Love, A.W.: Some highlights in reflector antenna development. Radio Sci. **11**(8–9), 671–684 (1976). https://doi.org/10.1029/RS011i008p00671
26. Lu, S., Zlatanov, D., Ding, X.: Approximation of cylindrical surfaces with deployable Bennett networks. J. Mech. Robot. **9**(2), 021001 (2017). https://doi.org/10.1115/1.4035801
27. Magisano, D., Corrado, A., Madeo, A., Garcea, G.: A reduced order model for nonlinear time history seismic analyzes of elasto-plastic 3D frame structures. Earthq. Eng. Struct. Dyn. **51**(9), 2055–2076 (2022). https://doi.org/10.1002/eqe.3653
28. Mahini, M.R., Moharrami, H., Cocchetti, G.: Elastoplastic analysis of frames composed of softening materials by restricted basis linear programming. Comput. Struct. **131**(1), 98–108 (2014). https://doi.org/10.1016/j.compstruc.2013.08.005
29. Maier, G., Giacomini, S., Paterlini, F.: Combined elastoplastic and limit analysis via restricted basis linear programming. Comput. Methods Appl. Mech. Eng. **19**(1), 21–48 (1979). https://doi.org/10.1016/0045-7825(79)90032-X
30. Maier, G., Pastor, J., Ponter, A.R.S., Weichert, D.: Direct methods of limit and shakedown analysis. In: Milne, I., Ritchie, R.O., Karihaloo, B. (eds.) Comprehensive Structural Integrity, Vol. 3, Chap. 12, pp. 637–684. Elsevier, Amsterdam (2003). https://doi.org/10.1016/B0-08-043749-4/03059-7
31. Maier, G.: Piecewise linearization of yield criteria in structural plasticity. Solid Mech. Arch. **2**(3), 239–281 (1976)
32. Meguro, A., Shintate, K., Usui, M., Tsujihata, A.: In-orbit deployment characteristics of large deployable antenna reflector onboard engineering test satellite VIII. Acta Astronaut. **65**(9–10), 1306–1316 (2009). https://doi.org/10.1016/j.actaastro.2009.03.052

33. Ozawa, S., Fujiwara, Y., Tsujihata, A.: Development of corotational formulated FEM for application to 30 m class large deployable reflector. IOP Conf. Ser.: Mater. Sci. Eng. **10**(1), 012006 (2010). https://doi.org/10.1088/1757-899X/10/1/012006
34. Ozawa, S., Tsujihata, A.: Lightweight design of 30 m class large deployable reflector for communication satellites. In: 52nd AIAA/ASME/ASCE/AHS/ASC Structures, Structural Dynamics and Materials Conference. American Institute of Aeronautics and Astronautics, Denver, Colorado (2011)
35. Puig, L., Barton, A., Rando, N.: A review on large deployable structures for astrophysics missions. Acta Astronaut. **67**(1–2), 12–26 (2010). https://doi.org/10.1016/j.actaastro.2010.02.021
36. Qi, X., Deng, Z., Li, B., Liu, R., Guo, H.: Design and optimization of large deployable mechanism constructed by Myard linkages. CEAS Space J. **5**(3–4), 147–155 (2013). https://doi.org/10.1007/s12567-013-0036-7
37. Qi, X., Huang, H., Li, B., Deng, Z.: A large ring deployable mechanism for space satellite antenna. Aerosp. Sci. Technol. **58**(11), 498–510 (2016). https://doi.org/10.1016/j.ast.2016.09.014
38. Semler, D., Tulintseff, A., Sorrell, R., Marshburn, J.: Design, integration, and deployment of the TerreStar 18–meter reflector. In: 28th AIAA International Communications Satellite Systems Conference. American Institute of Aeronautics and Astronautics, Anaheim, California (2010)
39. Shi, C., Guo, H., Zheng, Z., Li, M., Liu, R., Deng, Z.: Conceptual configuration synthesis and topology structure analysis of double-layer hoop deployable antenna unit. Mech. Mach. Theory **129**(11), 232–260 (2018). https://doi.org/10.1016/j.mechmachtheory.2018.08.005
40. Soliman, H.E., Stoumbos, T., Rebold, J.: FEM Correlation of sandwich composite antenna structure subject to thermal distortion. In: 55th AIAA/ASMe/ASCE/AHS/SC Structures, Structural Dynamics, and Materials Conference. American Institute of Aeronautics and Astronautics, National Harbor, Maryland (2014)
41. Talvik, I.: Finite element modelling of cable networks with flexible supports. Comput. Struct. **79**(26–28), 2443–2450 (2001). https://doi.org/10.1016/S0045-7949(01)00077-3
42. Terada, M., Bludworth, N., Moore, J., Sullivan, J.: Deployable reflector system for satellite applications. In: SBMO/IEEE MTT-S International Conference on Microwave and Optoelectronics. IEEE, Brasilia, Brazil (2005)
43. Thomson, M.: AstroMesh deployable reflectors for Ku and Ka band commercial satellites. In: 20th AIAA International Communication Satellite Systems Conference and Exhibit. TRW Astro Aerospace, American Institute of Aeronautics and Astronautics, Montreal, Quebec, Canada (2002). https://doi.org/10.2514/6.2002-2032
44. Tibert, A.G., Pellegrino, S.: Deployable tensegrity reflectors for small satellites. J. Spacecr. Rocket. **39**(5), 701–709 (2002). https://doi.org/10.2514/2.3867
45. Zhang, P., Lu, M., Hwang, K.: A mathematical programming algorithm for limit analysis. Acta Mech. Solida Sin. **7**(3), 267–274 (1991). https://doi.org/10.1007/BF02487595

An Introduction to the Probabilistic Linear Matching Method Framework for Structural Integrity Assessment Under Uncertain Design Conditions

Xiaoxiao Wang and Haofeng Chen

Abstract The novel probabilistic Linear Matching Method (pLMM) framework is developed by extending the current direct method, the Linear Matching Method (LMM), to deal with the probabilistic structural integrity assessment for engineering components under uncertain operating conditions. The pLMM framework covers several physics-based failure evaluation modules related to cyclic loads at elevated temperatures, including shakedown analysis, ratcheting analysis, low cycle fatigue (LCF) analysis and creep-fatigue analysis. To further improve the prediction efficiency, artificial neural network (ANN) technology is employed to build the data-driven surrogate relationship between the design parameters and the key responses regarding specified failure behaviour, with a series of probabilistic evaluation boundaries and assessment diagrams of engineering structures established to describe the uncertainty of the structural resistance. The reliability analysis techniques are involved as well, by which the failure probability is estimated considering the randomness of engineering problems. The pLMM framework is conducive to getting rid of the excessive dependence on the conventional safety factor with conspicuous conservativeness during risk management, enhancing the robustness of critical infrastructure.

1 Introduction

Structural integrity assessment provides a fundamental investigation of structural resistance against a certain failure mechanism by a series of deterministic experimental analyses and numerical simulations. To further reflect the uncertainty in the

X. Wang
Department of Mechanical & Aerospace Engineering, University of Strathclyde, James Weir Building, 75 Montrose Street, Glasgow G1 1XJ, UK

H. Chen (✉)
School of Mechanical and Power Engineering, East China University of Science and Technology, Shanghai 200237, China
e-mail: haofeng.chen@strath.ac.uk

G. Garcea and D. Weichert (eds.), *Direct Methods for Limit State of Materials and Structures*, Lecture Notes in Applied and Computational Mechanics 101,
https://doi.org/10.1007/978-3-031-29122-7_4

material properties, geometric parameters, boundary conditions and loads conditions of engineering infrastructures, the most commonly used strategy is to set an appropriate safety factor which is able to consider the adequate redundancy in the design process. [1–3]. Inevitably, this experience-based arbitrary decision-making may result in unreasonable conservativeness introduced due to the limited statistical information on the important structural responses. The feasible probabilistic structural integrity assessment technology contains two main prerequisites: the concise but accurate deterministic evaluation approach and the proper probabilistic analysis method, and integrating the two sections into a unified frame system is a challenging engineering problem to which enormous endeavour has been devoted in academia.

The case study of the reliability-based shakedown analysis for a high-pressure chamber is given out by Andrzej [4], where the shakedown analysis procedure (CYCLONE), Response Surface Method (RSM) and Monte Carlo Simulation (MCS) sampling method were adopted to address the randomness of the cyclic inner pressure and radius. Staat and Heitzer proposed a stochastic FEA procedure that is dedicated to implementing probabilistic shakedown analysis, and the benchmarks of the central holed plate, pipe junction and plate with mismatched weld and a crack were elaborated [5–7] and verified by related analytical studies.

The reduced-order model technique, as well as the surrogate model, is applicable to implicitly expressing the estimation of structural fatigue life [8, 9] under complicated operating environments with fewer computational resources. For instance, the artificial neural network (ANN) is widely applied to reliability-based fatigue evaluations [10], and the feedforward backpropagation multilayer perceptron (MLP) neural network was employed by Durodola [11], where the non-linear effect of mean stress on the fatigue life was examined in the probabilistic fatigue analysis. Another application was reported by Ref. [12], with the probabilistic fatigue damage of subsea pipe derived by the dynamic Bayesian network. A similar analysis scheme was also extended to investigate the probabilistic fatigue crack growth and propagation [13].

Concerning the components under the creep regime, Zhang [14] provided comparative research, where the abilities to predict the creep-fatigue life of 316 stainless steel of three neural networks were compared, revealing the superiority of physics-informed neural network for providing the creep-fatigue life approximation with better fitting quality. The machine learning-based creep-fatigue life prediction scheme of low-alloy steel 42CrMo4 specimen was constructed in Ref. [15], and the ANN-based damage model by the long short-term memory network and gated recurrent unit neural network was proved to be suitable for creep-fatigue life estimation under complicated conditions with non-linearly changing temperatures and mechanical strain rates.

Therefore, the plausible probabilistic analysis framework for structural integrity should satisfy the prerequisites in terms of three different levels. Firstly, the deterministic structural integrity evaluation program should keep a good balance between computational accuracy and efficiency. Next, the effective prediction model should be able to capture the pertinent nonlinear relationship between the design parameters and output, without depending on unaffordable computational resources. Finally, the universal probabilistic analysis tool should be applicable for solving the structure-level problems, providing comprehensive statistical information and the failure risk.

The LMM procedures have been proven to be able to deal with a wide range of structural integrity assessments (as shown in Fig. 1) by adopting a series of linear calculation approaches to precisely match the structural non-linear responses, including shakedown analysis [16], ratcheting analysis [17], low cycle fatigue (LCF) analysis [18] and creep-fatigue analysis [19]. The LMM shakedown analysis procedure is developed to calculate the structural shakedown limit leveraging the time-independent residual stress [20], according to the upper bound shakedown theory. When tackling the load conditions leading to alternating plasticity, the structural plastic response and the related Low Cycle Fatigue (LCF) evaluations are able to be implemented by the Direct Steady Cycle Analysis (DSCA) procedure [21], with the varying residual stress determined iteratively. Based on the time-dependent residual stress field by the DSCA procedure, the LMM ratcheting analysis procedure is dedicated to assessing the structural resistance to progressive plastic behaviour, where the limit of the additional constant load condition is acquired [18]. In order to further take the creep effect into consideration during LCF analysis for high-temperature components, the latest extended Direct Steady Cycle Analysis (eDSCA) procedure [19], can analyse the creep-fatigue interaction in terms of both stress–strain response level and the damage level, providing an accurate prediction of the lifetime to creep-fatigue crack initiation.

In this study, the deterministic analysis procedures are extended to address the probabilistic assessment for engineering components under cyclic load conditions and elevated temperature, and the recent research benchmarks and engineering applications by means of the proposed pLMM methodology are delivered.

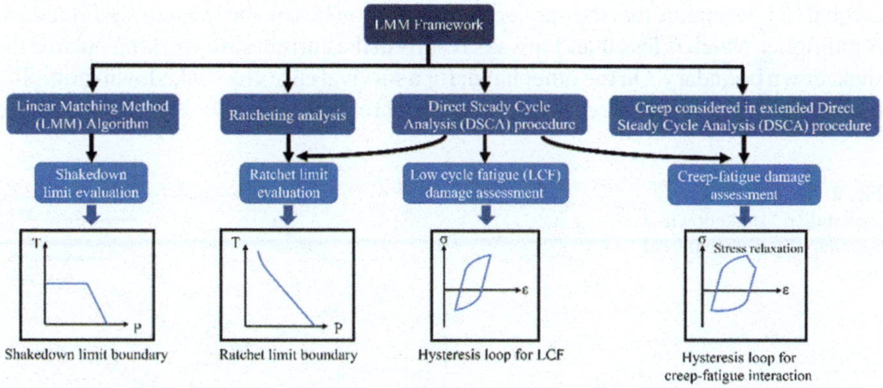

Fig. 1 Linear Matching Method (LMM) framework for structural integrity assessment regarding different failure mechanisms

2 Probabilistic Shakedown Analysis Under the pLMM Framework

2.1 *Probabilistic Shakedown Boundary and the Implementation of Reliability-Based Shakedown Analysis*

Establishment of the probabilistic shakedown boundary. The probabilistic shakedown boundary is conducive to visualizing the shakedown boundary affected by the uncertainty in design parameters, which can be built by implementing MCS samplings to estimate the statistical distribution of the LMM shakedown multipliers. As shown in Fig. 2, with the load ratio between Load 1 and Load 2, the envelope of probabilistic shakedown boundary and the associated statistical significance are displayed. And the failure risk against shakedown conditions raised by load conditions lying on the boundary is equal to the probability of the occurrence of the shakedown boundary.

Reliability-based shakedown analysis procedure. In order to perform the reliability-based shakedown evaluation and further examine the structural failure probability against shakedown conditions, a simplified governing function, the shakedown limit state indicator function (SLSIF), is expressed by Eq. (1) below.

$$G(\boldsymbol{X}) = \lambda(\boldsymbol{X}) - 1 \begin{cases} < 0, \text{ Failed} \\ = 0, \text{ Limit state} \\ > 0, \text{ Survival} \end{cases} \tag{1}$$

Here, the LMM shakedown multiplier λ functions as a state indicator, if the multiplier is equal to 1, it implies that the applied load condition lies on the shakedown boundary. A multiplier which is less than 1 always results in the current case working outside the shakedown boundary. On the other hand, for a survival case, the shakedown multiplier should be strictly guaranteed larger than the threshold value, $\lambda = 1$. The proposed

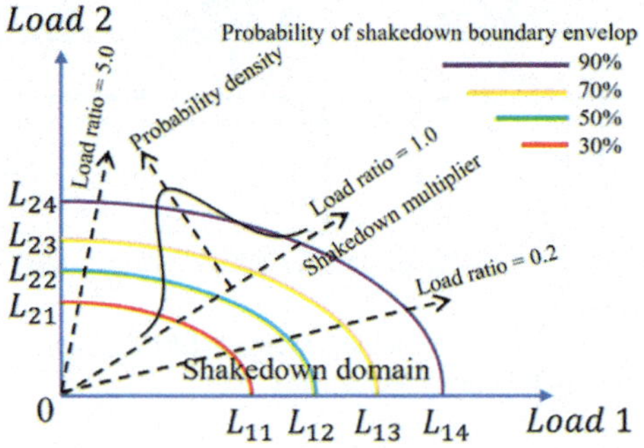

Fig. 2 Scheme of probabilistic shakedown boundary by different load ratios

SLSIF is then utilized directly by the First Order Reliability Method (FORM) [22] to acquire the structural failure probability and shakedown reliability index.

Description of deterministic FEA model. The benchmark of probabilistic shakedown analysis is demonstrated by the holed plate displayed in Fig. 3, which is discretized by the 20-node quadratic brick element C3D20R with reduced integration. And the ratio between the hole diameter D and the length L of the plate is 0.2, and the ratio between the depth d of the plate and the length L is equal to 0.05. Through the mesh convergence check, the total element number and minimum element size are determined to be 721 and 1.25 mm. Besides, Young's modulus E and the Poisson's ratio ν of the elastic-perfect plastic (EPP) material are assumed to be 2×10^5MPa and 0.3 respectively, and the material yield strength σ_y is defined to be 200MPa. The shakedown analysis is implemented considering the biaxial tensile load shown in Fig. 3, where the horizontal component P_1 is a constant uniform load and the vertical component P_2 is cyclic one. Detailed cyclic load pattern is depicted in Fig. 4.

Probabilistic shakedown boundary of the holed plate. Considering two random variables, the diameter of the central hole D and the material yield strength σ_y, the probabilistic shakedown boundaries of the holed plate are plotted in Fig. 5. This probabilistic shakedown assessment diagram is constructed by a series of random variables λ_i, and the statistical distributions of the LMM shakedown limit multiplier with different load ratios are displayed in Fig. 6. It is worth noting that the actual shakedown boundary should occur stochastically inside the envelope surrounded by a certain curve with corresponding probability.

Reliability-based shakedown analysis for the holed plate. According to the proposed shakedown limit state indicator function (SLSIF), the gradient vector of the SLSIF is generated during each iteration by calculating the convergent numerical partial derivatives of each random variable at the current design point. In the reliability-based shakedown analysis, there are two typical examples, with the

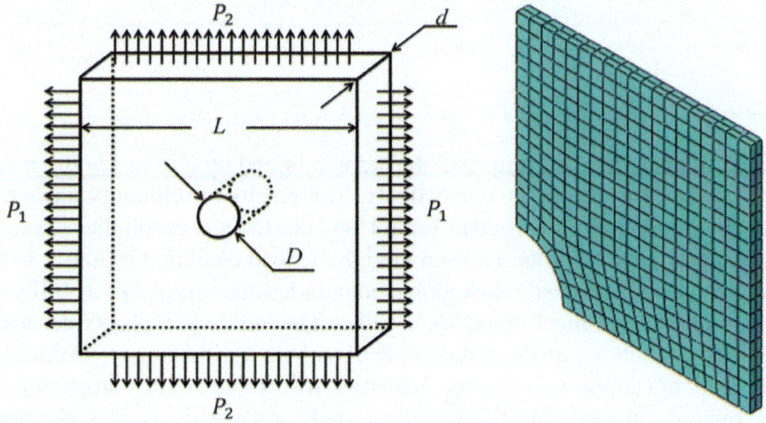

Fig. 3 FEA model of the central holed plate

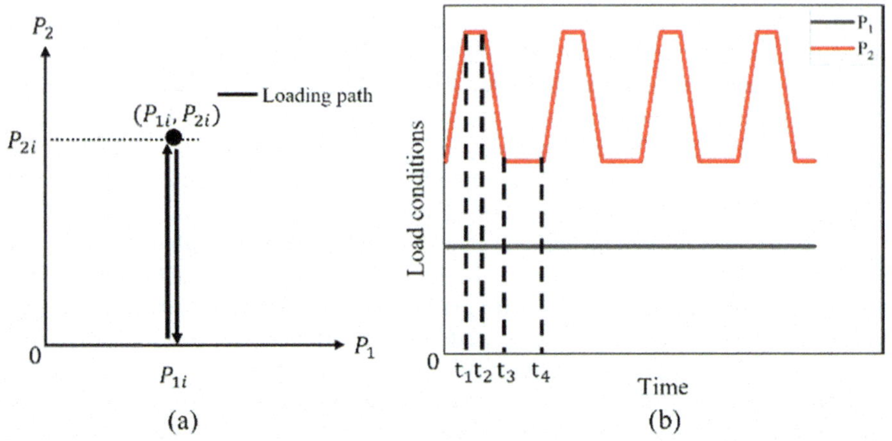

Fig. 4 Description of the time-dependent load conditions: **a** cyclic mode of the applied load path; **b** detailed spectrum of the applied load conditions

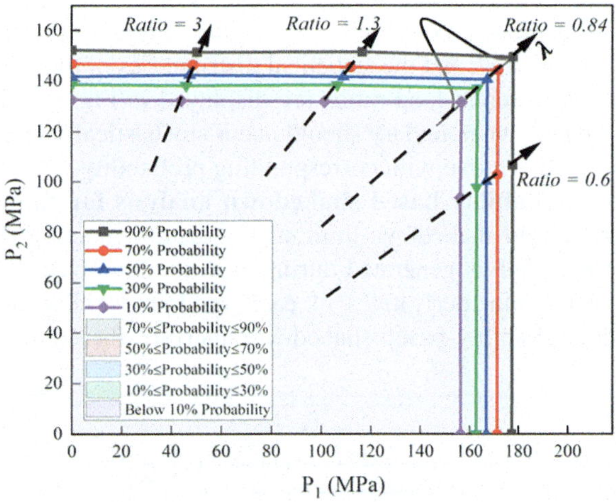

Fig. 5 Probabilistic shakedown boundaries of the central holed plate under predefined load path

load conditions on the deterministic shakedown boundary and inside the boundary involved (as shown in Fig. 7), respectively. The results of reliability analysis with different load conditions are listed in Table 1, and due to the uncertainties of geometric dimension and material property, even though the load condition (points 1 in Fig. 7) is located on the deterministic shakedown boundaries, the survival probability is only 70.45%, which is still much lower than 100%. The failure probability derived from pLMM-based reliability analysis is compared with the verifications with direct MCS, with all random variables remaining consistent with Table 1. The comparison shows that the results of the proposed probabilistic shakedown analysis are consistent with

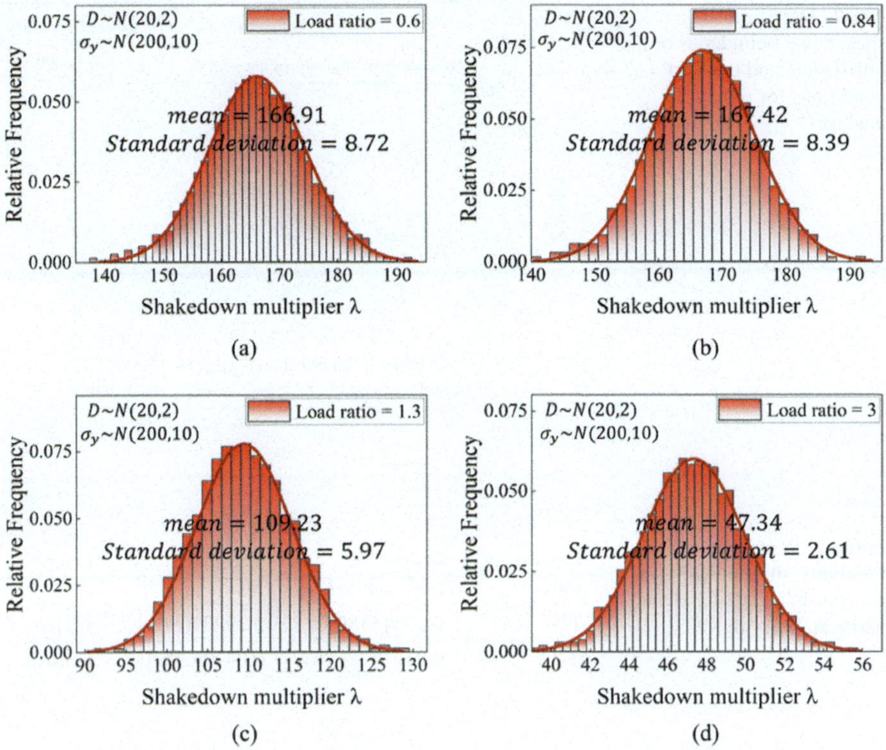

Fig. 6 Probabilistic distribution of LMM shakedown limit multiplier with different load ratios: **a** ratio = 0.6; **b** ratio = 0.84; **c** ratio = 1.3; **d** ratio = 3

the exact solutions provided by MCS (3,000 sampling points involved), with the relative error controlled within an acceptable range.

3 Probabilistic Low Cycle Fatigue and Ratcheting Analysis Under pLMM Framework

3.1 Linear Matching Method-Driven Neural Network (LDNN) for LCF Life and Ratchet Limit Predictions

The Linear Matching Method-driven neural network (LDNN) is built and employed as the multi-layer perceptron (MLP) for modelling and prediction [23], and the general network structure contains three layers: the input layer, the hidden layer and the output layer, as displayed in Fig. 8.

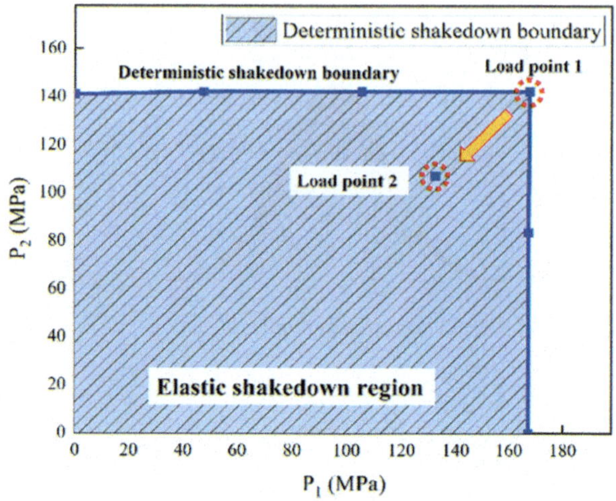

Fig. 7 Deterministic shakedown boundaries of the holed plate and the load conditions for reliability analyses

Table 1 Results of reliability-shakedown analysis by pLMM and the verifications with MCS

Load point index	Failure probability P_f	Verification by MCS	Error %
1	0.2954518037	0.2990909091	1.2167
2	0.04034325841	0.042962963	6.0976

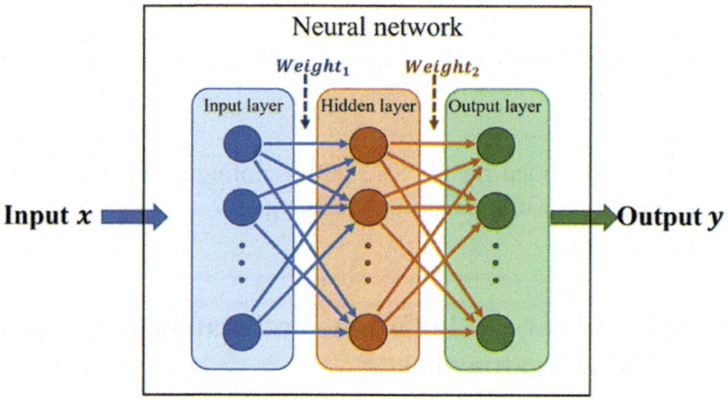

Fig. 8 General structure of a three-layer artificial neural network (ANN)

The data flow commences with the input data vector x passed in through the input layer to the hidden layer, during which the weight of each parameter is also added. Next, in the middle or hidden layer, the ellipsoidal basis function (EBF) [24] existing in each neuron plays the key role in the activation function, leading to the non-linear mapping process which is dependent on the summation of input with weights and bias. For implementing LCF lifetime and ratchet limit modelling, the interpolating functions are formulated by Eqs. (2) and (3), where $\phi_i(x)$ is the basis function of EBF neural network [24], α_i and α_{N+1} are the weight and bias terms, respectively, and the L_f and λ represent the approximations of LCF life and ratchet limit multiplier.

$$L_f(\boldsymbol{x}) = \sum\nolimits_{i=1}^{N} \alpha_i \phi_i(\boldsymbol{x}), \text{ for LCF lifetime modelling} \tag{2}$$

$$\lambda(\boldsymbol{x}) = \sum\nolimits_{i=1}^{N} \alpha_i \phi_i(\boldsymbol{x}) + \alpha_{N+1}, \text{ for ratchet limit modelling} \tag{3}$$

The working flow of LDNN-based probabilistic LCF and ratcheting analyses under the pLMM framework is illustrated by three main steps included:

Step 1. Generating the training dataset by Latin Hypercube Sampling and processing the input dataset with LMM to acquire the structural response;

Step 2. Training [25] and testing the LMM-driven neural network (LDNN);

Step 3. Performing the probabilistic LCF or ratcheting analysis with surrogate models to predict the distribution of key output and the failure probability.

3.2 Benchmark of Probabilistic Low Cycle Fatigue Analysis

Description of the FEA model. To illustrate the applicability of the proposed probabilistic analysis scheme for LCF life assessment, a benchmark of the elbow pipe bend is investigated, and detailed descriptions of the geometry are given out in Fig. 9. Firstly, the structure of the elbow pipe bend is characterized by the dimension parameters in Table 2. The parameters R_O, t_n, R and L denote the outer radius of elbow pipe, nominal pipe wall thickness, bending radius and straight length, respectively. Besides, the bending characteristic, h, is defined as $Rt/{r_m}^2$, where r_m refers to the nominal mean radius of the pipe. The FEA model is discretized by ABAQUS, with the 20-node quadratic brick element C3D20R adopted, which keeps a good balance of numerical precision and computational efficiency by reduced integration technology. And inside the black dotted box, the elbow zone is refined so as to satisfy the requirements of mesh convergence and to capture the prominent stress gradient around this local region, with 4,760 elements created in total.

Adopted material properties and boundary conditions (BCs). The elbow pipe bend is made of austenitic stainless steel 316L, and the temperature-dependent material properties are provided [26] in Table 3, including Young's modulus E, Poisson's ratio ν, the average coefficient of linear thermal expansion α_m and thermal conductivity k. In addition, the temperature-dependent cyclic stress–strain relationship is described by Eqs. (4) and (5) in Ramberg–Osgood (R-O) form, where ε_{ta} is the total true strain amplitude, σ_a is the total true stress amplitude, $\overline{E}$ is the multi-axial Young's modulus, and K and n are the strength coefficient and plastic hardening parameters controlling the cyclic responses.

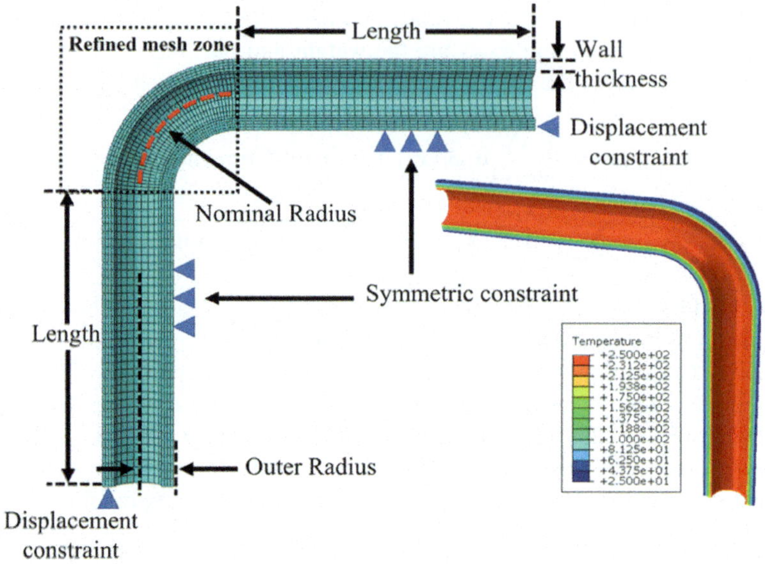

Fig. 9 Descriptions of the geometry and FEA model of the elbow pipe component for probabilistic LCF life and ratchet limit analyses

Table 2 Geometric dimensions of the elbow pipe bend

Outer radius $R_O(mm)$	Wall thickness $t_n(mm)$	Bending radius $R(mm)$	Length $L(mm)$	Bending characteristic h
180	60	500	1500	1.33

Table 3 Adopted temperature-dependent material properties of austenitic stainless steel 316L

Temperature (°C)	20	100	200	300	400
Young's modulus $E\left(10^3\text{MPa}\right)$	200	193	185	176	168
Poisson's ratio ν	0.3				
Coefficient of thermal expansion $\alpha_m\left(10^{-6}/°\text{C}\right)$	15.3	15.9	16.6	17.2	17.8
Thermal conductivity $k(\text{W/mm}\cdot\text{K})$	0.01428	0.01548	0.01698	0.01849	0.01999
K (MPa)	2286	2082	1860	1650	1650
n	0.351	0.339	0.325	0.31	0.31

$$\varepsilon_{ta} = \frac{\sigma_a}{\overline{E}} + \left(\frac{\sigma_a}{K}\right)^{\frac{1}{n}} \tag{4}$$

$$\overline{E} = \frac{3E}{2(1+v)} \tag{5}$$

The half model of the elbow pipe bend is created with the symmetric BCs exerted on the symmetric surfaces, and the displacement constraints are also applied along with the horizontal and vertical directions at each end of the elbow pipe, which are shown in Fig. 9. Besides, on the inner surfaces lies the constant inner pressure, and between the inner and outer surfaces of the pipe wall, there exists the cyclic temperature difference.

Deterministic ratchet limit boundary and alternating plastic response region. By means of the deterministic LMM procedures, the limit boundaries of the elbow pipe bend are established in Fig. 10, by which the structural responses are divided into two different sections: the ratcheting region and the reverse plasticity region. And the typical failure mechanisms in terms of LCF crack initiation and progressive plastic collapse (ratcheting) are compared in Fig. 11. Here, the horizontal and vertical coordinates are normalized by the limit load of the elbow pipe bend, 112.4MPa, and the applied reference temperature difference, 225°C, respectively.

Probabilistic LCF lifetime prediction by LDNN. Considering the uncertain design parameters, including the Material parameters *K* and *n*, Nominal thickness t_n, constant inner pressure P/P_0 and cyclic temperature difference $\Delta T/\Delta T_0$, the statistical distribution of the elbow pipe LCF life, visualized by the plots in Fig. 12, is investigated by employing the proposed LDNN-based surrogate model with 120 neuron pathways inside the hidden layer. Under the existence of random variables, including material property parameters (*K* and *n*), elbow pipe thickness and the inner pressure, the LCF life of elbow pipe bend tends to show the Lognormal distribution.

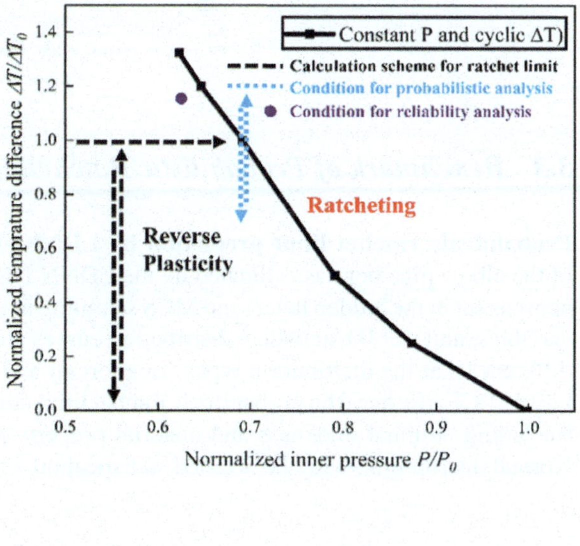

Fig. 10 Deterministic ratchet limit boundary and the reverse plasticity load region for the elbow pipe bend

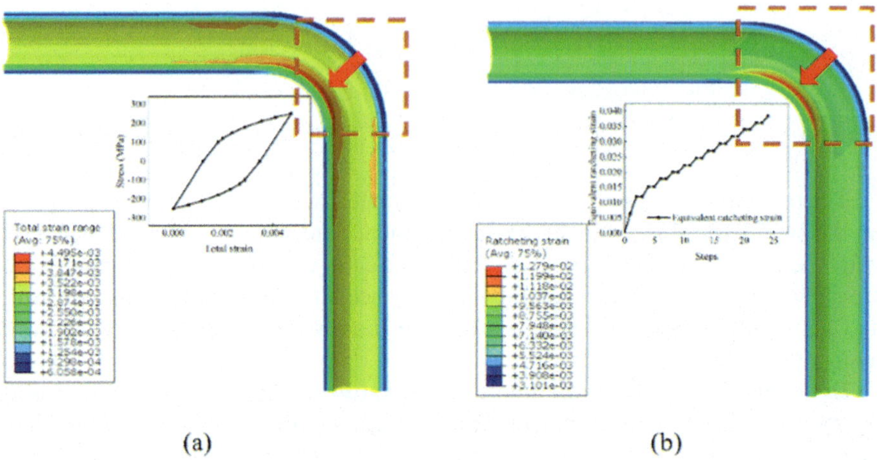

Fig. 11 Typical failure mechanisms in terms of LCF crack initiation and ratcheting for the elbow pipe bend

Fig. 12 Statistical distribution of LCF life of the elbow pipe

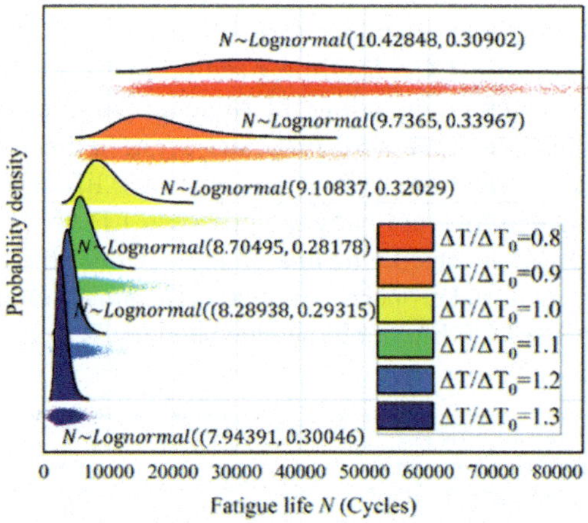

3.3 Benchmark of Probabilistic Ratcheting Analysis

Probabilistic ratchet limit prediction by LDNN. The probabilistic ratchet limit of the elbow pipe bend is estimated by the LDNN-based surrogate model (with 100 neurons set in the hidden layer) and MCS sampling in terms of the predefined random variables, and the 3D statistical distribution curves are plotted in Fig. 13. Definitely different from the distribution types (Lognormal and Weibull) to depict the probabilistic LCF life, here the probabilistic ratchet limit considering the random variables (including nominal thickness and material property parameters K and n) obey the Normal distribution, with the detailed statistical information provided in Table 4. As the level of cyclic temperature difference gradually increases, the mean values of the ratchet limit decrease from 79.04 to 69.42 MPa, whereas the standard deviations, accounting for the dispersion degree of the random variable, reduce slightly and remain stable at high-temperature conditions (for the cases where the normalized temperature difference ranges from 1.1 to 1.3).

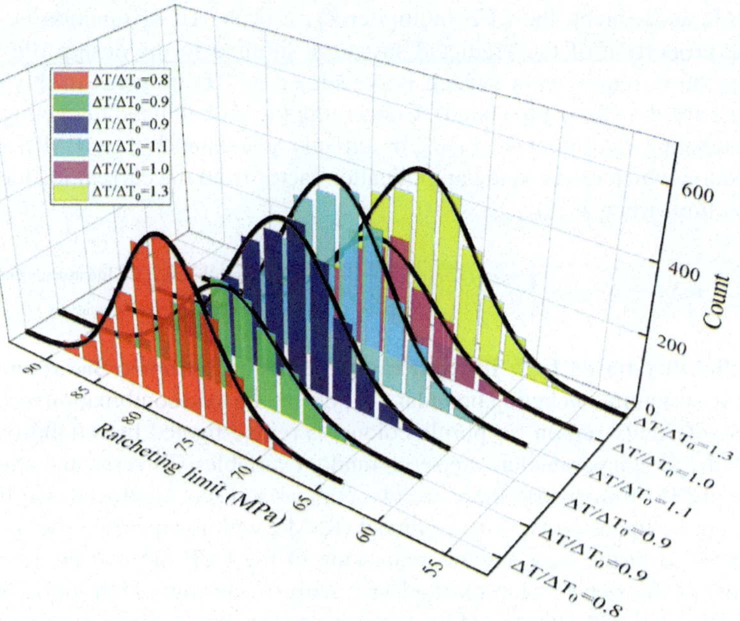

Fig. 13 Statistical distribution of ratchet limit of the elbow pipe under constant inner pressure and various cyclic temperature differences

Table 4 Statistical distribution information of ratchet limit of the elbow pipe under constant inner pressure and various cyclic temperature differences

$\Delta T/\Delta T_0$	Statistical distribution of ratchet limit
0.9	$N \sim Normal$ (76.446, 4.450)
1.0	$N \sim Normal$ (74.162, 4.564)
1.1	$N \sim Normal$ (72.389, 4.667)
1.2	$N \sim Normal$ (70.7137, 4.753)
1.3	$N \sim Normal$(69.416, 4.742)

3.4 Reliability-Based LCF and Ratchet Analyses for the Elbow Pipe Bend

Definition of the unified limit state indicator (ULSI) function. Under the pLMM framework, the unified limit state indicator (ULSI) function is proposed to simplify the subsequent reliability analysis with a concise form by Eq. (6). For the probabilistic LCF life assessment, the LCF multiplier λ_{LCF} in the ULSI function is expressed by the proportion of the predicted structural lifetime to the design life to satisfy the operation requirement (which is predefined as 2,000 cycles in this reliability analysis for the elbow pipe bend). Concerning the probabilistic ratcheting analysis, the ratcheting multiplier $\lambda_{ratcheting}$ is directly generated by the LMM ratcheting procedure, which refers to the amplification factor from the current load level to the ratchet limit when $\lambda_{ratcheting} > 1.0$.

$$\mathbf{G}(X) = \lambda(X) - \mathbf{1} = \begin{cases} \lambda_{LCF}(X) - 1 = \frac{LCF(X)}{L_{design}} - 1, & \text{for probabilistic LCF life assessment} \\ \lambda_{ratcheting}(X) - 1, & \text{for probabilistic ratcheting analysis} \end{cases} \quad (6)$$

Reliability-based LCF life and ratchet limit evaluations. Based on the deterministic structural ratcheting boundary in Fig. 10, the load combination located in the reverse plasticity region (in purple colour) is reinvestigated by reliability analysis, where the design parameters are set as random variables. To verify the effectiveness of the pLMM framework, there are two comparison sets employed: the first one is given out by the response surface model (RSM), with the quartic order polynomial leveraged to fit the least-squares regression of the LCF life and the ratchet limit. Another is the step-by-step elastoplastic analysis, during which the LCF life and ratcheting state are determined by either the cyclic plastic strain range or the accumulative plastic strain, respectively. The failure probability regarding a certain cyclic plastic response is calculated via the direct MCS.

The reliability analysis results of the proposed LDNN surrogate model and the RSM-based method are shown in Table 5, with the failure probability compared to the detailed MCS. It can be observed that with the lowest training points involved, the failure probability of the proposed LDNN-based approach is much closer to validation than the RSM-based method for the probabilistic fatigue and ratcheting evaluations. The error generated by the RSM-based method is mainly caused by insufficient training data, which means to guarantee adequate estimation quality, much more training points should be prepared and input during the fitting process of the RSM model.

Table 5 Results of reliability analyses for LCF and ratcheting failures by LDNN surrogate model, RSM model and MCS validation

Failure mode	LDNN-based method		RSM-based method		Verification of MCS with elastoplastic analysis	
	P_f	N_{LDNN}	P_f	N_{RSM}	P_f	N_{MCS}
LCF	0.1048	120	0.1079	250	0.0819	4452
Ratcheting	0.6635	100	0.6802	220	0.6679	4351

Noting that N_{LDNN} and N_{RSM} indicate the gross number of training points used in the training process of the LDNN and RSM surrogate model respectively, and N_{MCS} refers to the total number of elastoplastic simulations during MSC

4 Probabilistic Creep-Fatigue Analysis Under pLMM Framework

4.1 Linear Matching Method-Driven Neural Network (LDNN) for Creep-Fatigue Life Prediction

The probabilistic creep-fatigue assessment under the pLMM framework is performed by using the Linear Matching Method-driven neural network (LDNN), where the basic interpolating function of creep-fatigue lifetime L is given out by Eq. (7). And the numerical implementation process of this proposed probabilistic creep-fatigue analysis is carried out by using the conjunction of ABAQUS and Isight.

$$L(x) = \sum_{i=1}^{N} \beta_i \varphi_i(x) + \beta_{N+1} \tag{7}$$

4.2 Benchmark of Probabilistic Creep-Fatigue Analysis

Random variables in the probabilistic creep-fatigue analysis. In this case study, the same FEA model, the elbow pipe bend operating under a high-temperature environment, is utilized to introduce the applicability of the proposed probabilistic creep-fatigue analysis under the pLMM framework when solving engineering problems. The uncertain design parameters cover the cyclic yield strength $R_{p0.2}(T)$, the creep Norton law coefficients A and n, Nominal thickness t_n, cyclic inner pressure P/P_0, the cyclic temperature difference $\Delta T/\Delta T_0$, and creep dwell time t.

Creep-fatigue failure mechanism of the elbow pipe. Based on the mean values of the design parameters, the typical failure mechanism of the elbow pipe component is plotted in Fig. 14. Due to the effect of geometric discontinuity at the elbow location, the inner surface of the intrados exhibits the maximum strength of fatigue damage and creep damage simultaneously.

Probabilistic creep-fatigue lifetime prediction. Considering the uncertainty in design parameters of the elbow pipe, the proposed probabilistic creep-fatigue analysis is able to measure the statistical distributions of the creep-fatigue lifetime, as well as the estimation of the statistical information by the proposed LDNN with 120 neurons in the middle layer, as plotted in Fig. 15. The creep-fatigue lifetime follows the log-normal distribution [27], with the logarithmic mean value and the logarithmic standard deviation also fitted.

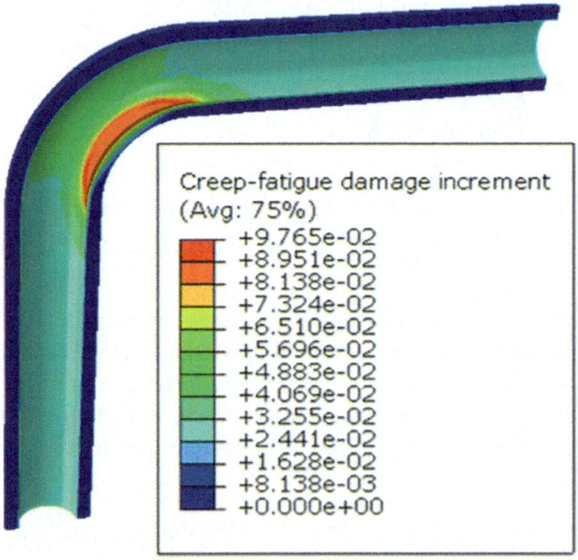

Fig. 14 Creep-fatigue damage increment per cycle of the elbow pipe under the mean values of the design conditions

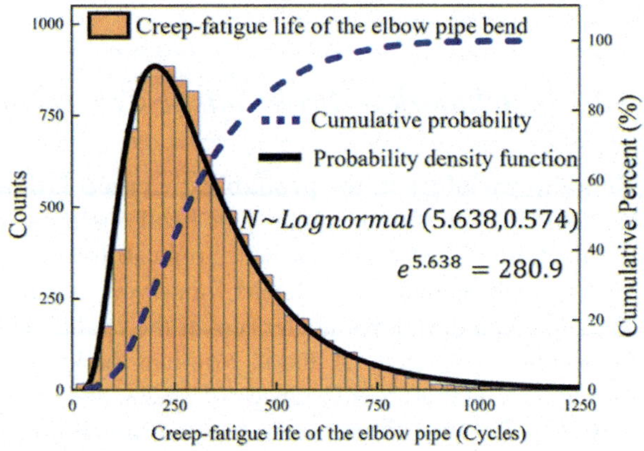

Fig. 15 Statistical distribution of the creep-fatigue responses: **a** damage increment of the elbow pipe; **b** lifetime of the elbow pipe

5 Conclusions

Traditional safety factors are dedicated to pursuing extremely safe designs under uncertain conditions with deterministic assessments, which, inevitably, is an obstacle to achieving precise design and risk management. In this study, a series of probabilistic structural integrity assessment strategies are elaborated under the probabilistic Linear Matching Method (pLMM) framework to consider the uncertainty of the design conditions, where different structural failure behaviours, including shakedown, ratcheting, low cycle fatigue (LCF), and creep-fatigue, are taken into consideration. Detailed benchmarks are also provided, showing the effectiveness and comprehensive applicability of the pLMM framework in engineering problems.

Acknowledgements The authors gratefully acknowledge the support from the National Natural Science Foundation of China (51828501 and 52150710540), the East China University of Science and Technology, and the University of Strathclyde during the course of this work.

References

1. Unfired pressure vessel standard EN 13445, Part 3: Design, Annex B Direct route for design by analysis. European Committee for Standardization (CEN) (2014)
2. Ainsworth RA e. R5: an assessment procedure for the high temperature response of structures. Procedure R5: Issue 3. Gloucester: UK: British Energy Generation Ltd. (2014)
3. ASME. ASME Boiler & Pressure Vessel Code, Section VIII Rules for Construction of Pressure Vessels Division 2, Alternative Rules. The American Society of Mechanical Engineers, New York (2015)
4. Siemaszko, A.: Computational methods for shakedown and limit reliability analysis. Inelastic Behaviour of Structures under Variable Repeated Loads, Springer, Berlin, pp. 345–359 (2002)
5. Heitzer, M., Staat, M.: Structural reliability analysis of elasto-plastic structures. Safety and Reliability AA Balkema, Rotterdam, Brookfield 1999;513:518
6. Staat, M., Heitzer, M.: Part VII probabilistic limit and shakedown problems. Numerical methods for limit and shakedown analysis, 217 (2003)
7. Staat, M.: Limit and shakedown analysis under uncertainty. Int. J. Comp. Meth.-Sing. **11**, 1343008 (2014)
8. Li, X.-Q., Bai, G.-C., Song, L.-K., Wen, J.: Fatigue reliability estimation framework for turbine rotor using multi-agent collaborative modeling. Structures **29**, 1967–1978 (2021)
9. Halfpenny, A., Chabod, A., Czapski, P., Aldred, J., Munson, K., Bonato, M.: Probabilistic fatigue and reliability simulation. Procedia Struct. Integr. **19**, 150–167 (2019)
10. Pierce, S.G., Worden, K., Bezazi, A.: Uncertainty analysis of a neural network used for fatigue lifetime prediction. Mech. Syst. Signal Process. **22**, 1395–1411 (2008)
11. Durodola, J.F., Ramachandra, S., Gerguri, S., Fellows, N.A.: Artificial neural network for random fatigue loading analysis including the effect of mean stress. Int. J. Fatigue **111**, 321–332 (2018)
12. Li, X., Zhang, Y., Abbassi, R., Khan, F., Chen, G.: Probabilistic fatigue failure assessment of free spanning subsea pipeline using dynamic Bayesian network. Ocean. Eng. **234**, 109323 (2021)
13. Mortazavi, S.N.S., Ince, A.: An artificial neural network modeling approach for short and long fatigue crack propagation. Comput. Mater. Sci. **185**, 109962 (2020)

14. Zhang, X.-C., Gong, J.-G., Xuan, F.-Z.: A physics-informed neural network for creep-fatigue life prediction of components at elevated temperatures. Eng. Fract. Mech. **258**, 108130 (2021)
15. Bartošák, M.: Using machine learning to predict lifetime under isothermal low-cycle fatigue and thermo-mechanical fatigue loading. Int. J. Fatigue **163**, 107067 (2022)
16. Chen, H., Chen, W., Li, T., Ure, J.: Shakedown analysis of a composite cylinder with a cross-hole. J. Press. Vessel. Technol. **133** (2011)
17. Chen, H., Ponter, A.R.S.: A direct method on the evaluation of ratchet limit. J. Press. Vessel. Technol. **132** (2010)
18. Chen, Y., Chen, H., Luan, W.: Shakedown, ratcheting and fatigue analysis of cathode coating in lithium-ion battery under steady charging-discharging process. J. Mech. Phys. Solids **150**, 104366 (2021)
19. Cho, N.-K., Wang, R.-Z., Ma, Z., Chen, H., Xuan, F.-Z.: Creep-fatigue endurance of a superheater tube plate under non-isothermal loading and multi-dwell condition. Int. J. Mech. Sci. **161**, 105048 (2019)
20. Wang, X., Ma, Z., Chen, H., Liu, Y., Shi, D., Yang, J.: Creep rupture limit analysis for engineering structures under high-temperature conditions. Int. J. Press. Vessel. Pip. **199**, 104763 (2022)
21. Wang, X., Yang, J., Chen, H., Ma, Z., Xuan, F.: Effect of constraint on cyclic plastic behaviours of cracked bodies and the establishment of unified constraint correlation. Eur. J. Mech.-A/Solids **97**, 104857 (2023)
22. Tran, T.N., Kreissig, R., Staat, M.: Probabilistic limit and shakedown analysis of thin plates and shells. Struct. Saf. **31**, 1–18 (2009)
23. Luo, J., Lin, S., Ni, J., Lei, M.: An improved fingerprint recognition algorithm using EBFNN. In: Proceedings—2nd International Conference on Genetic and Evolutionary Computing, WGEC 2008 (2008)
24. Mak, M.W., Li, C.K.: Elliptical basis function networks and radial basis function networks for speaker verification: a comparative study. In: IJCNN'99 International Joint Conference on Neural Networks Proceedings (Cat. No. 99CH36339), vol. 5, pp. 3034–3039 (1999)
25. Griffiths, G.W., Płociniczak, Ł, Schiesser, W.E.: Analysis of cornea curvature using radial basis functions—part I: methodology. Comput. Biol. Med. **77**, 274–284 (2016)
26. R.M.: Design and Construction rules for Nuclear Power Generating Stations, AFCEN, France (2015)
27. Wang, R., Liu, X., Hu, D., Meng, F., Li, D., Li, B.: Zone-based reliability analysis on fatigue life of GH720Li turbine disk concerning uncertainty quantification. Aerosp. Sci. Technol. **70**, 300–309 (2017)

Peak Load Prediction of Human Bone Proximal Femur: Sensitivity to Tissues Strength and Geometry

Aurora Angela Pisano and Paolo Fuschi

Abstract In a peculiar structure like the human proximal femur, the relevant literature oriented to describe the bone mechanical behavior gives a large variety of analysis methods, models and approaches and this due to many factors such as: age, life-style, ethnicity, gender, diseases in progress, etc. The main challenge is indeed the great variability of the input data/parameters to be used in the mechanical analysis. In this paper a sensitivity analysis is performed on a numerical predictive method for the evaluation of the collapse/limit load of the human proximal femur recently proposed by the authors. The influence of the cortical and trabecular thicknesses and strengths on the numerical findings is investigated. The results obtained show a great sensitivity of the predicted ultimate load to the strengths of bone tissues rather than to thicknesses.

Keywords Sensitivity analysis · Human proximal femur · FE-based limit analysis · *In silico* tests results

1 Introduction

The fracture of the human proximal femur is a very common pathology, it is known that every year hundreds of thousands people are affected worldwide by this pathology, especially the elderly, and this number is expected growing in the future aging industrialized societies. This explains the huge number of literature studies, both experimental and theoretical, which try to relate experimentally detectable quantities with the risk of rupture of the femur as well as the construction of mechanical models for predicting such rupture.

A. Angela Pisano (✉) · P. Fuschi
Department PAU, University Mediterranea of Reggio Calabria, 89124 Reggio Calabria, Italy
e-mail: paolo.fuschi@unirc.it

P. Fuschi
e-mail: aurora.pisano@unirc.it

G. Garcea and D. Weichert (eds.), *Direct Methods for Limit State of Materials and Structures*, Lecture Notes in Applied and Computational Mechanics 101,
https://doi.org/10.1007/978-3-031-29122-7_5

As well known, the risk of spontaneous or fall induced fractures of the femur is mainly associated to the presence of osteoporosis. Therefore, the clinical assessment of such risk is very often based on the measurement of bone mineral density (BMD) performed by dual-energy X-ray absorptiometry (DXA). It has been also shown that the BMD measures are not always reliable for the purpose and that better predictions of femoral strength can be obtained using subject-specific finite element (FE) models based on Computer Tomography (CT) images. The literature on this subject is truly vast so here reference is made to a recent review paper [1] and references therein, correlating experimental and clinical evidence of BMD and/or CT-finite element analysis with strength and fracture risk of human proximal femur.

Significant efforts have also been made in recent years in defining damage and fracture propagation models framed in the context of the extended finite element method (XFEM) [2, 3] and of the partition of unity finite element method (PUFEM) [4]. These models can be very complex, especially in 3D, and they are often not free from numerical problems and mesh dependence. These approaches to the problem, which try to simulate the initiation and propagation of the fracture until the bone breaks, are with no doubts interesting, but certainly deserve further studies.

On the other hand, the importance of the problem and its economic impact on public health, pushed researchers to find faster and reliable methods to predict the proximal femur collapse so to use this information in clinical decisions. The method promoted in this article, which belongs to a recent research program by the authors [5], can be framed within the latter faster predictive approaches, and applies the *theory of limit analysis* in the context of human bones.

The theory of limit analysis allows to determine the peak load or plastic collapse load of a structural element, without following the evolutive load history that has determined it, then it does not describe post-elastic phenomena, damage or fracture propagation, going rather to the direct determination of the ultimate load. A significant simplification offered by the limit analysis numerical procedure applied in this paper is that it is based only on elastic analyses of the proximal femur model. In particular, we refer to a numerical procedure, known as Elastic Compensation Method (ECM), able to define a *lower bound* to the collapse load. The ECM [6, 7] makes use of *sequences of linear elastic analyses* within which the elastic moduli of the "materials" the bone is made with are systematically varied to simulate the process of stress redistribution arising within the bone just before its collapse. The robustness of the method has already been experienced by the authors in several applications of engineering interest and for structures made up of composites, [8–10], or reinforced concrete [11–13] and a first promising attempt has also been made for "bone materials" [5, 14].

The application of the ECM to the human proximal femur essentially requires the definition of a *mechanical model* (geometry, thicknesses, loading and boundary conditions) together with *admissible stress domains* (yield surfaces) for the macro-constituent materials, namely the cortical and trabecular tissues. In the present paper attention will be focused on those parameters that more than others can influence the evaluation of the limit load via the promoted approach. Precisely, referring to a simplified mechanical model of the human proximal femur, a sensitivity analysis is carried out by varying both the cortical thickness, often indicated in the literature as a crucial parameter [15], (the trabecular thickness varying consequently), and the strength values entering the Tsai-Wu-type admissibility stress domain hereafter adopted for both cortical and trabecular tissues. The numerical results obtained show a great sensitivity of the predicted collapse load value on the material strengths rather than on the thickness of bone tissues.

2 Simplified Mechanical Model of Human Proximal Femur

As declared in the introductory section, the main object of the present paper is to study the influence on the limit load values of the human proximal femur exerted by the cortical thickness and the strengths of cortical and trabecular tissues.
To achieve the above goal, first of all, it is necessary to set up a mechanical model. A simplified (smoothed and regular) shape of the bone, already applied by the authors with success in [5], is adopted in order to make easier the implementation of a numerical FE model.

The geometry of the bone is assumed as in Fig. 1a, which represents a median section of a 3D model (the one used in the analyses) containing the shaft and neck axes. With reference to a Cartesian orthogonal system, the drawn section lies in

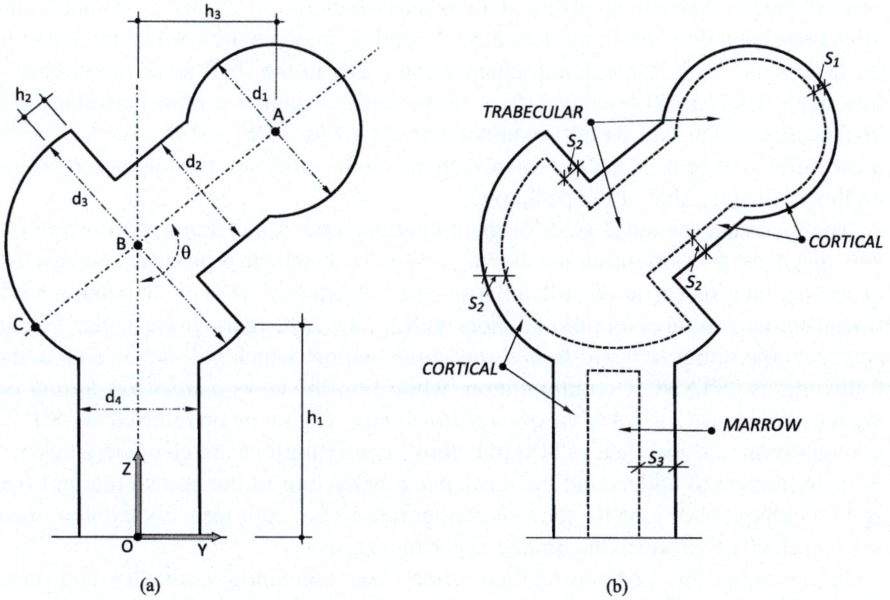

Fig. 1 Geometries of the (simplified) human proximal femur: **a** Characteristic outer dimensions; **b** Inner dimensions

Table 1 Geometrical data of the simplified proximal femur model

Dimension		Reference
d_1 [mm]	51.60	Holzet al. 2009 [16]
d_2	37.40	Holzet al. 2009 [16]
d_3	65.64	Michelotti and Clark 1999 [17]
d_4	35.40	Michelotti and Clark 1999 [17]
h_1	80.00	Dall'Ara et al. 2013 [18]
h_2	5.30	Michelotti and Clark 1999 [17]
h_3	44.00	Michelotti and Clark 1999 [17]
AC	90.26	Michelotti and Clark 1999 [17]
θ [grad]	130	Yang et al. 2014 [19]

the plane (Y, Z), with Y oriented along the radial or medio-lateral direction and Z coincident with the axial or inferior-superior shaft axis; eventually the X axis, orthogonal to such plane, is oriented in the posterior-anterior direction. The values of the dimensions indicated in Fig. 1a are reported in Table 1, together with the references from which they are borrowed.

Figure 1b depicts the internal geometry of the proximal femur with an external part made of cortical tissue and an internal part made of trabecular and marrow tissues. Unlike what was assumed by the authors in [5], the thickness of the cortical tissue is here considered variable in different areas into which the simplified proximal femur model has been divided; in particular S_1, S_2 and S_3 indicate the cortical thickness in the head zone, in the neck and trochanter zones and in the shaft zones, respectively (see again Fig. 1b). Following [15], it is possible to define a ratio between these thicknesses which can be approximately assumed as $S_1/S_2 = 0.5$, $S_1/S_3 = 0.25$, these ratios will be used later on to investigate on the effects of thicknesses variation on the predicted value of the peak load.

The mechanical model used for the numerical analyses is defined following the two mechanical configurations shown in Fig. 2a, b which reproduce the in-vitro experimental tests carried on till collapse by Dall'Ara et al. [18] on 36 pairs of fresh frozen human proximal femurs of donors with age 76 ± 12 years. In particular, Fig. 2a considers the femur in a *one-legged standing* position, which will be referred in the following as "STANCE" configuration, while Fig. 2b shows a *sideways falling on the postero-lateral aspect of the greater trochanter*, from now on referred as "SIDE" configuration. As specified in [18], the above configurations are characteristics and of main interest to understand the mechanical behaviour of the human femur. Load and boundary conditions for the two configurations are appropriately fixed in order to numerically reproduce the quoted experimental tests.

In particular, the load was applied in the plane containing both neck and proximal shaft axes with an angle of 20° and 60° from the proximal shaft axis for the STANCE and SIDE configurations, respectively. In order to simulate the real action of the testing machine, the loads are actually applied as distributed loads on a cap

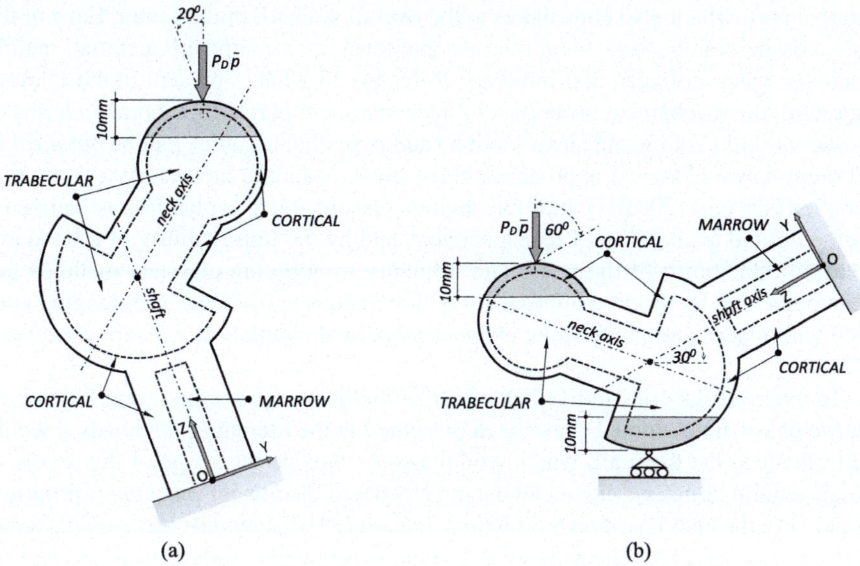

Fig. 2 Mechanical model, boundary and loading conditions, of the human proximal femur in: **a** STANCE configuration; **b** SIDE configuration

of the femoral head which extends itself to a depth of 10 mm. These caps, for both configurations, are highlighted in Fig. 2 and correspond to the grey parts under the drawn resultant concentrated loads. The same type of representation was used for the constraints in the SIDE configuration, where the roller constraint is extended to all the grey part (a zone of 10 mm depth) drawn above the schematized symbol of the roller.

3 Constitutive Assumptions: Tsai-Wu-Type Yield Surface

The constitutive characterization of the human proximal femur—to be meant in this context as the definition of the material mechanical properties of cortical and trabecular tissues and of their mutual dependence under loading—is a challenging problem, and this not only for the inherent complex structure of the human bones, but also because it depends on several factors, such as age, gender, geographical location and ethnic group. In spite of this, in the last decades, a wide number of researchers have faced this problem with the aim to set up numerical models able to forecast the bone mechanical response to assigned loads in different configurations.

It is well known that, at a macroscopic level, the proximal femur is made of two different types of tissues such as the *cortical tissue*, which constitutes the external and most resistant part of the bone, and the *trabecular* tissue, which constitutes its internal part, which also contributes to the overall strength of the femur. Each of the above tissues can be viewed, at microscopic level, as a composite material, mainly made by water, collagen and minerals. Referring to what is present in the relevant literature, the mechanical properties of the constituent parts of the bone, in terms of elastic moduli (Young and shear moduli) and material strengths, can be obtained in different ways. Classical approaches make use of standard laboratory tests on real samples (see e.g. [20, 21]), but, for obvious reasons, their application is hampered by the limited availability of real specimens and by the impossibility of a standardized reproducibility of the tests. The difficulty in applying classical methods has been overcome by modern approaches which make use of diagnostic imaging coupled with macro and micro finite element numerical simulation; see, among others, [22–29].

To interpret the constitutive behaviour of the human proximal femur beyond its elastic phase, many models have been presented in the literature, obviously it would be difficult to list them all, which would also be outside the scope of this study. In broad general terms, we can recall the models based on *damage mechanics* proposed in [30–33], the ones based on *fracture mechanics*, [34], the models based on *plasticity* (see e.g. [35] and references therein) and the more recent models proposed by [36, 37], in which a continuum *plasticity-damage* formulation is *enriched by nonlocal and/or gradient terms*. Moreover, there are specific models for trabecular and cortical tissues. The former have been proposed by [38–41], while the latter have been addressed in [42–45]. A general quadratic yield criterion has been also proposed by Zysset and Co-Workers [35]. This criterion adapts its shape from an ellipsoid to a cylinder or a cone surface so being able to fit, at every hierarchical level, the different behavior of the human bone tissues.

For the analysis method followed in this article, it is also essential to underline that *bones exhibit a plastic behaviour*, accompanied by dissipative processes, observable both at micro and macro scale. In particular, at micro scale, the plastic deformation are due to the breaking of hydrogen bonds within single collagen molecules followed by the breaking of bonds and intermolecular sliding within collagen fibrils, see e.g. [46–48]. This intermolecular sliding produces energy dissipation and represents the mechanism of plastic deformation. At macro scale, plastic deformation and energy dissipation are due to fibrillar sliding. Eventually, it is well accepted, that bones exhibit a plastic behaviour and than undergo plastic deformations; indeed the ductility of bones is a positive characteristic, it helps to absorbs impacts rendering them fracture resistant, [49].

Keeping in mind that bone is definitely a composite material, in the follow reference is made to one of the more accredited criterion for composites, that is the Tsai-Wu criterion, [50]. The Tsai-Wu criterion is hereafter used to interpret the constitutive mechanical behavior of both cortical and trabecular tissues. Indeed, the experimental stress-strain diagrams of [47, 51] justify the further hypothesis of perfect plasticity, i.e. the Tsai-Wu surface is assumed as a yield surface which plays the role of plastic potential. The same choice was made by the authors in a previous article, [5], of which this study is a further step forward.

In the principal stress space the Tsai-Wu-type yield surface writes

$$G_{11}\sigma_{11} + G_{22}\sigma_{22} + G_{33}\sigma_{33} + F_{1111}\sigma_{11}^2 + F_{2222}\sigma_{22}^2 + F_{3333}\sigma_{33}^2 + 2F_{1122}\sigma_{11}\sigma_{22} + 2F_{1133}\sigma_{11}\sigma_{33} + 2F_{2233}\sigma_{22}\sigma_{33} = 1 \quad (1)$$

with the stresses coefficients, G_{ii}, F_{iiii}, F_{iijj}, given by

$$G_{ii} = \frac{1}{\sigma_i^+} - \frac{1}{\sigma_i^-} \qquad F_{iiii} = \frac{1}{\sigma_i^+\sigma_i^-} F_{iijj} = \frac{1}{2}\left(\frac{1}{\sigma_i^+\sigma_i^-} + \frac{1}{\sigma_j^+\sigma_j^-} - \frac{1}{\sigma_{ij}^2}\right) \quad (2)$$

In Eq. (1), σ_{11}, σ_{22}, and σ_{33} are principal stresses, while in Eq. (2) σ_i^+, σ_i^-, σ_{ij} are the moduli of ultimate strengths in tension, compression and shear, respectively, with $i, j = 1, 2, 3$ referring to direction and plane of orthotropy. Moreover, the repeated indices in Eq. (2) do not imply summation. To be secure that the Tsai-Wu-type yield surface has an ellipsoidal shape, containing the stress space origin, also the condition $F_{iiii}\, F_{jjjj} - F_{iijj}^2 \geq 0$ has to be satisfied. The yield surface presented above is then quadratic and convex, it follows the standard transformation rules assuring invariance and symmetry, moreover it takes into account stress or strain components interaction.

A typical representation of the Tsai-Wu-type yield surface in principal stresses space is the one drawn in Fig. 3. The shape is that of an ellipsoid whose inclination

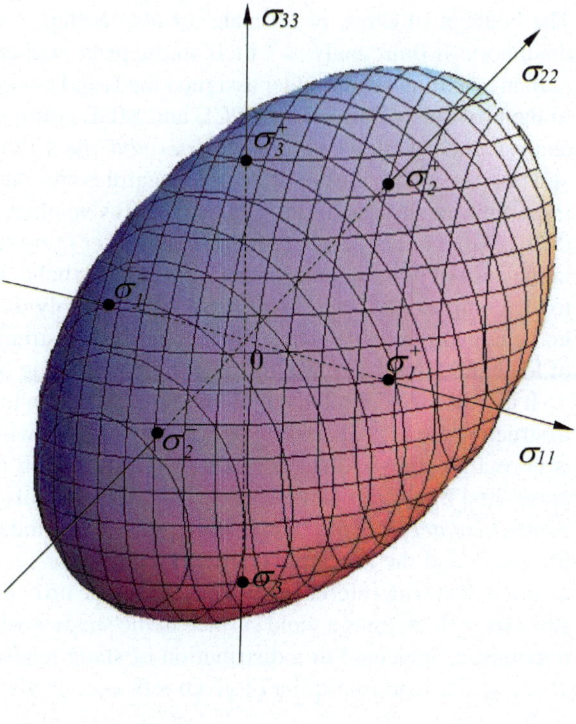

Fig. 3 Typical Tsai-Wu surface in principal stress space

with respect to the axes varies with the ultimate strengths values σ_i^+, σ_i^-, σ_{ij}, which are obviously different for cortical and trabecular tissues, so defining two distinct Tsai-Wu-type yield surfaces, one for the cortical and another one for the trabecular tissue. The strength values, for different fixed thicknesses of bone tissues, will also be used to carry on the sensitivity analysis.

At closure of this section, it is worth to remark that the ultimate strengths, different for the trabecular and the cortical tissues, will be assumed in the shape $\sigma_1^+ \equiv \sigma_2^+, \sigma_1^- \equiv \sigma_2^-, \sigma_{13} \equiv \sigma_{23}$, so adopting for both tissues a *transversally isotropic* behavior. Marrow will instead be assumed isotropic. Eventually, the values of the *elastic* parameters will be borrowed from the literature referring to a transversally isotropic cortical tissue and to an orthotropic trabecular tissue. The latter choices however *do not affect* the results because, as known, the limit load does not depend from the elastic behavior of the structure. The great variation of the local elastic properties in terms of their values (see e.g. [56, 57] for quantitative information) is indeed a drawback for analyses whose results are affected by the evolutive phenomena preceding the collapse. Limit analysis overcomes this drawback and the assumed values of the elastic constants are here chosen just to initialize the applied numerical procedure whose details are given in the next section.

4 Limit Load Prediction of Human Proximal Femur

The concept of limit, or ultimate, or also collapse, load of a structure belongs to the theory of limit analysis which, in the promoted approach, has been applied to a human proximal femur under assigned loads and boundary conditions corresponding to the previously defined STANCE and SIDE configurations. As well known, limit analysis theory allows to *directly determine the ultimate load* of a generic structure and this, as said at closure of Sect. 3, regardless its elastic behavior and, what is most important, regardless all the evolutive diffusive phenomena (damage, plasticity and the like) that occur before collapse. This latter property is particularly advantageous in the case of human femur, whose post elastic behaviour is not easy to describe due to the complexity of the "bone material". To apply the limit analysis theory only the knowledge of a yield surface (depending on the strength values of the material) and of load and boundary conditions, to which the bone is subjected, are required.

It is worth to remind that limit analysis theory, in its classical formulation, applies to structures made of perfectly plastic materials for which the plastic strain rate vector is normal to the yield surface. This latter condition is known as normality rule (or associated flow rule) and the materials that meet this requirement are referred to as *standard materials*, [52]. In this framework, two fundamental theorems hold, namely the *static* and the *kinematic* theorem, which, for sake of completeness, are briefly recalled next with reference to a structure made up of a material whose yield function, say $f(\boldsymbol{\sigma})= 0$, defines a yield surface in the shape given in Fig. 3. The structure is, by hypothesis, subjected to a distribution of static loads expressed as $\mathbf{P} = P\bar{\mathbf{P}}$, where P is a scalar load multiplier of given reference loads $\bar{\mathbf{P}}$.

The static theorem of limit analysis states that: "the exact limit load multiplier, say P_L,—i.e. the multiplier which gives the limit load $\mathbf{P}_L = P_L\bar{\mathbf{P}}$—is the largest one among all the possible *lower bound* multipliers, say P_{LB}, corresponding to statically and plastically admissible stress fields $\boldsymbol{\sigma}$, that is stress fields in equilibrium with the applied loads $P_{LB}\bar{\mathbf{P}}$ and satisfying the condition $f(\boldsymbol{\sigma}) \leq 0$". It is $P_{LB} \leq P_L$ and, for this reason, the theorem is also known as *lower bound* theorem giving rise to the so-called static approach to the determination of the limit load. The static approach considers only the equilibrium and the yield condition, no matter the kinematics at incipient collapse state.

The kinematic theorem of limit analysis states that: "the exact collapse load multiplier P_L is the smallest one among all the possible *upper bound* multipliers, say P_{UB}, corresponding to the set of all kinematically and plastically admissible velocity fields, say $\dot{\mathbf{u}}$". The displacement rates $\dot{\mathbf{u}}$ are compatible with the plastic strain rates $\dot{\boldsymbol{\varepsilon}}^p$ at collapse and the latter obey the normality rule, i.e., $\dot{\boldsymbol{\varepsilon}}^p = \dot{\lambda}\partial f/\partial\boldsymbol{\sigma}$ with $\dot{\lambda} > 0$ where $f(\boldsymbol{\sigma}) = 0$ and $\dot{\lambda} = 0$ otherwise. It is $P_{UB} \geq P_L$ and, for this reason, the theorem is also known as *upper bound* theorem giving rise to the so-called kinematic approach to the determination of the limit load. The kinematic approach considers only displacement and strain rate fields (which define a collapse mechanism), no matter the stress distribution.

For standard perfectly plastic materials the lower and the upper bound multipliers *coincide* with the exact limit load multiplier, for nonstandard materials the above two theorems can be rephrased enabling one *to bracket* the collapse load in a direct manner through the so-called "Radenkovic approach" for nonstandard limit analysis, see again [52]. During the last decades, many different procedures, have been developed for computing the limit load of real structural problems, see e.g. the contributions in the monographs [53, 54]. Among these procedures it must be included the one referred in this study and summarized in the next Section, known as Elastic Compensation Method (ECM). The ECM leads to the evaluation of a lower bound to the collapse load in the spirit of a static approach to limit analysis.

4.1 The Elastic Compensation Method for the Evaluation of a P_{LB}

The ECM it is a *FE-based iterative procedure* which evaluates the maximum multiplier P_{LB} of the applied loads $\mathbf{P}$ for which the stress distribution in the analyzed structure (the proximal femur in this context) is statically and plastically admissible. With reference to Fig. 3, the geometrical interpretation of such circumstance is that all the stress points, representing the stress state within the proximal femur under study, are located *inside* or, at least, *on* the yield surface. The ECM procedure is known in the literature and the authors have already successfully applied and extended it in several contexts, namely to structures made up of materials such as, steel, composites, concrete and also bone tissues (see e.g. [5, 9, 13]). Therefore, the Reader can find details on the numerical procedure by consulting the quoted papers.

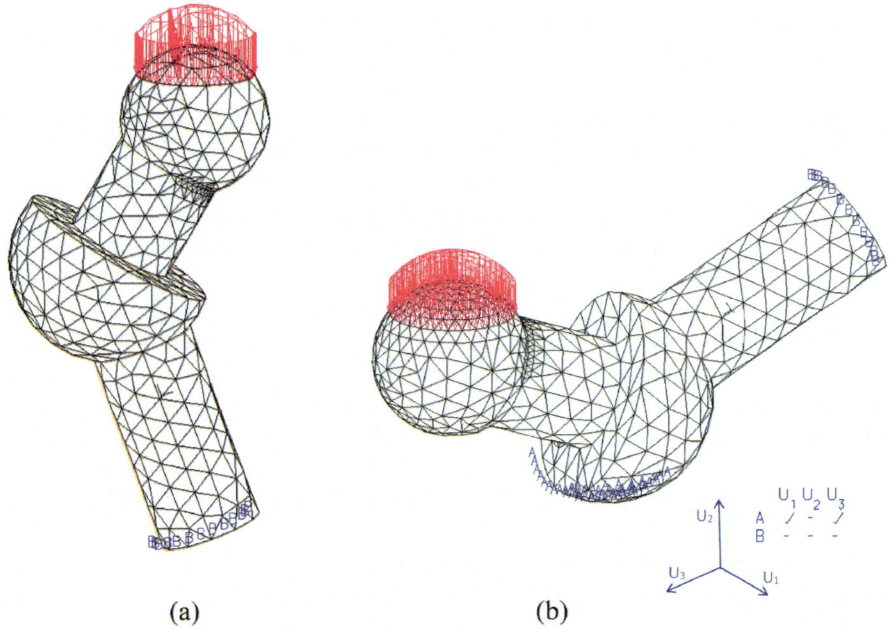

Fig. 4 Meshes, applied nodal loads and constraints in the FE models of the proximal femur: **a** STANCE configuration; **b** SIDE configuration

For completeness, the three main steps of the ECM, based on FE elastic analysis of the proximal femur, and the rationale on which the method is based are recalled hereafter.

Preliminarily, in Fig. 4a, b are reported the two meshes of 10-nodes 3D-solid tetrahedral finite elements associated to the mechanical models given in Fig. 2a, b, respectively. For both configurations the reference loads, acting on the cap of the femoral head (refer again to Fig. 2a, b), are applied (as point loads) on the FE nodes falling within the cap. The nodes marked with B are the fully constrained ones at the section where the shaft of the bone specimen was cut and clamped to the testing device. The nodes marked with A, on the greater trochanter of SIDE configuration, are those where the vertical displacement (DOF U_2) is not allowed being free the horizontal displacements (DOFs U_1 and U_3) so coming true the roller constraint applied during the experimental test (see Fig. 2b). Knowing that the output of an FE analysis gives, among other results, the principal stresses at each Gauss points (GPs) of each element; at each GP it can be easily verified the position of the corresponding stress point with respect to the assumed yield surface. Indeed, for the application of the ECM, principal stresses are evaluated *at element level*, by averaging their values between all the GPs within the same element. Eventually, stress points inside or on the yield surface are representative of stress states statically and plastically admissible, while stress points outside the yield surface are not admissible. The procedure then acts on these latter not admissible stresses, if any, and tries to redistribute them in such a way to obtain admissible stresses in the whole FE mesh of the bone. Because yield domains for cortical and trabecular tissues are different, the above check of admissibility for the stresses is made separately in the different tissues of the proximal femur.

- ECM-step #1. An FE model of the human proximal femur is built starting from its mechanical model in terms of geometry, boundary and loading conditions as well as elastic material parameters. In particular, the acting loads are defined in the shape $P_D \bar{p}_i$ where $\bar{p}_i$ are given applied *basic loads* specified in modulus, direction and versus and P_D denotes a scalar *design* load multiplier which is fixed as specified next. To start, a first multiplier for the loads, say $P_D^{(1)}$, is fixed and a *first sequence* of FE elastic analyses is carried on, using the adopted FE model.
- ECM-step #2. The ECM acts *by reducing the elastic parameters* of the elements for which the stress points, representative of the element stress state, are not admissible, and this to bring them back into, or onto, the yield domain [5]. It should be emphasized that, being in the elastic field, a reduction of the elastic parameters produces a proportional reduction of the stresses. At this stage, a new FE elastic analysis, or iteration, is performed within the current sequence, i.e. on keeping fixed the acting loads, $P_D^{(1)} \bar{p}_i$, but with the updated elastic moduli, and this with the aim of re-distribute the stresses within the whole mesh. The iterations continue until all the element stresses, for the loads $P_D^{(1)} \bar{p}_i$, just reach or are below their corresponding yield values.
- ECM-step #3. Further sequences of elastic analyses are carried on, (i.e. $P_D^{(2)} \bar{p}_i$, $P_D^{(3)} \bar{p}_i$, …), each one with an increased value of the initial load multiplier P_D, (i.e. $P_D^{(i+1)} > P_D^{(i)}$), and the ECM stress redistribution procedure repeated till further load increases do not allow all the stress points to be brought below or onto the yield surface. The greater value of P_D for which the stresses can be redistributed gives the searched (maximum) lower bound multiplier P_{LB} and, eventually, the limit load $P_{LB} \bar{p}_i$.

5 Sensitivity Analysis

This section reports, for both STANCE and SIDE configurations, the results of a sensitivity analysis carried out by considering different thicknesses for the cortical tissue (the trabecular's thickness, as said, varies consequently) and, for each fixed thickness value, by considering different strength values of cortical and trabecular tissues. The choice of the cortical thickness values and of the cortical and trabecular strength values is based on the reasoning reported below.

Concerning the choice of the cortical thickness values, reference is made to a study carried on by [55] on hundreds of patients. This study found that the values of the thickness (lateral and medial) of the cortical, measured at 5 cm below the less

Table 2 Cortical thicknesses of the proximal femur models

Model N.	S_3 [mm]	S_2 [mm]	S_1 [mm]
GM_1	5.7	2.85	1.425
GM_2	7.0	3.5	1.75
GM_3	8.3	4.15	2.075

trochanter, are equal to 7 ± 1.3 mm. Therefore, three geometric models, named GM_1, GM_2, GM_3, are considered for which the thickness of the cortical in the shaft (S_3) is assumed 5.7, 7 and 8.3 mm, respectively. Moreover, the cortical thickness in the head (S_1) and in the neck and trochanter areas (S_2) are varied as specified in Sect. 2, that is using the ratios $S_1/S_3 = 0.25$ and $S_1/S_2 = 0.5$. All the cortical thicknesses used in the three geometric models are given in Table 2.

On the other hand, the ultimate strength values for cortical and trabecular tissues are shown in Table 3, together with the referenced papers and the choices adopted to fix some of the strength values. Such values are function of the apparent density values, [56]. Three pairs of apparent density values are considered, corresponding to the minimum, average and maximum values of apparent density in cortical and in trabecular. These three pairs of apparent density values lead to the definition of three different materials, hereinafter referred to as Mat.1, Mat.2 and Mat3. Precisely, Mat.1 is obtained by assuming $\rho_{cortical} = 1.5\,\mathrm{g/cm^3}$ and $\rho_{trabecular} = 0.1\,\mathrm{g/cm^3}$; Mat.2 by assuming $\rho_{cortical} = 1.75\,\mathrm{g/cm^3}$ and $\rho_{trabecular} = 0.4\,\mathrm{g/cm^3}$; Mat.3 by assuming $\rho_{cortical} = 2.0\,\mathrm{g/cm^3}$ and $\rho_{trabecular} = 0.7\,\mathrm{g/cm^3}$. Obviously other combinations of values for $\rho_{cortical}$ and $\rho_{trabecular}$ would be possible, however it is likely that the material strengths assumed above correspond to a femur in condition of major fragility, medium resistant and higher resistant.

The values of the elastic moduli are also function of the apparent density ρ, see [57] and they are reported in Table 4 for the set of three materials considered. It should be pointed out that: for all the examined cases, the marrow has been assumed isotropic, homogeneous and incompressible with $E = 20$ MPa and $\nu = 0.499$, after [34]. As said, all the elastic parameters (for cortical, trabecular and marrow) are assigned only at the beginning of the first sequence of analyses, i.e. to initialize the procedure. They will remain unchanged within the elements that remain in the elastic regime (this is always the case in the marrow), while they will be updated within the elements of cortical and trabecular tissues when stress redistribution has to be performed. It is worth to remark once again that, as assured by limit analysis theory, their real values do not affect the value of the lower bound multiplier.

The limit analysis of the proximal femur via ECM was then carried on 18 times, 9 for the STANCE configuration, 9 for the SIDE one. Each of the three geometric models was discretized in finite elements by using 3D-solid tetrahedral elements each having 10 nodes and 17 Gauss points per element, see again Fig. 4a, b. Starting from a reference load equal to 1000 N, the P_{LB} values obtained, are drawn in Figs. 5 and 6.

Table 3 Strengths values [MPa] for cortical and trabecular tissues

Strength	Cortical	Trabecular	Cortical	Trabecular
	Mat.1		Mat.2	
$\sigma_1^- = \sigma_2^-$	68.249(1)	0.913(1)	86.137(1)	6.099(1)
σ_3^-	155.164(1)	0.526(1)	207.324(1)	7.220(1)
$\sigma_1^+ = \sigma_2^+$	23.887(2)	0.602(3)	30.148(2)	4.025(3)
σ_3^+	93.098(4)	0.347(5)	124.394(4)	4.765(5)
$\sigma_{13} = \sigma_{23}$	53.50(6)	0.315(7)	53.50(6)	4.332(7)
σ_{12}	67.80(6)	0.548(8)	67.80(6)	3.659(8)
	Mat.3			
$\sigma_1^- = \sigma_2^-$	105.380(1)	13.128(1)		
σ_3^-	266.486(1)	20.792(1)		
$\sigma_1^+ = \sigma_2^+$	36.883(2)	8.664(3)		
σ_3^+	159.892(4)	13.722(5)		
$\sigma_{13} = \sigma_{23}$	53.50(6)	12.475(7)		
σ_{12}	67.80(6)	7.877(8)		

(1) after [56, 58, 59]; (2) after [34, 60] with $\sigma_1^+ = \sigma_2^+ = 35\%\sigma_1^-$; (3) after [21] with $\sigma_1^+ = \sigma_2^+ = 66\%\sigma_1^-$; (4) after [34, 60] with $\sigma_3^+ = 60\%\sigma_3^-$; (5) after [21] with $\sigma_3^+ = 66\%\sigma_3^-$; (6) after [61]; (7) after [62] with $\sigma_{13} = \sigma_{23} = 0.6\sigma_3^-$; (8) after [62] with $\sigma_{12} = 0.6\sigma_1^-$

Table 4 Elastic constants for cortical and trabecular bone[a]

Cortical	Mat.1	Mat.2	Mat.3
E_1 [MPa]	3852.988	5484.106	7445.735
E_2	3852.988	5484.106	7445.735
E_3	6759.629	9621.239	13062.689
G_{12} [MPa]	1351.926	1924.248	2612.538
G_{23}	1960.292	2790.159	3788.179
G_{13}	1960.292	2790.159	3788.179
ν_{12}	0.425	0.425	0.425
ν_{23}	0.370	0.370	0.370
ν_{13}	0.370	0.370	0.370
Trabecular	Mat.1	Mat.2	Mat.3
E_1 [MPa]	4.884	137.954	531.445
E_2	8.462	229.293	868.602
E_3	13.698	327.636	1180.180
G_{12} [MPa]	2.049	68.369	281.673
G_{23}	2.686	81.309	322.116
G_{13}	4.073	125.018	498.056
ν_{12}	0.332	0.293	0.279
ν_{23}	0.164	0.149	0.143
ν_{13}	0.202	0.162	0.148

[a] After [57]

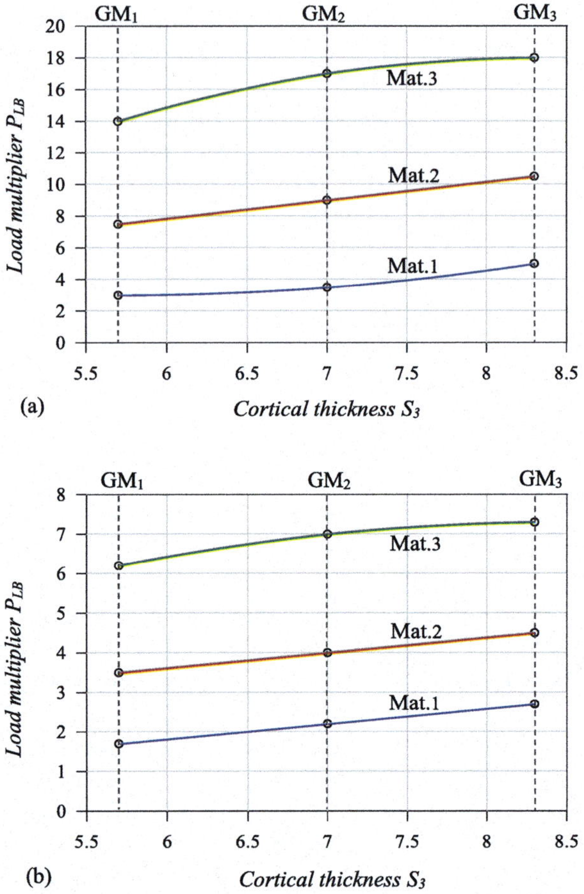

Fig. 5 P_{LB} load multiplier of the collapse load for the proximal femur versus cortical thickness S_3 in mm. Each curve refers to a fixed set of strengths for the cortical and trabecular tissues. **a** STANCE configuration; **b** SIDE configuration

Figure 5a, b show the load multiplier P_{LB} of the collapse load for the proximal femur versus the cortical thickness S_3 (to which S_2 and S_1 are associated). Each triplet of vertical dots identifies the computed P_{LB} values for each of the three geometrical models examined, namely: GM_1 with cortical thickness $S_3 = 5.7$ mm; GM_2 with cortical thickness $S_3 = 7$ mm; GM_3 with cortical thickness $S_3 = 8.3$ mm. Each dot within the triplet corresponds to a fixed set of strengths for the cortical and trabecular tissues, namely Mat.1, Mat.2 and Mat.3 of Tables 3 and 4. The curves interpolating the obtained findings for fixed material are aimed to show the sensitivity of the predicted load multipliers P_{LB} to the variation of the cortical thickness.

Figure 6a, b show the load multiplier P_{LB} of the collapse load for the proximal femur versus the apparent density of cortical tissue $\rho_{cortical}$, assumed as plotting parameter identifying the three different set of strengths for cortical and trabecular tissues used in the analyses and shown in Tables 3 and 4. Each triplet of vertical dots identifies the computed P_{LB} values for each of the three materials examined,

Fig. 6 P_{LB} load multiplier of the collapse load for the proximal femur versus the apparent density of cortical tissue $\rho_{cortical}$, assumed as plotting parameter identifying the three different set of strengths for cortical and trabecular tissues used in the analyses. Each curve refers to a fixed geometrical model of the proximal femur. **a** STANCE configuration; **b** SIDE configuration

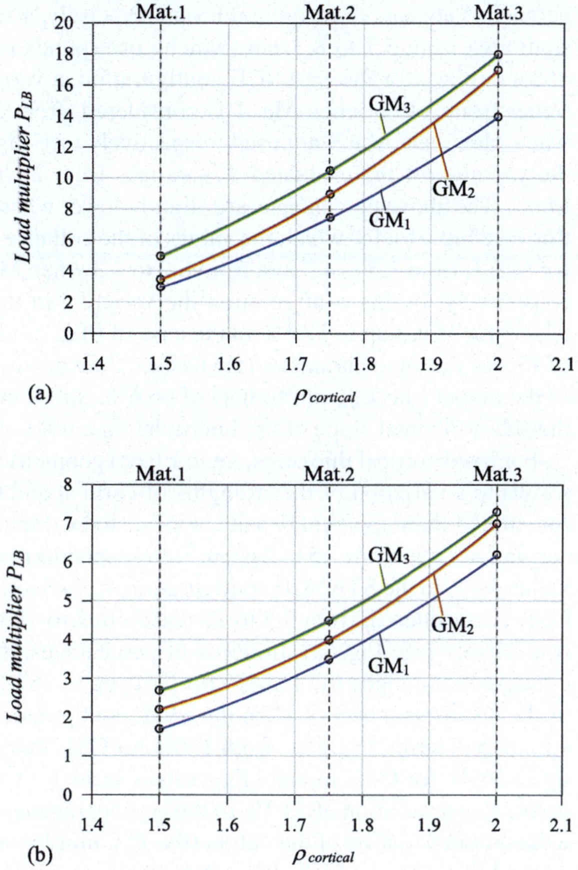

namely: Mat.1 with $\rho_{cortical} = 1.5\,\text{g/cm}^3$ and $\rho_{trabecular} = 0.1\,\text{g/cm}^3$; Mat.2 with $\rho_{cortical} = 1.75\,\text{g/cm}^3$ and $\rho_{trabecular} = 0.4\,\text{g/cm}^3$; Mat.3 with $\rho_{cortical} = 2.0\,\text{g/cm}^3$ and $\rho_{trabecular} = 0.7\,\text{g/cm}^3$. Each dot within the triplette corresponds to a fixed geometrical model of the proximal femur, namely GM_1, GM_2 and GM_3 of Table 2. The curves interpolating the obtained findings for fixed geometrical model are aimed to show the sensitivity of the predicted load multipliers P_{LB} to the variation of the strengths in the cortical and trabecular tissues.

6 Discussion

By inspection of Figs. 5 and 6 the following can be deduced. For fixed strength values of the bone tissues, i.e. for fixed mechanical characteristics of the bone tissues, the sensitivity to a variation of the cortical's thickness (variation of the proximal femur geometry) appears to be quite relevant. Precisely, a variation of the thickness in the shaft area from 5.7 to 8.3 mm (which corresponds to a variation of approximately 45%) implies, for the STANCE configuration, a variation of the lower bound multiplier from 3 to 5 when Mat.1 is considered, from 7.5 to 10.5 and from 14 to 18 when Mat.2 and Mat.3 are used, respectively (see Fig. 5a). In terms of percentages, the variation of the evaluated P_{LB} ranges from 28.6% for Mat.3 up to 66.6% for Mat.1. The above percentages are slightly lower when dealing with SIDE configuration (see Fig. 5b), for which the values of the collapse multiplier vary from 1.7 to 2.7 for Mat.1, from 3.5 to 4.5 and from 6.2 to 7.3 when Mat.2 and Mat.3 are considered, respectively. In this configuration the variation in the calculated P_{LB} ranges from 17.7% for Mat.3 up to 58.8% in the case of Mat.1. It is worth noting that a variation of 45% of the input parameter (the tissues' thicknesses) implies a maximum variation of the output (the P_{LB} multiplier) of 66.6%. Such circumstance is highlighted from the sub-horizontal slope of the interpolating curves of Fig. 5a, b.

For fixed cortical thickness, i.e. for fixed geometrical model, the sensitivity of the results to a variation of the strengths of cortical and trabecular tissues, plotted as a function of their apparent density, appears to be dramatically relevant. As in fact, by varying $\rho_{cortical}$ from 1.5 to 2 g/cm^3 (corresponding to a variation of 33.3%) the P_{LB} values vary, in the STANCE configuration, from 3 to 14 when the geometrical model GM_1 is considered, from 3.5 to 17 and from 5 to 18 when GM_2 and GM_3 are used, respectively (see Fig. 6a). In terms of percentages, the variation of the P_{LB} in this configuration ranges from 260% for GM_3 up to 367% for GM_1. An high variability of the results also occurs when the SIDE configuration is examined, in this case the variation in terms P_{LB} goes from 170% for GM_3 model (P_{LB} values from 2.7 to 7.3) up to 265% for GM_1 model (P_{LB} values from 1.7 to 6.2), (see Fig. 6b). It is worth noting that a variation of 33.3% of the input parameter (the tissues' strengths) implies a maximum variation of the output (the P_{LB} multiplier) of 367%. Such circumstance is highlighted by the high slope of the interpolating curves of Fig. 6a, b.

An accurate evaluation of the bone tissues strengths seems to be of utmost importance for a reliable prediction of the collapse load. Indeed, the strengths of cortical and trabecular tissues determine the shape of the Tsai-Wu-type yield domain used in the numerical procedure. It is this domain which discriminates between admissible and not admissible stresses so determining the P_{LB} value. On the other hand, a precise definition of the cortical thickness seems indeed to affect less the results of the numerical simulations.

The obtained results furnish an important information on the parameters to be taken into account for a good prediction. To this concern it is worth noting that: (i) the parameters necessary to the ECM are very few with respect to those required by a reliable post-elastic numerical analysis which implies a definition of the evolutive response beyond the elastic limit so introducing many other constitutive assumptions and other related material parameters; (ii) the FE analyses carried on within the ECM are really of predictive type. They are not used for a retrofitting of experimental tests to calibrate a numerical model. They are used to redistribute stresses, which is the key idea of the ECM.

Remark The ECM is able to reproduce the experimental data and, in fact, Fig. 6a, b provide, for the *average value* of $\rho_{cortical}$, corresponding to Mat.2, values of the collapse loads very close to each other and to the ones given in [18] and this *for all the three geometrical models* taken into account. More precisely, in the STANCE configuration the experimental collapse load value is 8.7 $\pm$ 2.9 kN and the one obtained from the numerical simulations (Fig. 6a) is 9 $\pm$ 1.5 kN, while in the SIDE configuration the experimental collapse load is 3.1 $\pm$ 1.1 kN and the one obtained applying the ECM procedure is 4 $\pm$ 0.5 kN (Fig. 6b). It should be remembered here that to obtain the collapse load, the scalar P_{LB} values have to be multiplied by the reference load which, as previously mentioned, was assumed equal to 1kN.

7 Concluding Remarks

A sensitivity study for the determination of the collapse load of a human proximal femur, by varying both the thickness of the cortical (the trabecular thickness varying consequently) and the strength values of cortical and trabecular tissues, has been presented. The numerical findings were carried out using a numerical predictive method applied on a geometrical simplified 3D model of the human proximal femur for assigned boundary and loading conditions. The adopted numerical method is based on the theory of limit analysis and is known in literature as Elastic Compensation Method. The proposed numerical approach allows to determine a lower bound to the collapse load and therefore provides a safety factor for the analyzed proximal femur. A limitation of the ECM is due to the fact that the acting forces are fixed on time, however there are no limits in considering different load conditions, which can simulate, for example, the action of muscles, as long as they are applied in a quasi-static equivalent manner.

The proposed numerical procedure provides a valuable tool for predicting the collapse load of the human proximal femur in a simple and relatively accurate manner. The accuracy of the results essentially depends on the accuracy of a few parameters entering the model, such as the cortical and trabecular strength values. The geometry seems indeed to have a weaker influence on the predicted collapse load. Finally, it is worth to remark that the procedure can be applied simultaneously to different materials that meet different yield criteria and this would allow, for example, the easy insertion in the mechanical model of a prostheses.

References

1. Bouxsein, M.L., Zysset, P., Glüer, C.C., McClung, M., Biver, E., Pierroz, D.D., Ferrari, S.L.: Perspectives on the non-invasive evaluation of femoral strength in the assessment of hip fracture risk. Osteoporos. Int. (2020). https://doi.org/10.1007/s00198-019-05195-0

2. Ali, A.A., Cristofolini, L., Schileo, E., Hu, H., Taddei, F., Kim, R.H., Rullkoetter, P.J., Laz, P.J.: Specimen-specific modeling of hip fracture pattern and repair. J. Biomech. **47**, 536–543 (2014)
3. Marco, M., Giner, E., Caeiro-Rey, J.R., Miguelez, M.H., Larrainzar-Garijo, R.: Numerical modelling of hip fracture patterns in human femur. Comput. Methods Programs Biomed. **173**, 67–75 (2019)
4. Gustafsson, A., Tognini, M., Bengtsson, F., Gasser, T.C., Isaksson, H., Grassi, L.: Subject-specific FE models of the human femur predict fracture path and bone strength under single-leg-stance loading. J. Mech. Behav. Biomed. Mater. **113**, 104118 (2021)
5. Pisano, A.A., Fuschi, P.: Limit analysis of human proximal femur. J. Mech. Behav. Biomed. Mater. **124**(6) (2021) https://doi.org/10.1016/j.jmbbm.2021.104844
6. Mackenzie, D., Boyle, J.T.: A method of estimating limit loads by iterative elastic analysis. I-Simple examples. Int. J. Press. Vessel. Pip. **53**(1), 77–95 (1993)
7. Mackenzie, D., Shi, J., Boyle, J.T.: Finite element modelling for limit analysis by the elastic compensation method. Comput. Struct. **51**(4), 403–410 (1994)
8. Pisano, A.A., Fuschi, P.: A numerical approach for limit analysis of orthotropic composite laminates. Int. J. Numer. Meth. Eng. **70**, 71–93 (2007)
9. Pisano, A.A., Fuschi, P., De Domenico, D.: A layered limit analysis of pinned-joint composite laminates: numerical versus experimental findings. Compos. B Eng. **43**, 940–952 (2012)
10. Pisano, A.A., Fuschi, P., De Domenico, D.: Peak load prediction of multi-pin joints FRP laminates by limit analysis. Compos. Struct. **96**, 763–772 (2013)
11. Pisano, A.A., Fuschi, P., De Domenico, D.: Peak loads and failure modes of steel-reinforced concrete beams: predictions by limit analysis. Eng. Struct. **56**, 477–488 (2013)
12. De Domenico, D., Pisano, A.A., Fuschi, P.: A FE-based limit analysis approach for concrete elements reinforced with FRP bars. Compos. Struct. **107**, 594–603 (2014)
13. Pisano, A.A., Fuschi, P., De Domenico, D.: Numerical limit analysis of steel-reinforced concrete walls and slabs. Comput. Struct. **160**, 42–55 (2015)
14. Pisano, A.A., Fuschi, P.: Evaluation of human bones load bearing capacity with the limit analysis theory. In: Pisano, A.A., Spiliopoulos, K.V., Weichert, D. (eds.) Direct Methods: Methodological Progress and Engineering Applications. Lecture Notes in Applied and Computational Mechanics, vol. 95, pp. 1–23. Springer International Publishing, Switzerland (2021). ISSN 1613-7736, ISBN 978-3-030-48833-8. https://doi.org/10.1007/978-3-030-48834-5_1
15. Väänänen, S.P., Grassid, L., Flivike, G., Jurvelina, J.S., Isaksson, H.: Generation of 3D shape, density, cortical thickness and finite element mesh of proximal femur from a DXA image. Med. Image Anal. **24**, 125–134 (2015)
16. Holzer, G., von Skrbensky, G., Holzer, L.A., Pichl, W.: Hip fractures and the contribution of cortical versus trabecular bone to femoral neck strength. J. Bone Miner. Res. **24**(3), 468–474 (2009)
17. Michelotti, J., Clark, J.: Femoral neck length and hip fracture risk. J. Bone Miner. Res. **14**(10), 1714–1720 (1999)
18. Dall'Ara, E., Luisier, B., Schmidt, R., Pretterklieber, M., Kainberger, F., Zysset, P., Pahr, D.: DXA predictions of human femoral mechanical properties depend on the load configuration. Med. Eng. Phys. **35**, 1564–1572 (2013)
19. Yang, Z., Jian, W., Zhi-han, L., Jun, X., Liang, Z., Ge, Y., Zhan-jun, S.: The geometry of the bone structure associated with total hip arthroplasty. PLoS One **9**(3), e91058 (2014). https://doi.org/10.1371/journal.pone.0091058
20. Beaupied, H., Lespessailles, E., Benhamou, C.-L.: Evaluation of macrostructural bone biomechanics. Jt. Bone Spine **74**(3), 233–239 (2007)
21. Rincón-Kohli, L., Zysset, P.K.: Multi-axial mechanical properties of human trabecular bone. Biomech. Model. Mechanobiol. **8**(3), 195–208 (2009)
22. Wolfram, U., Wilke, H.J., Zysset, P.K.: Valid micro finite element models of vertebral trabecular bone can be obtained using tissue properties measured with nanoindentation under wet conditions. J. Biomech. **43**(9), 1731–1737 (2010)

23. Nawathe, S., Akhlaghpour, H., Bouxsein, M.L., Keaveny, T.M.: Microstructural failure mechanism in the human proximal femur for sideway fall loading. J. Bone Miner. Res. **29**(2), 507–515 (2014)
24. Taghizadeh, E., Reyes, M., Zysset, P., Latypova, A., Terrier, A., Büchler, P.: Biomechanical role of bone anisotropy estimated on clinical CT scans by image registration. Ann. Biomed. Eng. **44**(8), 2505–2517 (2016)
25. Mirzaali, M.J., Schwiedrzik, J.J., Thaiwichai, S., Best, J.P., Michler, J., Zysset, P.K., Wolfram, U.: Mechanical properties of cortical bone and their relationships with age, gender, composition and microindentation properties in the elderly. Bone **93**, 196–211 (2016)
26. Schwiedrzik, J.J., Mirzaali, M.J., Thaiwichai, S., Best, J.P., Michler, J., Zysset, P.K., Wolfram, U.: Response to the commentary on mechanical properties of cortical bone and their relationships with age, gender, composition and microindentation properties in the elderly. Bone **105**, 312–314 (2017)
27. Musy, S.N., Maquer, G., Panyasantisuk, J., Wandel, J., Zysset, P.K.: Not only stiffness, but also yield strength of the trabecular structure determined by nonlinear μFE is best predicted by bone volume fraction and fabric tensor. J. Mech. Behav. Biomed. Mater. **65**, 808–813 (2017)
28. Guenoun, D., Pithioux, M., Souplet, J.-C., Guis, S., Le Coroller, T., Fouré, A., Pauly, V., Mattei, J.-P., Bernard, M., Guye, M., Chabrand, P., Champsaur, P., Bendahan, D.: Assessment of proximal femur microarchitecture using ultra-high field MRI at 7 Tesla. Diagn. Interv. Imaging **101**, 45–53 (2020)
29. Rajapakse, C.S., Farid, A.R., Kargilis, D.C., et al.: MRI-based assessment of proximal femur strength compared to mechanical testing. Bone (2020). https://doi.org/10.1016/j.bone.2020.115227
30. Zysset, P.K., Curnier, A.: A 3D damage model for trabecular bone based on fabric tensors. J. Biomech. **29**(12), 1549–1558 (1996)
31. Fondrk, M.T., Bahniuk, E.H., Davy, D.T.: A damage model for nonlinear tensile behavior of cortical bone. J. Biomech. Eng. **121**(5), 533–541 (1999)
32. Keaveny, T.M., Morgan, E.F., Niebur, G.L., Yeh, O.C.: Biomechanics of trabecular bone. Annu. Rev. Biomed. Eng. **3**(1), 307–333 (2001)
33. Garcia, D., Zysset, P.K., Charlebois, M., Curnier, A.: A three-dimensional elastic plastic damage constitutive law for bone tissue. Biomech. Model. Mechanobiol. **8**(2), 149–165 (2009)
34. Doblaré, M., García, J.M., Gómez, M.J.: Modelling bone tissue fracture and healing: a review. Eng. Fract. Mech. **71**, 1809–1840 (2004)
35. Schwiedrzik, J.J., Wolfram, U., Zysset, P.K.: A generalized anisotropic quadric yield criterion and its application to bone tissue at multiple length scales. Biomech. Model. Mechanobiol. **12**(6), 1155–1168 (2013)
36. Charlebois, M., Jirásek, M., Zysset, P.K.: A nonlocal constitutive model for trabecular bone softening in compression. Biomech. Model. Mechanobiol. **9**(5), 597–611 (2010)
37. Hosseini, H.S., Horák, M., Zysset, P.K., Jirásek, M.: An over-nonlocal implicit gradient-enhanced damage-plastic model for trabecular bone under large compressive strains. Int. J. Numer. Methods Biomed. Eng. **31**(11), 1–32 (2015)
38. Zysset, P.K.: A review of morphology-elasticity relationships in human trabecular bone: theories and experiments. J. Biomech. **36**, 1469–1485 (2003)
39. Bayraktar, H.H., Gupta, A., Kwon, R.Y., Papadopoulos, P., Keaveny, T.M.: The modified super-ellipsoid yield criterion for human trabecular bone. J. Biomech. Eng. **126**(6), 677–684 (2004)
40. Tellache, M., Pithioux, M., Chabrand, P., Hochard, C.: Femoral neck fracture prediction by anisotropic yield criteria. Rev. Europénne de Mécanique Numérique/Eur. J. Comput. Mech., Hermès/Paris: Lavoisier **18**(1), 33–41 (2009)
41. Wolfram, U., Gross, T., Pahr, D.H., Schwiedrzik, J., Wilke, H.J., Zysset, P.K.: Fabric-based Tsai-Wu yield criteria for vertebral trabecular bone in stress and strain space. J. Mech. Behav. Biomed. Mater. **15**, 218–228 (2012)
42. Hayes, W.C., Wright, T.M.: An empirical strength theory for compact bone. Fracture **3**, 1173–1179 (1977)

43. Cezayirlioglu, H., Bahniuk, E., Davy, D.T., Heiple, G.: Anisotropic yield behavior of bone under combined axial force and torque. J. Biomech. **18**(1), 61–69 (1985)
44. Carnelli, D., Gastaldi, D., Sassi, V., Contro, R., Ortiz, C., Vena, P.: A finite element model for direction-dependent mechanical response to nanoindentation of cortical bone allowing for anisotropic post-yield behavior of the tissue. J. Biomech. Eng. **132**, 081008-1-10 (2010)
45. Carnelli, D., Lucchini, R., Ponzoni, M., Contro, R., Vena, P.: Nanoindentation testing and finite element simulations of cortical bone allowing for anisotropic elastic and inelastic mechanical response. J. Biomech. **44**(10), 1852–1858 (2011)
46. Fratzl, P., Gupta, H.S., Paschalis, E.P., Roschger, P.: Structure and mechanical quality of the collagen-mineral nano-composite in bone. J. Mater. Chem. **14**, 2115–2123 (2004)
47. Gupta, H.S., Fratzl, P., Kerschnitzki, M., Benecke, G., Wagermaier, W., Kirchner, H.O.K.: Evidence for an elementary process in bone plasticity with an activation entalphy of 1 eV. J. R. Soc. **4**, 277–282 (2007)
48. Ritchie, R.O., Buehler, M.J., Hansma, P.: Plasticity and toughness in bone. Phys. Today **62**(6), 41–47 (2009)
49. Schwiedrzik, J.J., Zysset, P.K.: An anisotropic elastic-viscoplastic damage model for bone tissue. Biomech. Model. Mechanobiol. **12**, 201–213 (2013)
50. Tsai, S.W., Wu, E.M.: A general theory of strength for anisotropic materials. J. Compos. Mater. **5**, 58–80 (1971)
51. Zimmermann, E.A., Schaible, E., Bale, H., Barth, H.D., Tang, S.Y., Reichert, P., Busse, B., Alliston, T., Ager, J.W., Ritchie, R.O.: Age-related changes in the plasticity and toughness of human cortical bone at multiple length scales. Proc. Natl. Acad. Sci. **108**(35), 14416–14421 (2011)
52. Lubliner, J.: Plasticity Theory. Macmillan Publishing Company, New York (1990)
53. Fuschi, P., Pisano, A.A, Weichert, D.: Direct methods for limit and shakedown analysis of structures: advanced computational algorithms and material modelling. Springer International Publishing, Switzerland (2014). ISBN: 978-3-319-12927-3. https://doi.org/10.1007/978-3-319-12928-0
54. Pisano, A.A., Spiliopoulos, K.V., Weichert, D.: Direct Methods: Methodological Progress and Engineering Applications. Lecture Notes in Applied and Computational Mechanics, vol. 95, Springer International Publishing, Switzerland (2021). ISSN 1613-7736, ISBN 978-3-030-48833-8. https://doi.org/10.1007/978-3-030-48834-5
55. Niimi, R., Kono, T., Nishihara, A., Hasegawa, M., Matsumine, A., Kono, T., Sudo, A.: Cortical thickness of the femur and long-term bisphosphonate use. J. Bone Miner. Res. **30**(2), 225–231 (2015)
56. Wirtz, D.C., Schiffers, N., Pandorf, T., Radermacher, K., Weichert, D., Forst, R.: Critical evaluation of known bone material properties to realize anisotropic FE-simulation of proximal femur. J. Biomech. **33**, 1325–1330 (2000)
57. San, Antonio T., Ciaccia, M., Müller-Karger, C., Casanova, E.: Orientation of orthotropic material properties in a femur FE model: a method based on the principal stresses directions. Med. Eng. Phys. **34**, 914–919 (2012)
58. Lotz, J.C., Gerhart, T.N., Hayes, W.C.: Mechanical properties of trabecular bone from the proximal femur: a quantitative CT study. J. Comput. Assist. Tomogr. **14**, 107–114 (1990)
59. Lotz, J.C., Gerhart, T.N., Hayes, W.C.: Mechanical properties of metaphyseal bone in the proximal femur. J. Biomech. **24**, 317–329 (1991)
60. Currey, J.: Cortical bone. In: Murphy, W., et al. (eds.) Handbook of Biomaterial Properties, Chapter A1. Springer Science + Business Media, New York (2016)
61. Turner, C.H., Wang, T., Burr, D.B.: Shear strength and fatigue properties of human cortical bone determined from pure shear tests. Calcif. Tissue Int. **69**, 373–378 (2001)
62. Sanyal, A., Gupta, A., Bayraktar, H.H., Kwon, R.Y., Keaveny, T.M.: Shear strength behavior of human trabecular bone. J. Biomech. **45**, 2513–2519 (2012)
63. Zysset, P.K., Dall'Ara, E., Varga, P., Pahr, D.H.: Finite element analysis for prediction of bone strength. BoneKEy Rep. **2**(386) (2013)

Graded Damage Solutions in One Dimension

Nunziante Valoroso

Abstract A regularized damage model is considered named *Graded damage* in which the gradient enhancement has the form of an explicit bound for the spatial gradient of damage. The key features of the proposed approach are demonstrated by computing the analytical solution of two problems that are one-parameter dependent. The first one is the classical one-dimensional damageable rod under tensile load, for which the hardening function is determined based on the equivalence with a given cohesive relationship. The second application is a mode-I delamination problem for which the cohesive law for the interface is formulated starting from the graded damage concept, i.e. by prescribing the shape of damage distribution within the cohesive process zone.

Keywords Damage mechanics · Regularization · Cohesive zone models

1 Introduction

Damage and Fracture Mechanics find their *raison d'être* in the need for predictive computations able to prevent catastrophic failure in engineering structures. The complexity of the physics of damage, which rules out any homogeneity of materials at the usual macroscopic scale of laboratory experiments, has led to many different modeling assumptions in Solid Mechanics, each of them resulting from a suitable trade-off between physical relevance at different scales and applicability to structural design [5]. Nonetheless, computations of failure mechanisms and ultimate load-carrying capacity of structures still stay as difficult tasks in civil and mechanical engineering owing to the intrinsic non-smoothness of damage and fracture phenomena [7].

Roughly speaking, one can categorize the computational approaches to failure into two families, the continuous and the discontinuous one, each of them with advantages and limitations. Discontinuous descriptions allow for jumps in the displacement

N. Valoroso (✉)
Dipartimento di Ingegneria, Università di Napoli Parthenope, Naples, Italy
e-mail: nunziante.valoroso@uniparthenope.it

G. Garcea and D. Weichert (eds.), *Direct Methods for Limit State of Materials and Structures*, Lecture Notes in Applied and Computational Mechanics 101,
https://doi.org/10.1007/978-3-031-29122-7_6

field, whereby one has to deal with changes in topology that are intrinsic to the representation of discrete cracks. In a finite element context this requires special elements with embedded discontinuities [3] or extended finite element formulations (X-FEM) either in the original setup of Belytschko and co-workers [17] or in the format of the so-called *Thick Level Set* model [18].

One can also include into the discontinuous family the cohesive zone models originating from the work of Barenblatt [4]. Initially motivated by the need to characterize stress states in the vicinity of cracks, in cohesive zone models one may speak e.g. of damage, delamination or de-cohesion to designate all those progressive phenomena preceding fracture, during which material separation is resisted by attractive forces that develop along an extended crack tip, i.e. the *cohesive process zone*.

In the cohesive zone approach crack progression is governed by an independent relationship between surface tractions and displacement jumps that incorporates typical fracture parameters, i.e. the cohesive strength and the fracture toughness. Under certain conditions the shape of the softening curve does also play a role in fracture predictions [1], but it is commonly believed to be less relevant compared to the other parameters. Anyway, classical implementations of the cohesive zone concept only allow for strong discontinuities along interfaces that pre-exist in the material before any loading, whereby in numerics use is made of degenerated (zero-thickness) finite elements that are placed along potential discontinuity surfaces [14].

A discrete crack representation closely reflects the physics of fracture but includes a number of difficulties, most of which are related to crack tracking. This partly motivates the continuous approach as a tool for modeling fracture starting from the strain localization stage; here no physical crack opening exist and fully damaged states can be understood as the smeared, diffuse representation of macro-fractures. In this context the basic idea consists of preserving the topology of the initial finite element mesh and to bring into the material model a concise information about material microstructure via a length scale parameter; the latter is used to introduce the necessary regularization that restores well-posedness of the problem either via a nonlocal integral approach after Pijaudier-Cabot and Bažant [21] or in the form of a gradient enhancement in the wake of the works of Peerlings et al. [8, 20]. We also note the family of *phase-field models* initiated from the regularized form of the variational theory of quasi-static fracture [6]; though starting from a different perspective, i.e. global energy minimization, these models end up with a field equation of diffusive type that is quite close to that of gradient damage models, see e.g. [15, 16] among others.

In all such cases averaging or differential operators come into play, whereby the constitutive equations are no longer defined at the local level but are established at the scale of the structural model. One may then conclude that continuous representations of discontinuities provide globally smoothed solutions through elements, whereas in usual local models stresses, strains and internal variables are all defined in a pointwise fashion that can be understood as generally discontinuous fields inside elements and across elements boundaries [2].

This chapter is concerned with a gradient-based continuum damage formulation named *Graded damage* [23], i.e. a Generalized Standard Model with convex constraints that admits the geometrical interpretation of the Thick Level Set approach of Moës et al. [18]. The variational structure of the model along with its directional convexity properties allow for an effective implementation based on convex programming in the spirit of *direct methods*, by alternating minimization with respect to displacements and damage and maximization with respect to the Lagrange multipliers that implicitly contain the information necessary to track the interphases between fully damaged regions and the sound material.

In particular, in the following the graded damage model is applied to two different one-dimensional problems for which a non-homogeneous solution is computed in closed form. The first problem is presented in Sect. 2; it is rather classical and refers to the rod under tensile load, whose interest lies in the fact that the relevant solution is considered to be representative of the response of a three-dimensional structure across a localization band. Moreover, this analytical solution is typically used to design the constitutive functions of a continuum damage formulation that render the response of the damageable rod identical to the one of an elastic bar in which a cohesive interface is the only source of dissipation, see e.g. [13, 25].

The second problem is discussed in Sect. 3; it is in a sense analogous to the one of the tensile rod and refers to a one-parameter-dependent delamination problem where the interface constitutive relationship is gradient-enhanced based on the graded damage concept. Worth noting is the fact that in cohesive models understood in the sense of Hillerborg [11] there is in principle no need for any regularization, whereby few attempts have been made so far to introduce gradients along a cohesive interface. However, the general consensus that nonlocal interactions may occur at the meso-scale level suggest that introduction of a material length scale into an interface model is a worthwhile attempt [12, 19].

2 The Damageable Rod

To begin with, consider the homogeneous elastic-damageable rod depicted in Fig. 1; the domain Ω occupied by the structure is the interval $[-L, L]$, body forces are neglected and loading is performed via an increasing elongation at the two ends of the bar.

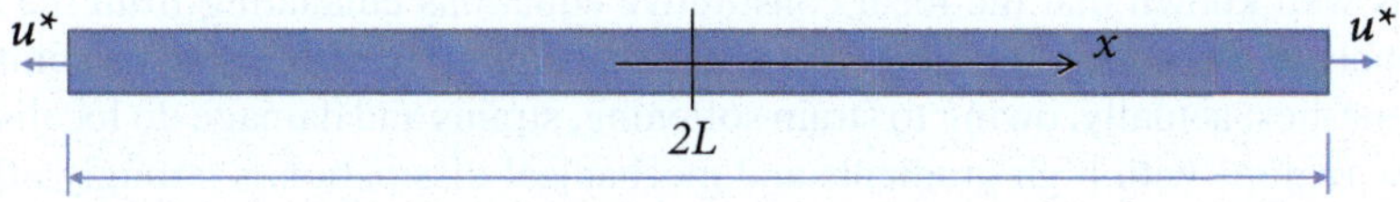

Fig. 1 One-dimensional rod. Model problem

In the present one-dimensional context the stored energy function reads:

$$\psi(u, d) = \frac{1}{2}\,\omega(d)\,E\left(\frac{\mathrm{d}u}{\mathrm{d}x}\right)^2 \tag{1}$$

where $E > 0$ is the elastic modulus, u is the axial displacement obeying the essential boundary conditions:

$$u(-L) = -u^\star, \quad u(L) = u^\star; \qquad u^\star > 0 \tag{2}$$

d is the damage variable and $\omega(d)$ is a monotonically decreasing function that accounts for material degradation.

Restricting attention to the class of Generalized Standard Materials, the local model is completed by prescribing a dissipation pseudo-potential [10]; for rate-independency, it must be positively homogeneous of degree-one with respect to the flux $\dot{d}$:

$$\varphi(\dot{d}) = Y_c(d)\,\dot{d} + \sqcup_{\Re^+}(\dot{d}) \tag{3}$$

In the above relationship $Y_c(d)$ is a positive convex function of the current damage state d, here considered as a parameter, and $\sqcup_{\Re^+}$ is the convex indicator of non-negative reals that enforces irreversibility:

$$\sqcup_{\Re^+}(d^*) = \begin{cases} 0 & \text{if } d^* \geq 0 \\ +\infty & \text{otherwise} \end{cases} \tag{4}$$

Likewise, damage is a constrained variable since it has to comply with the physical bounds:

$$0 \leq d \leq 1 \tag{5}$$

This can be accounted for in the present formulation either via an indicator function of the admissibility domain or using a smoothed version of it, say $g_1(d)$, with the relevant Karush-Kuhn-Tucker conditions

$$g_1(d) \leq 0; \qquad \gamma_1 \geq 0; \qquad \gamma_1\, g_1(d) = 0 \tag{6}$$

It is well known that the local constitutive equations emanating from the above potentials will produce non-objective numerical solutions with respect to finite element meshes; actually, owing to strain softening, strains and damage do localise into narrow regions with high gradients and mechanical dissipation is strongly affected by mesh refinements. Objectivity can be restored by appealing to a nonlocal formulation, i.e. introducing spatial interactions into the constitutive equations to provide a suitable localization limiter [5]. To this end, in the graded damage model an explicit nonlocal constraint acting on the damage gradient is prescribed via the following [23]:

$$g_2(d) = ||\nabla d|| - f(d) \tag{7}$$

where the bounding function $f(d) > 0$ may be arbitrarily nonlinear provided that it is concave; in addition, the following complementarity conditions apply:

$$g_2(d) \leq 0\,; \qquad \gamma_2 \geq 0\,; \quad \gamma_2\, g_2(d) = 0 \tag{8}$$

that characterize the gradient constraint (7) as non-dissipative.

As a direct consequence of nonlocality, the thermodynamic potentials are functionals of the state variables u and d, here understood as fields [9]. In particular, the internal energy of the rod is the Lagrangian:

$$\mathcal{E}(\boldsymbol{\varepsilon}, d, \gamma_i) = \int_\Omega \psi(\boldsymbol{\varepsilon}, d)\, \mathrm{d}x + \int_\Omega [\gamma_1\, g_1(d) + \gamma_2\, g_2(d)]\, \mathrm{d}x \tag{9}$$

where ψ is the local stored energy function defined by Eq. (1); likewise, a global pseudo-potential of dissipation is obtained by integrating the dissipation function (3) over the physical domain Ω:

$$\mathcal{D}(\dot{d}) = \int_\Omega \varphi(\dot{d})\, \mathrm{d}x \tag{10}$$

The forces work-conjugate to the axial strain and damage are the Cauchy stress and the energy release rate. In particular, the former is obtained along with the equilibrium equation by zeroing the first variation of the potential energy (9) with respect to the displacement u as

$$\sigma = \omega(d)\, E\, \frac{\mathrm{d}u}{\mathrm{d}x} \tag{11}$$

For the ensuing developments the gradient constraint is taken as:

$$g_2(d) = \left|\frac{\mathrm{d}d}{\mathrm{d}x}\right| - \frac{1}{l_c} \leq 0 \tag{12}$$

whence results a piece-wise linear distribution of damage along the rod. The energy release rate is a variational derivative and includes a nonlocal term originating from (12) plus two boundary conditions [16]. In particular, the first variation of the functional (9) with respect to damage followed by integration by parts yields:

$$\frac{\partial \mathcal{E}}{\partial d}\delta d^* = \left[-\int_\Omega G\; \mathrm{d}x + \left[\!\left[\gamma_2\, \mathrm{sign}\left(\frac{\mathrm{d}d}{\mathrm{d}x}\right)\right]\!\right]_S + \left[\gamma_2\, \mathrm{sign}\left(\frac{\mathrm{d}d}{\mathrm{d}x}\right)\right]_{\partial\Omega}\right] \delta d^* \tag{13}$$

where the domain term G reads:

$$G = -\frac{\partial \psi}{\partial d} - \gamma_1 \frac{\mathrm{d}g_1}{\mathrm{d}d} + \frac{\mathrm{d}}{\mathrm{d}x}\left(\gamma_2 \,\mathrm{sign}\left(\frac{\mathrm{d}d}{\mathrm{d}x}\right)\right) \tag{14}$$

S being the set of possible discontinuity points for the damage gradient. For non-dissipative internal discontinuities the (internal) jump relationships give:

$$\gamma_2^+ = \gamma_2^- = 0 \tag{15}$$

whereas the (external) natural boundary conditions read:

$$\gamma_2(x) = 0; \quad x \in \partial\Omega \tag{16}$$

which allow for non-zero damage derivatives on the outer boundary.

Damage evolution is governed by the normality rule:

$$G - Y_c(d) \leq 0, \qquad \dot{d} \geq 0, \qquad (G - Y_c(d))\,\dot{d} = 0 \tag{17}$$

that follows from the Biot-like subdifferential inclusion:

$$-\frac{\partial \mathcal{E}}{\partial d} \in \partial \mathcal{D}(\dot{d}) \tag{18}$$

During initial loading the displacement $u^\star$ increases up to the elastic limit:

$$u_{el} = \sqrt{\frac{2\,Y_c(0)}{-\omega'(0)\,E}}\,L \tag{19}$$

and the unique response is the homogeneous elastic one:

$$u^\star = \frac{\sigma\,L}{E} \tag{20}$$

Once damage has started to grow and its spatial distribution $d(x)$ is known, the relationship between the constant stress and the prescribed displacement can be made explicit as:

$$u^\star = \frac{\sigma}{2\,E}\int_\Omega \omega^{-1}(d(x))\,\mathrm{d}x \tag{21}$$

Solutions beyond the elastic limit are associated with the initiation and growth of defects and can be either homogeneous or localized; in both cases one can use a parametrization in terms of the maximum damage level $d_m \leq 1$.

For the homogeneous inelastic case, damage evolution requires the local strain energy to increase everywhere in the bar; this can be expressed as

$$\left(\frac{Y_c(d)}{-\omega'(d)} \right)' > 0 \tag{22}$$

where use is made of the local limit condition emanating from (17) and the prime denote differentiation with respect to the driving variable d. The above inequality is equivalent to:

$$Y_c(d)\,\omega''(d) - Y_c'(d)\,\omega'(d) > 0 \tag{23}$$

that is a necessary requirement for local stability. Condition (23) is suggested in [13] along with an additional strain softening condition, whereby the complementary elastic energy should decrease with damage, that is:

$$\left(\frac{\omega^2(d)\,Y_c(d)}{-\omega'(d)} \right)' < 0 \tag{24}$$

whereby one has

$$\left[Y_c'(d)\,\omega^2(d) + Y_c(d)\,2\,\omega(d)\omega'(d) \right] \omega'(d) - Y_c(d)\omega^2(d)\,\omega''(d) > 0 \tag{25}$$

Without loss of generality, for non-homogeneous damage we assume that strain localization associated with one single defect initiates at point $x = 0$ immediately after the initial elastic limit (19) has been attained; the study can therefore be limited to half of the bar on account of the symmetry of (12).

The complementarity conditions (8) imply that the multiplier γ_2 can be non-zero only where the nonlocal constraint (12) is met with the equality. In this case the damage field reads:

$$d(x) = d(0) - \frac{x}{l_c} \tag{26}$$

and the constraint set coincides with the interval $[0, l_m]$, being

$$l_m = l_c\,d(0) = l_c\,d_m \leq l_c \tag{27}$$

the half-width of the localization band, see also Fig. 2.

For damage evolution ($\dot{d} > 0$) one has from (17) the differential problem:

$$Y(d(x)) - \frac{\mathrm{d}\gamma_2}{\mathrm{d}x} = Y_c\,(d(x)) \tag{28}$$

where the local damage-driving force reads:

$$Y = -\frac{\partial \psi}{\partial d} = \frac{-\omega'(d)}{\omega^2(d)} \frac{\sigma^2}{2\,E} \tag{29}$$

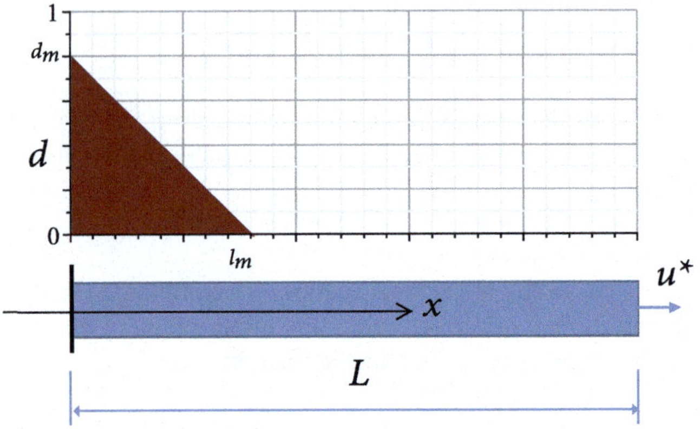

Fig. 2 One-dimensional rod. Damage distribution for localized solution

Relationship (28) is a first-order differential equation subject to two boundary conditions; the first one allows to compute the (uniform) stress σ as a function of the driving variable d_m and the second one is needed to set the integration constant for the Lagrange multiplier field γ_2. The latter is certainly nihil either on the boundary of the active constraint set defined by $g_2(d) = 0$, either where the gradient of damage is discontinuous, or on the outer boundary of the domain, where condition (16) holds. Therefore, integration of (28) between 0 and l_m, which correspond to two discontinuity points for the damage gradient, provides the *averaged limit condition* [23]:

$$\int_0^{l_m} Y(d(x))\,\mathrm{d}x = \int_0^{l_m} Y_c\,(d(x))\,\mathrm{d}x \tag{30}$$

The integrals are computed via u-substitution in the form

$$\int_0^{l_m} y(d(x))\,\mathrm{d}x = -l_c \int_{d(0)}^{d(l_m)} y(d)\,\mathrm{d}d \tag{31}$$

and one obtains the stress as a function of the maximum damage level d_m as:

$$\sigma(d_m) = \left[\frac{2\,E}{\omega^{-1}(d_m) - 1}\,H(d_m)\right]^{\frac{1}{2}} \tag{32}$$

where $H(d_m)$ is the definite integral:

$$H(d_m) = \int_0^{d_m} Y_c(d)\,\mathrm{d}d \tag{33}$$

and the integrand $Y_c(d)$ is the constitutive function, which can in turn be determined in a way consistent with a cohesive model. To this end re-write Eq. (21) for half of the bar and split the integral into two parts, respectively accounting for damage behaviour and a purely elastic response:

$$u^\star = \frac{\sigma}{E}\left[\int_0^{l_m} \left[\omega^{-1}(d(x)) - 1\right] \mathrm{d}x + L\right] = \frac{1}{2}\, w + \frac{\sigma\, L}{E} \tag{34}$$

In the above equation w is the apparent opening displacement across the localization band; it can be expressed in terms of the chosen parametrization as:

$$w(d_m) = \frac{2\,\sigma\,(d_m)}{E}\, l_c\, F(d_m) \tag{35}$$

where the non-dimensional term $F(d_m)$ reads:

$$F(d_m) = \int_0^{d_m} \left(\omega^{-1}(d) - 1\right)\, \mathrm{d}d \tag{36}$$

and depends only upon the assumed form of the degradation function $\omega(d)$.

Relationships (32) and (35) are used to determine the constitutive function $Y_c(d)$ by requiring that the macroscopic response of the damageable rod be equivalent to that of an elastic bar in which the localization band is replaced by a cohesive interface of given properties. In particular, we consider the linear softening law depicted in Fig. 3, whose analytical expression reads:

$$\sigma = \sigma_c\left(1 - \frac{\sigma_c}{2\,G_c}\, w\right) \tag{37}$$

where σ_c and G_c respectively denote the peak stress and the fracture energy.

Substitution of (35) into (37) and solution for $\sigma\,(d_m)$ provides the stress as a function of the maximum damage d_m as:

$$\sigma\,(d_m) = \frac{\sigma_c}{1 + \lambda\, F(d_m)} \tag{38}$$

Fig. 3 Linear softening function

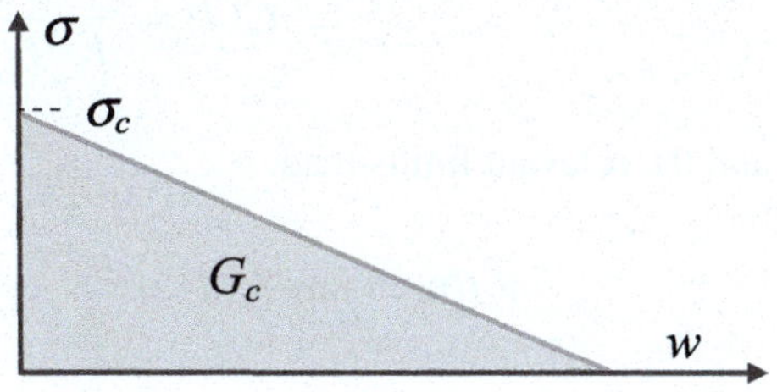

where λ is the non-dimensional parameter:

$$\lambda = \frac{l_c}{l_{coh}} \tag{39}$$

expressing the ratio between the characteristic length l_c of the graded damage model and the length scale l_{coh} of the cohesive zone [11]:

$$l_{coh} = \frac{E\,G_c}{\sigma_c^2} \tag{40}$$

For a given d_m, the value $H(d_m)$ of the integral (33) follows from substitution of (38) into (32) as:

$$H(d_m) = \frac{\omega^{-1}(d_m) - 1}{2\,E} \left(\frac{\sigma_c}{1 + \lambda\, F(d_m)} \right)^2 \tag{41}$$

Motivated by stability arguments that are being illustrated later on, we choose for the degradation function the quadratic expression:

$$\omega = (1 - d)^2 \tag{42}$$

whereby one has from (36):

$$F(d_m) = \frac{d_m^2}{1 - d_m} \tag{43}$$

while (32) provides the following expression for the (uniform) stress:

$$\sigma(d_m) = \sigma_c \, \frac{1 - d_m}{\lambda\, d_m^2 + 1 - d_m} \tag{44}$$

with the limits

$$\lim_{d_m \to 0} \sigma(d_m) = \sigma_c; \qquad \lim_{d_m \to 1} \sigma(d_m) = 0 \tag{45}$$

The constitutive function $Y_c(d)$ is computed by differentiation of (41) as:

$$Y_c(d) = \frac{\sigma_c^2}{E} \, \frac{1 + \lambda\, d^2 (d - 3)}{\left(\lambda\, d^2 + 1 - d\right)^3} \tag{46}$$

and the relevant limits read:

$$Y_c(0) = \lim_{d \to 0} Y_c(d) = \frac{\sigma_c^2}{E}; \qquad \lim_{d \to 1} Y_c(d) = \frac{\sigma_c^2}{E} \, \frac{1 - 2\,\lambda}{\lambda^3} \tag{47}$$

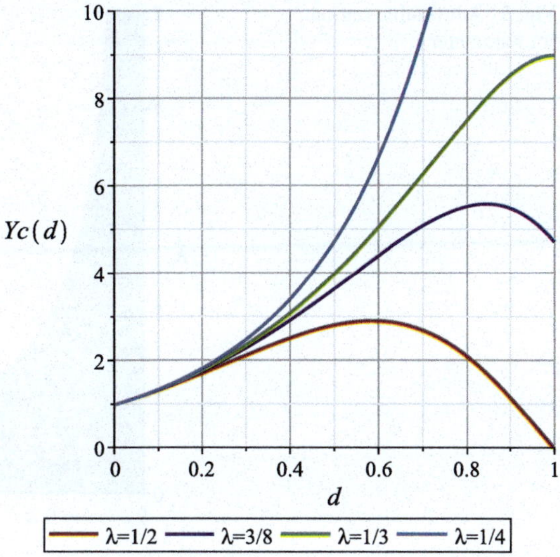

Fig. 4 The constitutive function Y_c consistent with linear softening

Figure 4 depicts the function $Y_c(d)$ normalized to $Y_c(0)$ for different values of the parameter λ, which determines the properties of the function $Y_c(d)$ itself. In practice, a *safe value* of λ to be used in numerical computations [23] can be taken in a way to comply with conditions (23) and (25).

For the case at hand the local stability requirement (23) provides:

$$\lambda < \frac{(d-2)\sqrt{d^4-4d^3+40d^2-72d+36}+d^3-4d^2-10d+12}{8d^4-36d^3+24d^2} \tag{48}$$

with limit

$$\lim_{d\to 1}\lambda = \frac{1}{2} \tag{49}$$

On the other hand, the strain softening condition (25) implies:

$$0 < \lambda < \frac{1+(1-d)^2}{2d} \tag{50}$$

The above relationships define the admissible region for λ that is shaded in Fig. 5. Clearly, any positive value of λ lower than 0.5 allows to fulfill both conditions for each $d \in [0, 1]$.

For a given damage distribution the relationship between the stress and the displacement $u^\star$ is obtained from Eq. (21) as:

$$\sigma(d_m) = \frac{E\,u^\star}{L}\,\frac{1-d_m}{(\beta\,d_m^2+1-d_m)} \tag{51}$$

Fig. 5 Admissible region for parameter λ

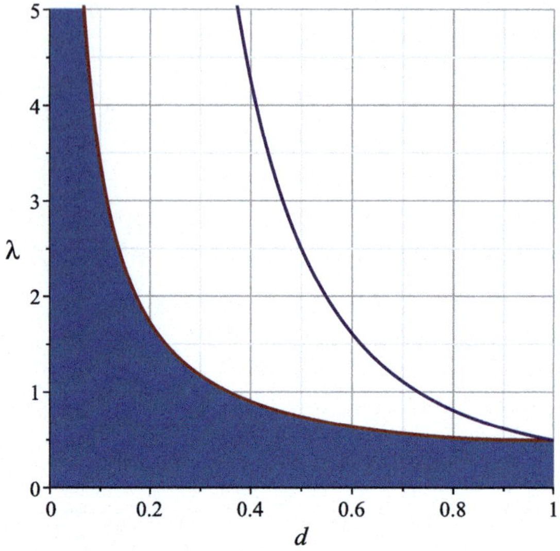

with

$$\beta = \frac{l_c}{L} \tag{52}$$

Evidently, in the present context knowledge of the damage distribution is equivalent to knowledge of the constraint set $[0, l_m]$, where the integral in Eq. (21) is non-trivial and the Lagrange multiplier γ_2 is non-zero.

Evaluation of the multiplier γ_2 amounts to compute the integral of the differential equation (28). To this end use is made of the chain rule as:

$$\frac{\mathrm{d}\gamma_2}{\mathrm{d}x} = \gamma_2'(d)\frac{\mathrm{d}d}{\mathrm{d}x} = -\frac{1}{l_c}\gamma_2'(d) \tag{53}$$

to get the integral as:

$$\frac{1}{l_c}\gamma_2(d) = \frac{-\sigma^2(d_m)}{2\,E\,\omega(d)} + \frac{\sigma_c^2}{E}\,\frac{d\,(2-d)}{2\,\left(\lambda\,d^2+1-d\right)^2} + C \tag{54}$$

The integration constant C is obtained using one of the two boundary conditions on γ_2, i.e. $\gamma_2(0) = \gamma_2(d_m) = 0$, that is:

$$C = \frac{\sigma^2(d_m)}{2\,E} \tag{55}$$

whereby one has:

$$\gamma_2(d) = \frac{-\sigma^2(d_m)\,l_c}{2\,E}\left(\omega^{-1}(d) - 1\right) + \frac{\sigma_c^2\,l_c}{E}\,\frac{d\,(2-d)}{2\left(\lambda\,d^2 + 1 - d\right)^2}$$
$$= \frac{l_c}{2\,E}\,\frac{d\,(2-d)}{(1-d)^2}\left[\sigma^2(d) - \sigma^2(d_m)\right] \tag{56}$$

Equation (56) describes the variation of the multiplier γ_2 within the active constraint interval and depends upon d_m, which is a fixed value, and $d = d(x)$, which is a function of the abscissa x along the bar. It is immediately recognized that the Lagrange multiplier γ_2 is zero either at $x = 0$, where $d = d_m$, and at $x = l_m$, where $d = 0$. Moreover, its maximum value within the interval occurs at point $\bar{x}$ where $Y(x) = Y_c\,(d(x))$ owing to the differential relationship (28).

A typical spatial distribution of the Lagrange multiplier $\gamma_2(x)$ within the symmetric localization band is depicted in Fig. 6 along with the local damage-driving force Y and the constitutive function Y_c .

The response of the elasto-damaging bar is clearly dependent upon the length scales l_c and l_{coh} via the non-dimensional parameter λ defined by (39) and from the geometric factor β given by (52).

Actually, by comparison of Eq. (44) with (51) one has:

$$u^\star(d_m) = \frac{\sigma_c\,L}{E}\,\frac{\beta\,d_m^2 + 1 - d_m}{\lambda\,d_m^2 + 1 - d_m} \tag{57}$$

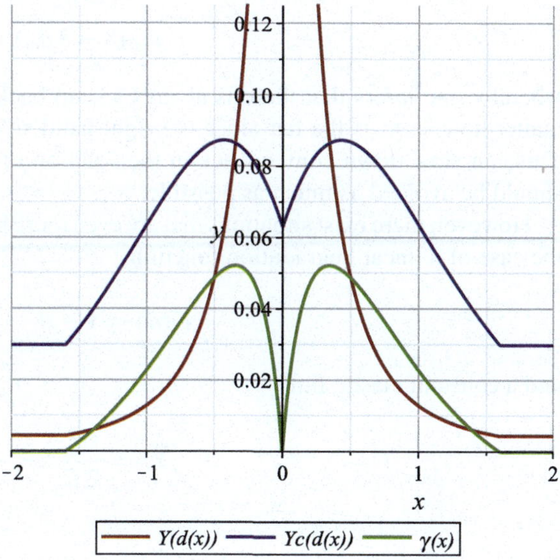

Fig. 6 One-dimensional rod. A typical spatial distribution of Y, Y_c and γ_2 corresponding to the symmetric localized solution with a single defect

This relationship allows to obtain the condition under which the response of the rod is stable under displacement control, i.e. it does not exhibit a snap-back. This requires the end-displacement $u^\star$ to be an increasing function of the maximum damage level d_m, that is:

$$\frac{\partial u^\star}{\partial d_m} = \frac{\sigma_c\,L}{E}\,\frac{(\beta-\lambda)(2\,d_m - d_m^2)}{\lambda\,d_m^2 + 1 - d_m} > 0 \tag{58}$$

whereby one obtains the condition that governs the stability of the response for the damaging tensile bar under displacement control

$$\beta > \lambda \;\Leftrightarrow\; L < l_{coh} \tag{59}$$

The stability condition strongly depends upon the expressions of the degradation function (42) and of the constitutive function (46). Actually, taking for Y_c the constant function, i.e. $Y_c(0)$ given by (47), and the quadratic degradation function (42) one obtains:

$$\sigma(d_m) = \frac{2\,\sigma_c\,(1-d_m)}{\sqrt{4-2\,d_m}} \tag{60}$$

in place of (44) and

$$u^\star(d_m) = \frac{\sigma_c\,L}{E}\,\frac{2(\beta\,d_m^2 + 1 - d_m)}{\sqrt{4-2\,d_m}} \tag{61}$$

that replaces (57). The stability condition now reads:

$$\beta > \frac{3-d_m}{d_m\,(8-3\,d_m)} \tag{62}$$

whereby one infers that there is always a snap back right after the elastic limit no matter how short is the bar since the right-hand side of (62) diverges for $d_m \to 0$. This can slow down convergence in the solution of a Finite Element problem and should be avoided as much as possible.

However, there exist situations that are even more harmful. In this respect, consider the case of a linear degradation function

$$\overline{\omega}(d) = 1 - d \tag{63}$$

and a constant elastic limit

$$\overline{Y}_c = \frac{\sigma_c^2}{2\,E} \tag{64}$$

In this case one obtains the stress and the end displacement as:

$$\sigma(d_m) = \sigma_c \sqrt{1 - d_m} \tag{65}$$

$$u^\star(d_m) = \frac{\sigma_c L}{E} \left(1 - \beta\, d_m - \beta \ln(1 - d_m)\right) \tag{66}$$

The stability condition now reads:

$$\beta > \frac{1}{\ln(1 - d_m) + 3\, d_m} \tag{67}$$

Clearly, the right-hand side of (67) diverges for both $d_m \to 0$ and for $d_m \to 1 - \exp(-3d_m) \simeq 0.94048$.

Figure 7 depicts the $\sigma - u^\star$ response of the bar for the different choices of the constitutive functions $\omega(d)$ and $Y_c(d)$ considered above, that is:

(i) quadratic degradation and non-constant limit $Y_c(d)$ that realizes the equivalence with linear softening;
(ii) quadratic degradation and constant limit $Y_c = \sigma_c^2/E$;
(iii) linear degradation and constant limit $Y_c = \sigma_c^2/2E$.

For all these cases the non-dimensional parameters respectively defined by (39) and (52) are such that $\lambda \leq 0.5$ and $\beta > \lambda$, which correspond to a rod that can be considered a short one.

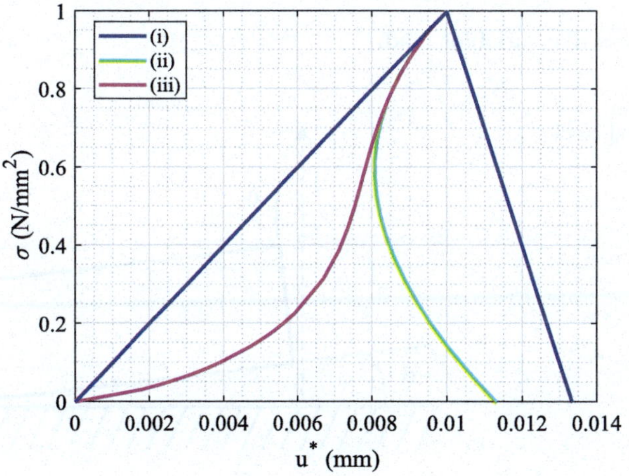

Fig. 7 One dimensional rod. Normalized stress-displacement responses obtained for different choices of the constitutive functions $\omega(d)$ and $Y_c(d)$

3 The Block with Cohesive Interface

Generally speaking, a cohesive law is a relationship between a displacement discontinuity vector, which is understood as the interface strain, and a surface traction vector playing the role of the stress. For the developments that follow attention will be restricted to mode-I opening; tractions and displacement jumps will then be normal to the interface while negative relative displacements will be left out for notational simplicity.

As a model problem consider the structure in Fig. 8, consisting of a rigid block connected to a fixed support via a damageable adhesive layer of negligible thickness [24]. A monotonic increase of the end-displacement δ produces a uniform rotation of the block but a non-uniform distribution of damage, which starts nucleating from the left edge with non-zero gradient.

Denoting by w the opening displacement across the interface, a stored energy function from which one can obtain a (local) cohesive law using a damage-based formulation reads [22]:

$$\tilde{\psi}(w, d) = \frac{1}{2}\,\omega(d)\,k\,w^2 \tag{68}$$

where k is the (undamaged) interface stiffness in tension.

For the problem at hand the kinematics of deformation is completely described by a single parameter, i.e. the rotation α, here assumed to be small in the usual sense. Therefore, one has the opening displacement:

$$w(x) = \alpha\,(L - x) \tag{69}$$

while the stress-like variables read:

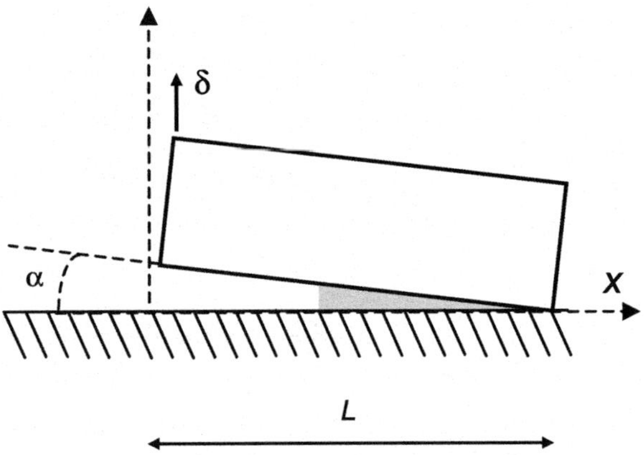

Fig. 8 The rigid-block problem

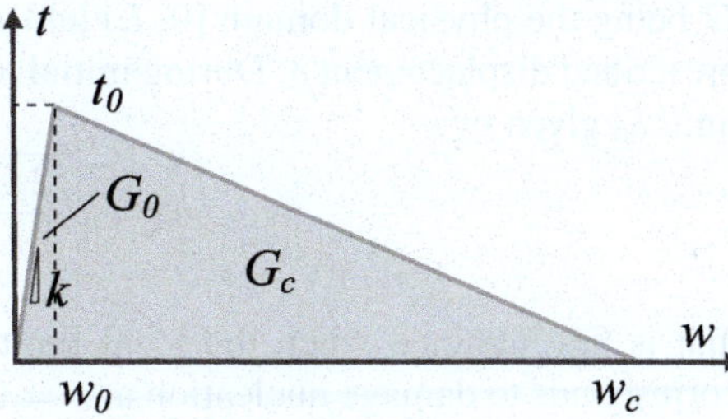

Fig. 9 Bilinear cohesive law

$$
t(x,d) = \frac{\partial \tilde{\psi}(w,d)}{\partial w} = \omega(d)\, k\, \alpha\, (L-x)
$$
$$
\tilde{Y}(x,d) = -\frac{\partial \tilde{\psi}(w,d)}{\partial d} = -\omega'(d)\, \frac{1}{2}\, k\, \alpha^2 (L-x)^2 \tag{70}
$$

The governing equations of a nonlocal interface model based on the graded damage concept are formally identical to those developed in Sect. 2 for the tensile rod with two minor modifications. In particular, the stored energy function (1) has to be replaced by (68), while the function $\tilde{Y}_c(d)$ is now directly prescribed, e.g. based on the shape of a chosen traction-separation relationship.

For instance, a constitutive function $\tilde{Y}_c(d)$ that yields the (local) bilinear cohesive law of Fig. 9 reads:

$$
\tilde{Y}_c(d) = \frac{-\omega'(d)\, G_0\, G_c^2}{\left[G_0 + (G_c - G_0)\, \omega(d)\right]^2} \tag{71}
$$

to which corresponds the work of separation:

$$
\int_0^{+\infty} \tilde{Y}_c(d)\, \dot{d}\; \mathrm{d}t = -\int_0^1 \tilde{Y}_c(\omega)\; \mathrm{d}\omega = G_c \tag{72}
$$

where G_0 and G_c respectively denote the initial energy threshold and the interface fracture toughness:

$$
G_0 = \frac{1}{2} k\, {w_0}^2; \qquad G_c = \frac{1}{2} k\, w_0\, w_c \tag{73}
$$

Without loss of generality, in the remainder we shall assume Eqs. (12) and (42) to hold; moreover, the length scale l_c of the nonlocal interface model is supposed to be greater than the width L of the block.

The equilibrium path of the structure can be traced using the balance of moments about the center of rotation:

$$
P\, L = \int_\Omega t(x,d)\, (L-x)\; \mathrm{d}x \tag{74}
$$

Ω being the physical domain $[0, L]$ and P the reaction force corresponding to the prescribed displacement δ. During initial loading the latter increases up to the elastic limit δ_0 given by:

$$\delta_0 = \sqrt{\frac{2\,G_0}{k}} = \alpha_0\,L \tag{75}$$

that is first attained when the local limit condition $\tilde{Y} = \tilde{Y}_c\,(0)$ is met. This state corresponds to damage nucleation at $x = 0$ and from this point onwards loading can be effectively parametrized in terms of the size $l_m > 0$ of the damaged portion of the domain of interest.

For $l_m \le l_c$ the length l_m does also coincide with the size of the active constraint set (i.e. where $g_2(d) = 0$) and the adopted parametrization is fully equivalent to the one given in terms of the maximum damage d_m defined as:

$$d_m = \min\left\{1, \frac{l_m}{l_c}\right\} \tag{76}$$

It is worth emphasizing that, unlike the case of the tensile rod with a single evolving defect, for the problem at hand the constraint set translates along the interface once the damage process zone has fully developed. In particular, this occurs when the size of the damaged region l_m equals the length scale l_c; to account for this case, the (piece-wise) linear damage function that is prescribed via the gradient constraint (12) is conveniently defined as:

$$d(x) = \max\left\{0, d_m - \frac{x - c}{l_c}\right\} \tag{77}$$

where the (finite) size of the fully damaged subdomain reads:

$$c = \max\left\{0, l_m - l_c\right\} \tag{78}$$

As discussed in Sect. 2, owing to gradient-dependence the normality rule yields the differential equation (28); the latter now admits two sets of boundary conditions for the opening angle α and the Lagrange multiplier γ_2.

For the rigid block problem the averaged limit condition reads:

$$\int_0^H \tilde{Y}(x, d(x))\,\mathrm{d}x = \int_0^H \tilde{Y}_c\,(d(x))\,\mathrm{d}x \tag{79}$$

where H denotes the size of the active process zone portion that is contained within the physical domain $[0, L]$:

$$H = \min\left\{l_m, L\right\} \tag{80}$$

With this notation in hand, balance of moments about the center of rotation can be expressed as:

$$P\,L = \int_c^H \omega(d) k\,\hat{\alpha}\,(L-x)^2\,\mathrm{d}x + \int_H^L k\,\hat{\alpha}\,(L-x)^2\,\mathrm{d}x \tag{81}$$

whereby one obtains the different branches of the equilibrium path by distinguishing the different possible cases for the integration limits of Eq. (79), which in turn take into account the boundary conditions (16) for the Lagrange multiplier field and the internal jump conditions (15), if any.

Phase 1. Linear elastic The initial linear elastic phase is purely local; the limit value for the displacement is given by (75), to which corresponds the reaction force:

$$P_0 = \sqrt{2\,G_0\,k}\,\frac{L}{3} \tag{82}$$

Phase 2. Damage nucleation and growth for $0 < l_m \leq L$ In this case use of the integral limit condition (79) allows to compute the opening angle α as a function of the length l_m (driving variable) as:

$$\hat{\alpha}_1 = \alpha_0 \sqrt{\frac{G_c\,l_c^2\,L^2\,(2\,l_c - l_m)}{A_1\,A_2}} \tag{83}$$

with

$$A_1 = \left[\frac{l_m^3}{6} - \frac{2}{3}(l_c + L)\,l_m^2 + (L + 2\,l_c)\,L\,l_m - 2\,L^2\,l_c\right] \tag{84}$$

$$A_2 = \left[(G_0 - G_c)(l_m^2 - 2\,l_c\,l_m) - G_c\,l_c^2\right] \tag{85}$$

and limits

$$\hat{\alpha}_1^0 = \lim_{l_m \to 0} \hat{\alpha}_1 = \alpha_0 \tag{86}$$

$$\hat{\alpha}_1^L = \lim_{l_m \to L} \hat{\alpha}_1 = \alpha_0 \sqrt{\frac{6\,G_c\,l_c^2\,(2\,l_c - L)}{(4\,l_c - 3\,L)\,[(l_c - L)^2\,G_c - (L - 2\,l_c)\,G_0\,L]}} \tag{87}$$

Obviously, the limit (86) coincides with the opening angle α_0 defined by (75) whereas the upper limit (87) marks the end of the domain of validity of relationship (83), to which corresponds by equilibrium the end reaction force:

$$\hat{P}_1 = \left[L^3\,l_c^2 + (l_m - 3\,l_c)\,l_m^2\,L^2 - \frac{(l_m - 4\,l_c)\,l_m^3\,L}{2} + \frac{(l_m - 5\,l_c)\,l_m^4}{10}\right]\frac{k\,\hat{\alpha}_1}{3\,l_c^2\,L} \tag{88}$$

Phase 3. Damage growth for $L \le l_m \le l_c$ When the driving variable l_m grows beyond the length L, the zero-damage boundary would be located outside the physical domain of the interface. In this case the boundary conditions for the Lagrange multiplier and the limit condition (79) yield an averaged equality over the entire domain $[0, L]$, whereby one obtains another nonlinear branch of the equilibrium path defined by the following:

$$\hat{\alpha}_2 = \alpha_0 \sqrt{\frac{6\, G_c^2\, l_c^4\, (L + 2\, l_c - 2\, l_m)}{B_1\, B_2\, B_3}} \tag{89}$$

with

$$B_1 = L + 4\, l_c - 4\, l_m \tag{90}$$

$$B_2 = (l_c - l_m)^2\, G_c - l_m\, (l_m - 2\, l_c)\, G_0 \tag{91}$$

$$B_3 = (L - l_m + l_c)^2\, G_c - (L - l_m)(L - l_m + 2\, l_c) G_0 \tag{92}$$

The corresponding value of the reaction force reads:

$$\hat{P}_2 = \frac{k\, L^2}{30\, l_c^2} \left[10\, l_c^2 + (5\, L - 20\, l_m)\, l_c + L^2 - 5\, L\, l_m + 10\, l_m^2 \right] \hat{\alpha}_2 \tag{93}$$

Phase 4. Crack propagation For $l_m > l_c$ there exists a fully damaged region of finite size c defined by (78); the latter can be taken as the driving variable for computing the last part of the equilibrium curve because in this case the non-trivial limit condition reduces to an averaged equality over the interval $[c, L]$ on account of the jump relationships (15).

The opening angle is now computed as:

$$\hat{\alpha}_3 = \frac{\alpha_0}{(L - c)^2} \sqrt{\frac{6\, G_c^2\, l_c^2\, L^2}{G_0 \left[G_c\, (L - c)^2 - G_0 (-L + c - l_c)(-L + c + l_c) \right]}} \tag{94}$$

whereas the equilibrium equation yields:

$$\hat{P}_3 = \frac{k\, (L - c)^5}{30\, l_c^2\, L}\, \hat{\alpha}_3 \tag{95}$$

As expected, the reaction force (95) converges to zero when the portion of the active damage process zone lying within the physical domain progressively shrinks and collapses to a point.

The complete equilibrium path for the block delamination problem is depicted in Fig. 10. The curve corresponds to a length scale $l_c = 6\,\text{mm}$ and to the data set of Table 1. The different colors on the plot are used to distinguish the four branches of the theoretical solution whereas the points highlighted on the load-deflection curve

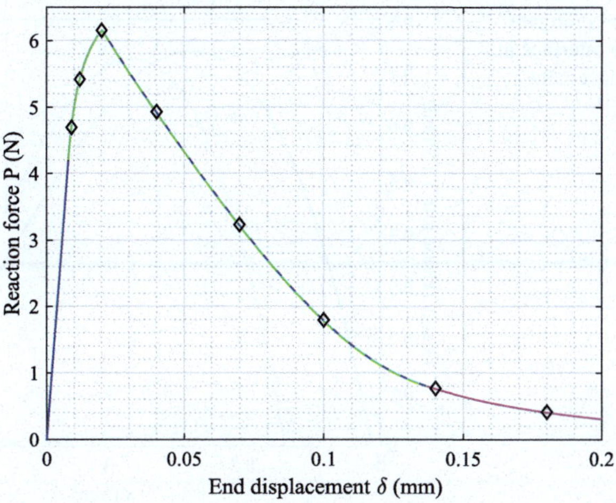

Fig. 10 Equilibrium curve for the rigid-block problem

Table 1 Data set for the rigid block problem

$L = 2\,\text{mm}$;	$k = 800\,\text{N/mm}^3$
$G_c = 0.25\,\text{N/mm}$;	$G_0 = 0.025\,\text{N/mm}$

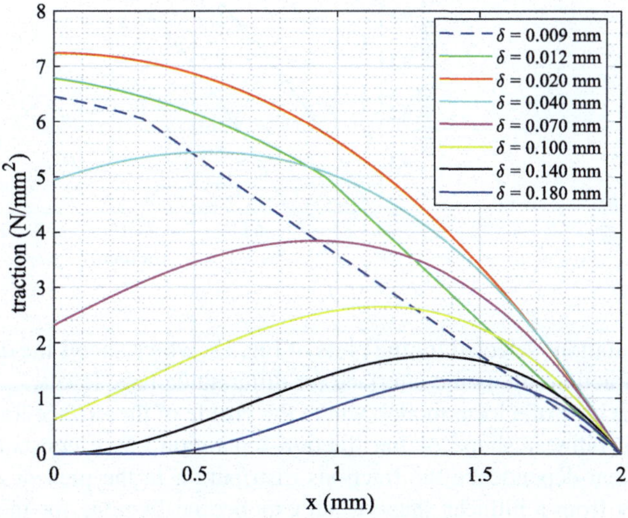

Fig. 11 Tractions distributions along the interface at varying end-displacement δ

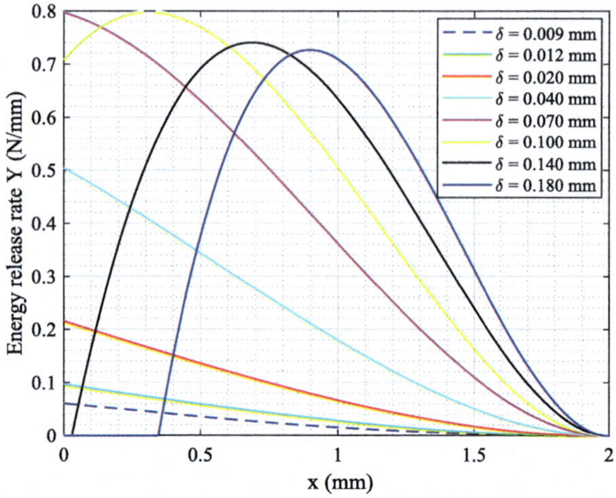

Fig. 12 Damage-conjugate force $\tilde{Y}$ along the interface at varying end-displacement δ

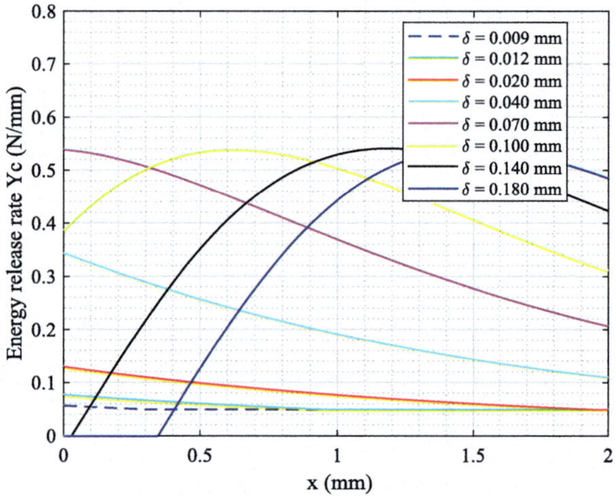

Fig. 13 Threshold function $\tilde{Y}_c$ along the interface at varying end-displacement δ

indicate the stations selected for the plots of Fig. 11, which shows the distribution of the surface tractions along the interface for different damage levels.

Unlike the case of a local model, where the profile of the surface tractions would replicate the bilinear shape of the traction-separation curve consequent to (71), due to gradient-dependence the tractions distribution in the present case changes continuously from a bilinear shape to an exponential-like one for increasing end-displacement δ.

Moreover, it is noted that the differential character of the constitutive relationship allows for point-wise values of the surface tractions higher than the peak stress $\sqrt{2k\,G_0}$ of the underlying local model. Likewise, the point-wise values of the damage-conjugate variable $\tilde{Y}$ can exceed those of the limit function $\tilde{Y}_c$ owing to

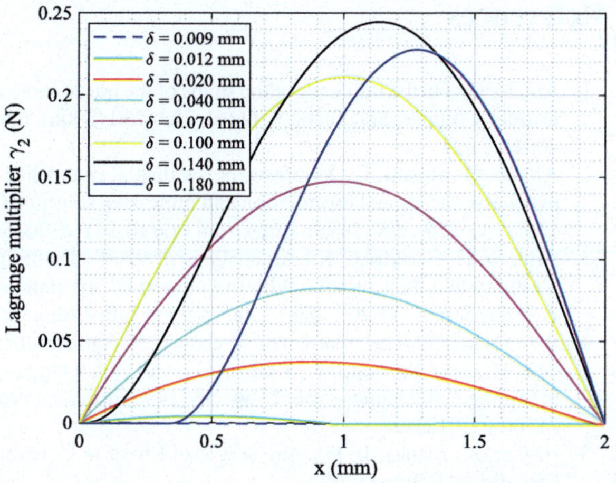

Fig. 14 Lagrange multiplier γ_2 along the interface at varying end-displacement δ

the averaged character of the limit condition. This behaviour is clearly put forward in Figs. 12 and 13, that have been plotted using the same scale to ease the comparison.

It is also worth noting that functions $\tilde{Y}$ and $\tilde{Y}_c$ do share only one common pointwise value over the physical domain $[0, L]$. This corresponds to the extremum point, which is indeed a maximum because of the obvious sign constraint, of the Lagrange multiplier field γ_2 that is used to enforce the nonlocal, gradient constraint, see e.g. Fig. 14.

4 Closure

Based on the graded damage formulation contributed in [23] two problems in one dimension have been discussed and the relevant analytical solutions computed to be used as a reference for finite element procedures.

In the tensile rod problem the uniform stress is obtained from the averaged limit condition and the constitutive function Y_c is determined in a way to produce a global response curve that is consistent with a cohesive zone model with linear softening.

On the other hand, in the block delamination problem the constitutive function Y_c is prescribed based on a local cohesive model while the integral limit condition provides the parameter governing the kinematics of deformation. In this case one obtains a global equilibrium curve that depends from the length scale since the effect of gradient-dependence is that of relaxing the surface tractions in that they are no longer constrained to the shape of the underlying local softening curve.

References

1. Alfano, G.: On the influence of the shape of the interface law on the application of cohesive-zone models. Compos. Sci. Technol. **66**(6), 723–730 (2006). https://doi.org/10.1016/j.compscitech.2004.12.024
2. Alfano, G., Rosati, L., Valoroso, N.: A displacement-like finite element model for J2 elastoplasticity: variational formulation and finite-step solution. Comput. Methods Appl. Mech. Eng. **155**(3–4), 325–358 (1998). https://doi.org/10.1016/S0045-7825(97)00171-0
3. Armero, F., Garikipati, K.: An analysis of strong discontinuities in multiplicative finite strain plasticity and their relation with the numerical simulation of strain localization in solids. Int. J. Solids Struct. **33**(20), 2863–2885 (1996). https://doi.org/10.1016/0020-7683(95)00257-X
4. Barenblatt, G.: The mathematical theory of equilibrium cracks in brittle fracture. In: Dryden, H., von Kármán, T., Kuerti, G., van den Dungen, F., Howarth, L. (eds.) Advances in Applied Mechanics, vol. 7, pp. 55–129. Elsevier (1962). https://doi.org/10.1016/S0065-2156(08)70121-2
5. Bažant, Z., Planas, J.: Fracture and Size Effect in Concrete and Other Quasibrittle Materials. CRC Press, London (1998)
6. Bourdin, B., Francfort, G., Marigo, J.J.: Numerical experiments in revisited brittle fracture. J. Mech. Phys. Solids **48**(4), 797–826 (2000). https://doi.org/10.1016/S0022-5096(99)00028-9
7. François, D., Pineau, A., Zaoui, A.: Mechanical Behaviour of Materials. Volume II: Fracture Mechanics and Damage. Solid Mechanics and Its Applications. Springer, Berlin (2013). https://doi.org/10.1007/978-94-007-4930-6
8. Geers, M., De Borst, R., Brekelmans, W., Peerlings, R.: Strain-based transient-gradient damage model for failure analyses. Comput. Methods Appl. Mech. Eng. **160**(1–2), 133–153 (1998). https://doi.org/10.1016/S0045-7825(98)80011-X
9. Germain, P., Nguyen, Q., Suquet, P.: Continuum thermodynamics. J. Appl. Mech. **50**(4b), 1010–1020 (1983). https://doi.org/10.1115/1.3167184
10. Halphen, B., Nguyen, Q.S.: Sur les matériaux standard généralisés. J. de Mécanique **14**, 39–63 (1975)
11. Hillerborg, A., Modéer, M., Petersson, P.E.: Analysis of crack formation and crack growth in concrete by means of fracture mechanics and finite elements. Cem. Concr. Res. **6**(6), 773–781 (1976). https://doi.org/10.1016/0008-8846(76)90007-7
12. Latifi, M., van der Meer, F., Sluys, L.: An interface thick level set model for simulating delamination in composites. Int. J. Numer. Meth. Eng. **111**(4), 303–324 (2017). https://doi.org/10.1002/nme.5463
13. Lorentz, E., Godard, V.: Gradient damage models: toward full-scale computations. Comput. Methods Appl. Mech. Eng. **200**(21–22), 1927–1944 (2011). https://doi.org/10.1016/j.cma.2010.06.025
14. Mi, Y., Crisfield, M.A., Davies, G.A.O., Hellweg, H.B.: Progressive delamination using interface elements. J. Compos. Mater. **32**(14), 1246–1272 (1998). https://doi.org/10.1177/002199839803201401
15. Miehe, C., Hofacker, M., Welschinger, F.: A phase field model for rate-independent crack propagation: robust algorithmic implementation based on operator splits. Comput. Methods Appl. Mech. Eng. **199**(45), 2765–2778 (2010). https://doi.org/10.1016/j.cma.2010.04.011
16. Miehe, C., Welschinger, F., Hofacker, M.: Thermodynamically consistent phase-field models of fracture: variational principles and multi-field FE implementations. Int. J. Numer. Meth. Eng. **83**(10), 1273–1311 (2010). https://doi.org/10.1002/nme.2861
17. Möes, N., Dolbow, J., Belytschko, T.: A finite element method for crack growth without remeshing. Int. J. Numer. Meth. Eng. **46**(1), 131–150 (1999). https://doi.org/10.1002/(SICI)1097-0207(19990910)46:1<131::AID-NME726>3.0.CO;2-J
18. Moës, N., Stolz, C., Bernard, P.E., Chevaugeon, N.: A level set based model for damage growth: the thick level set approach. Int. J. Numer. Meth. Eng. **86**(3), 358–380 (2011). https://doi.org/10.1002/nme.3069

19. Nguyen, T., Yvonnet, J., Zhu, Q.Z., Bornert, M., Chateau, C.: A phase-field method for computational modeling of interfacial damage interacting with crack propagation in realistic microstructures obtained by microtomography. Comput. Methods Appl. Mech. Eng. **312**, 567–595 (2016). https://doi.org/10.1016/j.cma.2015.10.007
20. Peerlings, R., De Borst, R., Brekelmans, W., De Vree, J.: Gradient enhanced damage for quasi-brittle materials. Int. J. Numer. Meth. Eng. **39**(19), 3391–3403 (1996). https://doi.org/10.1002/(SICI)1097-0207(19961015)39:19<3391::AID-NME7>3.0.CO;2-D
21. Pijaudier-Cabot, G., Bažant, Z.: Nonlocal damage theory. J. Eng. Mech. **113**(10), 1512–1533 (1987). https://doi.org/10.1061/(ASCE)0733-9399(1987)113:10(1512)
22. Valoroso, N., Champaney, L.: A damage-mechanics-based approach for modelling decohesion in adhesively bonded assemblies. Eng. Fract. Mech. **73**(18), 2774–2801 (2006). https://doi.org/10.1016/j.engfracmech.2006.04.029
23. Valoroso, N., Stolz, C.: Graded damage in quasi-brittle solids. Int. J. Numer. Meth. Eng. **123**(11), 2467–2498 (2022). https://doi.org/10.1002/nme.6947
24. Volokh, K.Y.: Comparison between cohesive zone models. Commun. Numer. Methods Eng. **20**(11), 845–856 (2004). https://doi.org/10.1002/cnm.717
25. Wu, J.Y.: A unified phase-field theory for the mechanics of damage and quasi-brittle failure. J. Mech. Phys. Solids **103**, 72–99 (2017). https://doi.org/10.1016/j.jmps.2017.03.015

Fatigue Strength Prediction of Nodular Cast Iron by Shakedown Analysis

Christian Gebhardt, Felix Weber, Tobias Sedlatschek, Alexander Bezold, and Christoph Broeckmann

Abstract A large number of heavy-section castings are used in current wind turbines. Considerable potential for innovation arises, on the one hand, from the use of new materials with different mechanical properties and, on the other hand, from exploiting strength gradients in the component which result from the microstructure of the material. In the present paper, therefore, the influence of the graphite morphology on the fatigue strength of high-silicon alloyed ductile cast iron (Si-DCI) is analyzed by shakedown analysis of real and synthetic microstructures. Finite element models are built from low magnification micrographs of Si-DCI by approximating the geometry of the precipitated graphite with B-splines. The fatigue strength is then predicted using shakedown analysis with Melan' static theorem. Simulation studies with synthetically generated microstructures show that the so-called nodularity can already be used as a suitable qualitative parameter for evaluating the influence of graphite morphology on the fatigue strength. A comparison of the results of shakedown analyses with experimentally measured fatigue strengths of Si-DCI shows good agreement. Due to the simplified modeling based on micrographs, the method may be used for component design, quality assurance and extrapolation of fatigue strength.

Keywords RVE · Micromechanical simulation · Shakedown analysis · B-Splines

1 Introduction

Hardly any mechanical failures of heavy-section castings in wind turbines are known [38], which implies a high safety potential in these components. According to the standard DIN EN ISO 1563 [16], a decisive cross-section needs to be specified in part design, which results in requirements for the microstructure and thus a mini-

C. Gebhardt (✉) · F. Weber · T. Sedlatschek · A. Bezold · C. Broeckmann
Institute for Materials Applications in Mechanical Engineering, RWTH Aachen University, Aachen, Germany
e-mail: c.gebhardt@iwm.rwth-aachen.de

G. Garcea and D. Weichert (eds.), *Direct Methods for Limit State of Materials and Structures*, Lecture Notes in Applied and Computational Mechanics 101,
https://doi.org/10.1007/978-3-031-29122-7_7

mum strength, often not defined by the standard but only by agreement of customer and foundry. As a result of complex geometries, however, castings in wind turbines exhibit locally different microstructures, which is manifested in various gradients, one of which is the graphite morphology [44]. In addition to the metallic matrix, shrinkage cavities and graphite nodules influence the strength of nodular cast iron. In particular, they have a crack-initiating effect under cyclic loading and are responsible for the scattering of the fatigue strength due to local variations in morphology, size and distribution of defects [18]. Large shrinkage cavities lower the fatigue strength and are therefore often modeled as cracks in the context of component design [30]. In general, the fatigue mechanisms of ductile cast iron cannot be generalized. Since ductile cast irons tolerate shrinkage up to a certain size according to the Kitagawa diagram, the design guideline [25] defines an elongation at fracture above which a damage tolerant design is required. In more brittle cast iron materials, the crack initiation phase tends to play a minor role and crack growth under a Mode I loading dominates the fatigue life [30], while the fatigue phenomenon in ductile materials is caused by dislocation motion [26]. The latter results in higher defect tolerance and an increased occurrence of non-propagating cracks [24]. In high silicon alloyed nodular cast irons (Si-DCI), the metallic matrix is almost exclusively ferritic. In the absence of larger shrinkage cavities, which can be locally avoided by cooling irons, the graphite morphology has a major influence on the fatigue strength in this material [24]. This complicates the derivation of microstructure-property relationships, which are, however, of great interest for the local dimensioning of the castings. Micromechanical models based on the real microstructure in the material provide a remedy. Beyond analytical and semi-analytical micromechanical models, representative volume elements (RVE) based on the finite element method are well established [8]. By applying numerical homogenization methods, they can be used to determine effective mechanical material properties - including the fatigue properties of a material. In the last decades, micromechanical modeling methods have been further developed which allow the simulation of graphite morphology in ductile iron castings using representative volume elements [5]. For ductile cast irons, both two- and three-dimensional models are well established, with two-dimensional models currently having advantages over three-dimensional ones in resolving the phase boundary. For example, the interface between graphite and matrix was approximated with Bezier curves in the literature [9]. The advantage is a smooth interface between graphite and ferrite, which avoids artificial stress peaks at the interface. Furthermore, multiphase elements are known, where the interface passes a finite element [20]. In the literature, the constitutive modeling of the metallic matrix of DCI is mainly performed on the basis of infinitesimal strain tensors and assuming the validity of the associated flow rule [35] as well as the von Mises yield surface [4]. Homogeneous and isotropic modeling of the metallic matrix, where the grain structure is explicitly modeled, are common. Several phenomenological hardening laws were applied for the matrix, such as the Voce's law [6], the Armstrong-Frederich law [35] and the Simo and Hughes law [19].

For DCI, it is common to inversely calibrate model parameters by comparison with macroscopic material tests. Kasvayee et al. [32] calibrated the model parameters with tensile tests. Andriollo et al. [5] calibrated the homogenized properties of the simulation to experimental cyclic material properties. In addition, characterization of the individual microstructural constituents by nanoindentation is frequently applied for DCI [3, 19]. Fernandino et al. determined the model parameters inversely by using FEM [19]. Andriollo et al. [2] used nanoindentation to characterize the mechanical property gradient within graphite nodules. However, nanoindentation of such soft phases is problematic because the results depend not only on the heterogeneity of the graphite nodules themselves, but also on the constitutive properties of underlying microstructural constituents. More recently, Andriollo et al. [6] prepared an equivalent matrix material that corresponded to the effective composition of the ferritic matrix. The model material is then characterized in macroscopic material tests. Gebhardt et al. [22] applied an alternative approach in which the plastic deformation of the ferritic matrix in Si-DCI was characterized by a combination of micromechanical modeling and full-field digital image correlation. If calibrated properly, effective material properties are obtained from the RVE through homogenization. Such approach fails if the microstructure is not statistically homogeneous which is particularly crucial for DCI since the microstructural constituents vary widely locally [6]. In the literature for DCI, numerical homogenization methods based on averaging techniques are largely known [10]. For purely elastic deformation, the effective elasticity tensor can be obtained immediately from the homogenized fields. If plastic deformation is simulated, a stress-strain curve is often plotted in the literature by using the homogenized fields, from which the effective properties are then directly determined [19]. Compared to the material properties under monotonous loading, microstructure-sensitive fatigue properties and microstructure-fatigue relationships are not sufficiently investigated in the current scientific literature [23]. Generally, in various modeling approaches, crack initiation and crack propagation were explicitly simulated for each loading cycle [31] which results in high computational costs. Consequently, these models may contribute to an understanding of the underlying fatigue mechanisms, but are unsuitable for generating large data sets. However, these are required for the derivation of microstructure-property relationships. Fatigue Indicator parameters (FIP) were developed based on this motivation [27]. Therein, only a few cycles of the cyclic loading are simulated explicitly, which still results in a high, but reduced, computational time. On the other hand, so-called direct methods offer an advantage, since they can efficiently predict a microstructure-sensitive fatigue strength based on RVEs by using the shakedown effect on the microscale [48]. Here, the limit state sought is the cyclic loading of the RVE at which the accumulation of plastic dissipation work is limited. Thereby, the time evolution of fatigue damage and cracking are not explicitly simulated. The shakedown state can be determined by mathematical optimization [41]. The advantage of this method is that the fatigue strength (shakedown limit) of a microstructure can be predicted very quickly without expensive incremental simulations [13]. The static shakedown theorem was applied to idealized material microstructures to determine optimal design parameters for fiber composites [14]. By using the average stress theorem, safe load spaces can be determined for RVEs [39]. Hachemi et al. [28] presented an approach to account for the debonding of defects in idealized material microstructures. Moreover, the static shakedown theorem was used for three-dimensional RVEs. However, the material structure consisted of only a single idealized inclusion [34]. You et al. [49] studied copper composites using incremental and direct methods. The results show comparable shakedown behavior with both methods. Shakedown analysis of cemented carbide microstructures were performed by Chen et al. [13]. The RVEs were based on SEM images or generated synthetically from statistical data. In more recent publications, the shakedown behavior at the grain level was also discussed. These models are based on the research of Dang Van and Papadopolous [12]. Auricchio et al. [7] derived a fatigue model for shape memory alloys from them. Gebhardt et al. performed shakedown analysis of nodular cast iron and, the results of image analysis and shakedown limits were correlated by using a simplified residual neural network [23]. Other than that, no systematic study of the influence of the microstructure on the fatigue strength of Si-DCI is known in the literature. Thus, this paper presents shakedown analyses on real and synthetically generated microstructures of Si-DCI.

2 Material

The basis of the investigations was the standardized material EN-GJS-500-14 which was cast in cuboids of size $100 \times 100 \times 220$ mm by sand casting. For details of process control and fabrication, the reader is referred to [21]. Starting from a reference alloy, the graphite morphology was modified by adding cerium and manipulating the magnesium pretreatment, resulting in three materials with spheroidal graphite, where the graphite nodule density increases from *S1* to *S3*, and *V* with vermicular graphite (see Table 1).

As a first approach to understand the influence of the graphite morphology on the fatigue limit, fatigue tests under axial loading were carried out in accordance with DIN 50100 [15] with constant stress amplitudes on modernized 100 kN Amsler HFP 422 high-frequency pulsators which were also partly published in [21, 24]. The cycle limit was 10^7 and a frequency drop of 5 Hz was used to detect the macroscopic crack. At least 20 specimens were tested at stress ratios of $R_\sigma = -1$, $R_\sigma = 0$, and $R_\sigma = -2.3$ for each S-N-curve, 15 of which were used in a stair-case procedure. An additional five specimens were used to estimate the transition region to infinite fatigue

Table 1 Chemical compositions of investigated alloys in wt.%

Alloy	*C*	*Si*	*Ce*	*Mg*	*Ni*	*P*	*Cr*	*Mn*	*Fe*
S1	2.87	3.85	<0.0033	0.028	0.028	0.077	0.045	0.172	bal.
S2	2.76	3.78	<0.0033	0.033	0.057	0.035	0.050	0.181	bal.
S3	2.82	3.95	0.047	0.042	0.037	0.036	0.057	0.202	bal.
V	2.95	3.95	<0.0033	0.022	0.027	0.233	0.048	0.175	bal.
Standard DIN EN 1563 [16]									
	–	>3.8	–	–	–	< 0.05	–	< 0.5	

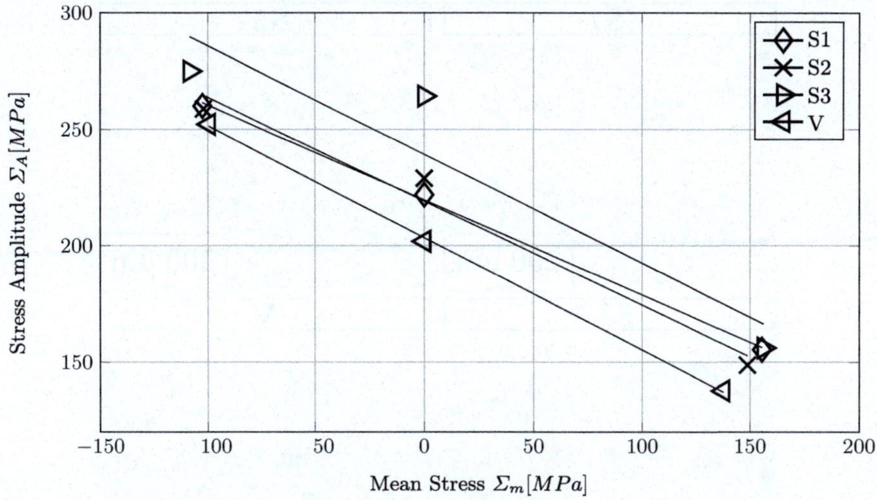

Fig. 1 Haigh diagrams of the investigated alloys of EN-GJS-500-14. Σ_A is the stress amplitude at 50% failure probability. *S1–S3* show spheroidal graphite with increasing nodule density, whereas *V* shows vermicular graphite

Table 2 Quantitative graphite analysis

Alloy	Nodularity [%]	Nodule density [1/mm^2]	Mean maximum feret diameter [μm]
S1	62.6	88.4	32.5
S2	64.0	75.4	30.8
S3	57.6	22.2	51.1
V	44.5	67.5	36.9

life by using a combined statistical analysis [33]. Figure 1 shows the 50% values of the stress amplitudes Σ_A over the mean stress Σ_m. Alloy group *V* exhibits the lowest fatigue strength. It is noteworthy that alloy group *S3* predominantly exhibits the highest fatigue strengths. Alloys of groups *S1* and *S2* are comparable.

For quantitative image analysis as well as for the generation of the micromechanical models, metallographic sections of the previously tested fatigue specimens were prepared for each S-N curve. At least ten microstructural images were taken for each specimen at 100× magnification and a pixel scale of 1.0234 μm. The results in Table 2 show that the nodularity according to ASTM E2567 [17] is lowest for the alloy group *V*. The alloy group *S3* had the largest graphite nodules, resulting in a lower nodule density. Figure 2 shows microstructural images representative of the different alloy groups.

The fatigue mechanisms of these particular alloys were published in [24]. It was shown that the graphite morphology influences the fatigue mechanisms locally. At spheroidal graphite mostly slip is observed, whereas non-speroidal graphite shows

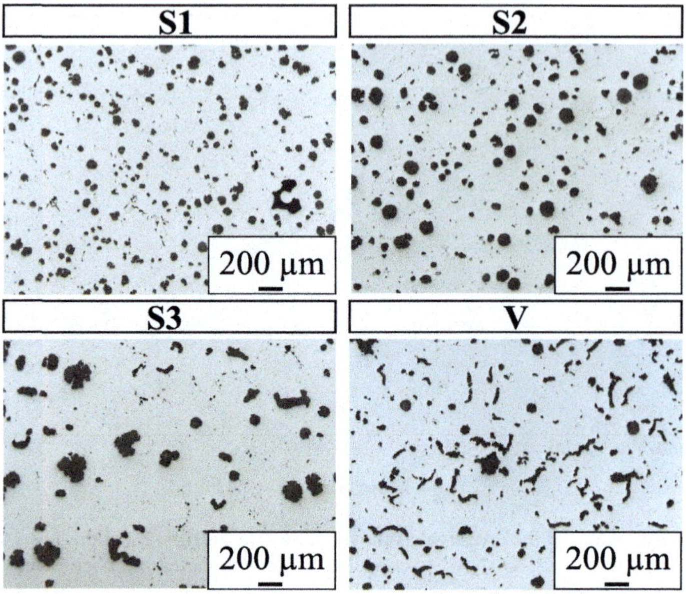

Fig. 2 Exemplary micrographs of the four different alloys investigated. Spheroidal graphite *S1–S3* and vermicular graphite *V*

early Mode I crack initiation. In the course of fatigue loading these fatigue mechanisms, the cracks become non-propagating and no further plastic slip accumulates at spheroidal graphite. Instead more plastic slip and cracks form at other graphite nodules. In a continuum mechanics sense, these microscale fatigue mechanisms can be considered as effective plastic deformation. In the following, therefore, these mechanisms are interpreted as shakedown phenomena at the mesoscale.

3 Methods

3.1 Generation of Finite Element Representative Volume Elements

Shakedown analyses were carried out on both real and synthetic micrographs of the cast iron grade EN-GJS-500-14. The analyses on synthetic micrographs were used to investigate the effect of graphite on the fatigue behavior, while the analyses of the real micrographs were used to compare the numerical results with experiments. To this end 254 real micrographs of the four investigated alloy groups *S1–S3* and *V* were taken from the fatigue specimens as mentioned above.

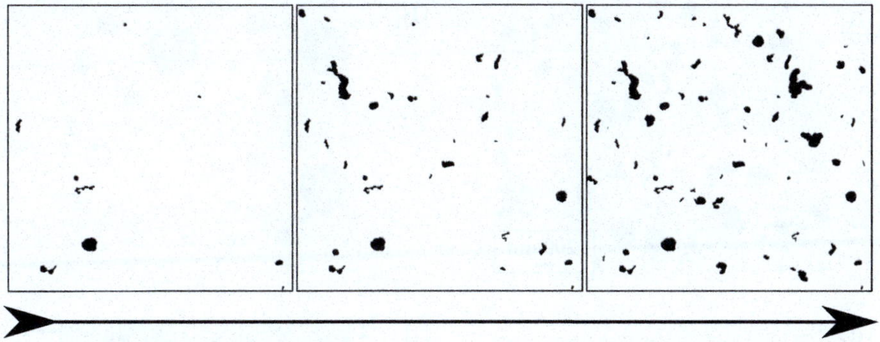

Fig. 3 Example of a synthetic RVEs built with the RSA algorithm. The graphite nodules are randomly projected into the RVE

The synthetic microstructures were constructed by using a so-called Random Sequential Adsorption algorithm (RSA) (see Fig. 3). According to DIN 945-4 [1], graphite nodules may be classified into classes I to VI depending on their shape as well as categories three to eight depending on their size. Velichko et al. [46] quantified the comparative images for graphite analysis given in the standard [1] by quantitative image analysis. These results were used in this work to classify all graphite nodules acquired from all 254 micrographs in a database. The synthetic RVEs were then generated by randomly projecting graphite nodules into an image size of 1 mm^2. After each iteration, the respective graphite morphology and graphite fraction as well as the nodule count were checked. In this way, selected graphite features were kept constant or within a maximum allowed relative deviation of 1%. The result of the algorithm is a binary micrograph. In a post-processing step, the graphite-ferrite interfaces were diluted by assigning to each pixel the mean of its $3 \cdot 3$ submatrix, which simplifies the construction of the B-spline around the graphite nodules, described in the next chapter.

The fatigue mechanisms identified and related to the graphite morphology require that the graphite geometry must be reconstructed from the micrographs as accurately as possible. Hence, approximated B-splines were used to model the graphite. Figure 4 shows the main steps of the implemented meshing method. The implementation was carried out in Python language and uses the open-source packages *sci-kit* [36], *sci-kit image* [45] and *open-cv* [11]. The details of the program code are published in [23] and are briefly outlined here.

Starting from the micrograph (Fig. 4a), the graphite contour is extracted by image analysis. The pixel coordinates of the respective contours serve as control points of a B-spline (Fig. 4b). A B-spline curve $\mathbf{C}(u)$ of the degree p is a sum of piecewise polynomials constructed by using a knot vector $\mathbf{U} = [u_0, \ldots, u_m]$ and has the control points $\{\mathbf{P}_i\} = \{\mathbf{P}_0, \ldots, \mathbf{P}_n\}$. Knot vector and control points are related by the following relation:

$$m = n + p + 1. \tag{1}$$

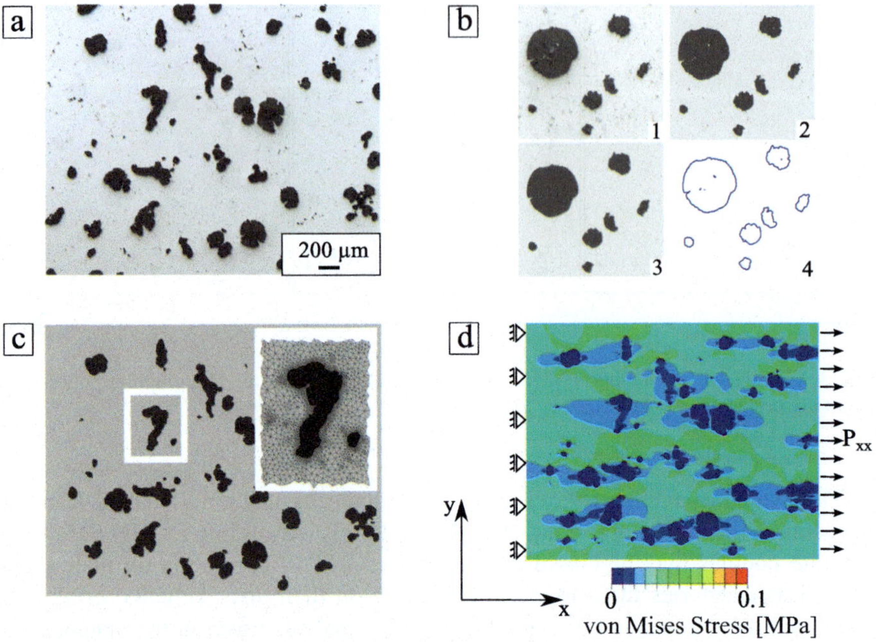

Fig. 4 Generation of the Finite Element Model: optical microscope image (**a**), image processing and B-spline modeling (**b**), meshing (**c**), exemplary linear elastic RVE simulation (**d**)

Here $m + 1$ is the number of knots and $n + 1$ is the number of control points. The B-spline curve is then

$$\mathbf{C}(u) = \sum_{i=0}^{n} N_{i,p}(u)\mathbf{P}_i. \tag{2}$$

with ith basis functions $N_{i,p}$, degree p and $\mathbf{P}_i$ the ith control point. The basis functions are defined over the knot vector $\mathbf{U}$ and are constructed according to the Cox-de-Boor formula [37] as follows:

$$N_{i,0}(u) = \begin{cases} 1 & \text{if } u_i \leq u \leq u_{i+1} \\ 0 & \text{otherwise} \end{cases},$$
$$N_{i,p}(u) = \frac{u - u_i}{u_{i+p} - u_i} N_{i,p-1}(u) + \frac{u_{i+p+1} - u}{u_{i+p+1} - u_{i+1}} N_{i+1,p-1}(u). \tag{3}$$

An approximation according to the algorithm of Piegl and Tiller [37] is performed to smooth the contour. The approximation of the graphite geometry from the discrete image data is based on the removal of knots. In a first step, an overdetermined B-spline curve is constructed which interpolates all discrete image data. As input parameters, besides the error tolerance ξ, the starting degree and the final degree of the B-spline curve are defined. From this, a B-spline curve $\mathbf{C}(u)$ with a knot vector $\mathbf{U}$ and a set of evaluation points $\{\bar{u}_k\}$ is constructed. The approximated curve results from a reduced knot vector of a B-spline and the resulting newly placed control points. Here, an approximation error ξ between the B-spline curve $\mathbf{C}(u)$ and the coordinates $\{\mathbf{D}_k\}$ of the graphite contour is tolerated. Estimated errors $\{\tilde{e}_k\}$ are used within the algorithm. After the least square fit, the exact errors $\{\bar{e}_k\}$ are determined as the shortest distance between the data points $\{\mathbf{D}_k\}$ and new evaluation points. After that, the resulting graphite geometries are slightly extruded in-direction to obtain a 2.5D RVE model that avoids the pure plane strain or stress state. The model is meshed by using the free software gmsh with wedge elements C3D6 from the abaqus element library having two gauss points and six nodes (Fig. 4c). Finally, the linear-elastic stress field is solved in Abaqus by using static uniform boundary conditions (Fig. 4d).

3.2 *Implementation of Melan's Static Shakedown Theorem*

Melan's static shakedown theorem was implemented with and without kinematic hardening. The discretized shakedown theorem here is based on the notation given by Simon et al. [41] and uses, among others, the open Python libraries NumPy [29] and SciPy [47]. The discretized shakedown theorem is given by

$$\begin{aligned}
&\text{maximize} && \alpha, \\
&\text{subjected to} && [C] \cdot \{\overline{\boldsymbol{\rho}}_i\} = \mathbf{0}, && \text{(4a)} \\
& && f(\alpha \boldsymbol{\sigma}^e_{ik} + \overline{\boldsymbol{\rho}}_i, \sigma_{Yi}) \leq \mathbf{0} && \text{(4b)} \\
& && i = 1, \ldots, NG; k = 1, \ldots, NV,
\end{aligned}$$

with the load factor α, the system matrix $[\boldsymbol{C}]$, the discretized residual stresses $\{\overline{\boldsymbol{\rho}}_i\}$, discretized elastic stresses $\boldsymbol{\sigma}^e_{ik}$ and yield limits σ_{Yi} for all Gauss point NG. The inequality constraints in Eq. 4b ensure that the stress states lie within the yield surface f. The loadspace $\mathcal{L}$ is defined by the load vertices NV (see Fig. 5). Therefore, for a purely pulsating loading, which is applied here, the number of vertices is two. The mesoscopic elastic stresses $\boldsymbol{\sigma}^e$ increase with increasing load space by a factor of α and are determined by using the FEM software Abaqus. They are a function of the applied loads $\boldsymbol{P}_k$. Figure 6 shows schematically the setup of the simulation for a two-dimensional RVE and the load point $P_{xx} > 0$.

Kinematic and isotropic hardening are considered in this work with the inequality constraints

$$f\left(\alpha \boldsymbol{\sigma}^e_{ik} + \overline{\boldsymbol{\rho}}_i - \overline{\boldsymbol{\pi}}_i, \sigma_{Yi}\right) \leq 0 \tag{5a}$$

$$f\left(\alpha \boldsymbol{\sigma}^e_{ik} + \overline{\boldsymbol{\rho}}_i, \sigma_{max,i}\right) \leq 0 \tag{5b}$$

$$i = 1, \ldots, NG; k = 1, \ldots, NV$$

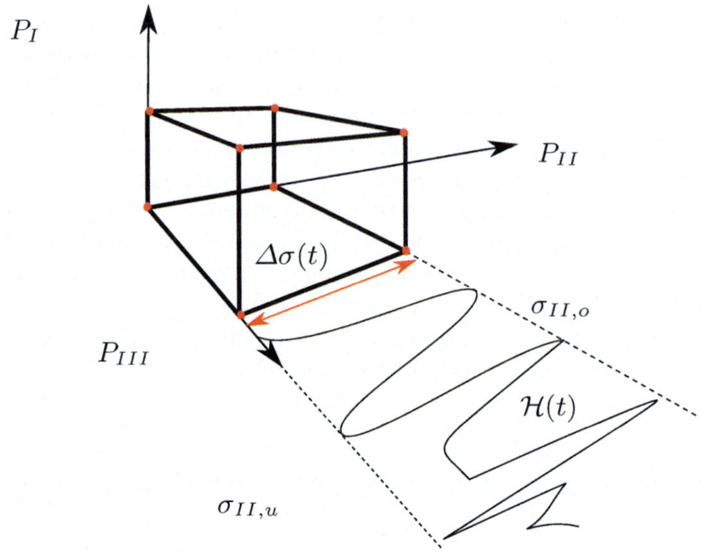

Fig. 5 Definition of the load space $\mathcal{L}$. The load considered in this work are two load vertices corresponding to a purely pulsating stress

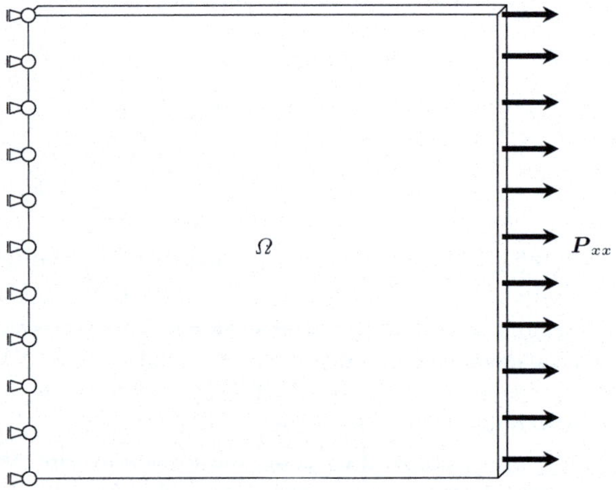

Fig. 6 Boundary conditions for the linear-elastic finite element RVE simulation

which limit the movement and expansion of the yield surface f by a second yield surface (two-surface model) [40]. For this purpose, the hardening stress $\bar{\boldsymbol{\pi}}$ is introduced. For the radius of the second yield surface, the tensile strength σ_{max} of the respective phase, i.e. graphite or ferrite, is used. The shakedown conditions are checked at the Gaussian points for each load vertex.

Equation 4a is the equilibrium condition of the discretized, time-independent residual stresses. The principle of virtual work for the time-independent residual stresses is

$$\int_V \delta \boldsymbol{\varepsilon}^T \overline{\boldsymbol{\rho}} \, dV = \int_V \delta \boldsymbol{u}^T \boldsymbol{B}^T \overline{\boldsymbol{\rho}} \, dV = \mathbf{0}. \tag{6}$$

The B-matrix $\boldsymbol{B}$ denotes the matrix of partial derivative shape functions $\boldsymbol{N}(\boldsymbol{x}) \in K^{NNE \times NDOF}$, where NNE is the number of nodes of the element type used and $NDOF$ is the number of degrees of freedom. Equation 6 is discretized as follows, where w_{ij} denotes the weights following from the numerical integration of the integrals:

$$\sum_{1}^{NE} \int_{Ve} \boldsymbol{B}^T \overline{\boldsymbol{\rho}} \, dV = \sum_{1}^{NE} \iiint_0^1 \boldsymbol{B}^T \overline{\boldsymbol{\rho}} \, |\boldsymbol{J}| \, d\xi \, d\eta \, d\zeta \tag{7a}$$

$$\sum_{i=1}^{NE} \sum_{j=1}^{NGE} w_{ij} \, |\boldsymbol{J}|_{ij} \, \boldsymbol{B}_{ij}^T \overline{\boldsymbol{\rho}}_{ij} = \sum_{1}^{NG} \boldsymbol{C}_i \overline{\boldsymbol{\rho}}_i. \tag{7b}$$

The sum of the matrices $\boldsymbol{C}_i$ per Gauss point symbolizes the assembled system matrix, whose structure is explained below. The Jacobi matrix $\boldsymbol{J} \in K^{NNE \times NDOF}$ reads

$$\boldsymbol{J} = \begin{bmatrix} \frac{\partial N_1}{\partial \xi}(\boldsymbol{\xi}) & \frac{\partial N_2}{\partial \xi}(\boldsymbol{\xi}) & \ldots & \frac{\partial N_{NNE}}{\partial \xi}(\boldsymbol{\xi}) \\ \frac{\partial N_1}{\partial \eta}(\boldsymbol{\xi}) & \frac{\partial N_2}{\partial \eta}(\boldsymbol{\xi}) & \ldots & \frac{\partial N_{NNE}}{\partial \eta}(\boldsymbol{\xi}) \\ \frac{\partial N_1}{\partial \zeta}(\boldsymbol{\xi}) & \frac{\partial N_2}{\partial \zeta}(\boldsymbol{\xi}) & \ldots & \frac{\partial N_{NNE}}{\partial \zeta}(\boldsymbol{\xi}) \end{bmatrix} \begin{bmatrix} x_1 & y_1 & z_1 \\ x_2 & y_2 & z_2 \\ \vdots & \vdots & \vdots \\ x_{NNE} & y_{NNE} & z_{NNE} \end{bmatrix}. \tag{8}$$

The shape functions in element coordinates are transformed into the global coordinate system by using the Jacobian matrix

$$\boldsymbol{N}(\boldsymbol{x}) = \begin{bmatrix} \frac{\partial N_1}{\partial \xi} & \frac{\partial N_1}{\partial \eta} & \frac{\partial N_1}{\partial \zeta} \\ \frac{\partial N_2}{\partial \xi} & \frac{\partial N_2}{\partial \eta} & \frac{\partial N_2}{\partial \zeta} \\ \vdots & \vdots & \vdots \\ \frac{\partial N_{NNE}}{\partial \xi} & \frac{\partial N_{NNE}}{\partial \eta} & \frac{\partial N_{NNE}}{\partial \zeta} \end{bmatrix} (\boldsymbol{J}^{-1})^{\boldsymbol{T}}. \tag{9}$$

The B-matrix in global coordinates is then

$$
\boldsymbol{B}(\boldsymbol{x}) = \begin{bmatrix} \frac{\partial N_1}{\partial x} & 0 & 0 & \dots & 0 \\ 0 & \frac{\partial N_1}{\partial y} & 0 & \dots & 0 \\ 0 & 0 & \frac{\partial N_1}{\partial z} & \dots & \frac{\partial N_{NNE}}{\partial z} \\ \frac{\partial N_1}{\partial y} & \frac{\partial N_1}{\partial x} & 0 & \dots & 0 \\ 0 & \frac{\partial N_1}{\partial z} & \frac{\partial N_1}{\partial y} & \dots & \frac{\partial N_{NNE}}{\partial y} \\ \frac{\partial N_1}{\partial z} & 0 & \frac{\partial N_1}{\partial x} & \dots & \frac{\partial N_{NNE}}{\partial x} \end{bmatrix} \tag{10}
$$

and has the dimensions $\boldsymbol{B} \in K^{6\times 3NNE}$. The matrices $\boldsymbol{B}$ are evaluated at the Gauss points. By multiplying with the Jacobi determinant and the respective weights, the submatrices per Gauss point $\boldsymbol{C}_{minor} \in K^{3NNE\times 6}$ correspondingly

$$
\boldsymbol{C}_{minor} = \boldsymbol{C}_i = det(\boldsymbol{J}_i)\, w_i\, \boldsymbol{B}_i^T \tag{11}
$$

are formed. The global system matrix $\boldsymbol{C} \in K^{3NN\times 6NG}$ is composed of submatrices of $\boldsymbol{C}_{minor}$, i.e. $\boldsymbol{C}_{minor,S} \in K^{3\times 6}$, so that $\boldsymbol{C}_{minor,S} = C_{i,j}$ with $i \in NN$, $j \in NG$. The system matrix follows by positioning the matrices $\boldsymbol{C}_{minor,S}$ according to

$$
\boldsymbol{C} = \begin{bmatrix} \boldsymbol{C}_{1,1} & \boldsymbol{C}_{1,2} & \dots & \boldsymbol{C}_{1,NG} \\ \boldsymbol{C}_{2,1} & \boldsymbol{C}_{2,2} & \dots & \boldsymbol{C}_{2,NG} \\ \vdots & \vdots & \ddots & \vdots \\ \boldsymbol{C}_{NN,1} & \boldsymbol{C}_{NN,2} & \dots & \boldsymbol{C}_{NN,NG} \end{bmatrix}. \tag{12}
$$

The commercial optimizer Gurobi is used to solve the discretized shakedown theorem. To determine the shakedown limit at the end of the optimization, the average stress theorem is applied, leading to

$$
\Sigma_\infty = \frac{1}{\Omega} \int_{\bar{\Omega}} \alpha \sigma^e d\Omega, \tag{13}
$$

where Ω is the RVE volume.

3.3 Model Calibration and Validation

In the RVE, graphite and ferrite were modelled homogeneously and isotropically. The elastic modulus of ferrite was chosen based on results from nanoindentations determined in [22]. Therein, graphite was also characterized, but the Young's moduli of graphite were highly scattered, which is also known from the literature [6]. The parameters of the graphite were therefore determined exclusively in subsequent

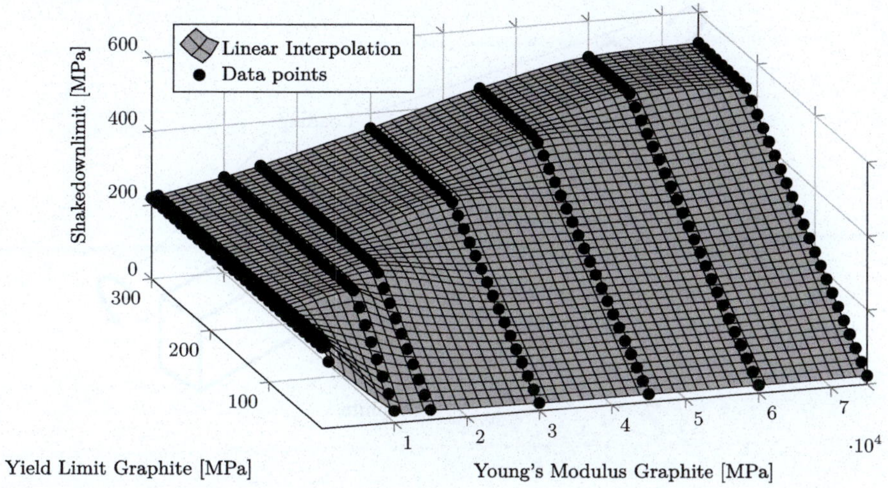

Fig. 7 Influence of elastic modulus and yield limit of graphite on the shakedown limit of Si-DCI

parameter studies. The parameters for the plastic deformation of the ferrite were identified by full-field inverse analysis published in [22]. A cost function based on simulated and experimental strains which were measured by digital image correlation was built. In order to investigate the effect of graphite on the shakedown limit, a single micrograph of the alloy group *S1* was meshed, where the yield limit of graphite $\sigma_{Y,0.2,Graphite}$ was varied from 10 to 300 MPa and the elastic modulus of graphite $E_{Graphite}$ was varied from 100 to 75000 MPa. The results in Fig. 7 show that the shakedown limit saturates with increasing values of the yield limit of the graphite. The onset of the saturation shifts to higher yield limits as the elastic modulus of graphite increases and is reached immediately for low elastic moduli of graphite. Therefore, the model parameters for the graphite must be chosen from the range where the shakedown limit is no longer affected by the yield limit of the graphite, otherwise the model parameters of the graphite would control the shakedown limit. Meanwhile, an increase in the elastic modulus from 10 to 15 GPa at a yield limit of 300 MPa results in the increase of the shakedown limit of 8.6%. In light of the fact that the elastic deformation of graphite cannot be measured reproducibly, a model error due to the experimental uncertainties can be expected.

The shakedown simulation of Si-DCI was validated by using flat bending fatigue tests. For this purpose, fatigue specimens of the alloy group *S1* were used. The failure probabilities corresponding to the experimental fatigue strengths were determined by Rossow's estimation function and the data were fitted by using a normal distribution function. By extrapolating the regression in Fig. 9, the fatigue strength amplitude $\sigma_{A,P0.1\%}$ was determined. Ten micrographs were taken and modeled in the highly stressed near-surface region, and the size of the RVE in each case was the entire image area of approximately $1250 \cdot 900$ pixels with a pixel size of $1.0234\,\mu\mathrm{m}/\mathrm{pixel}$ (see Fig. 8). Shakedown analyses were then performed on the RVEs. Figure 9 shows

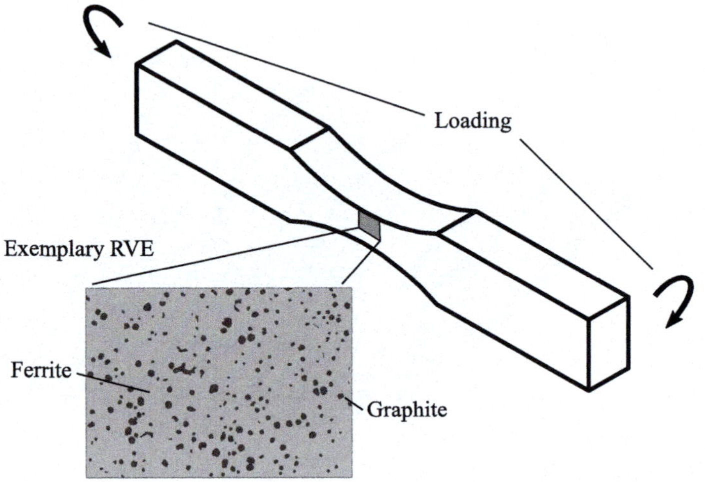

Fig. 8 Position of the RVE in the bending fatigue specimen for validation

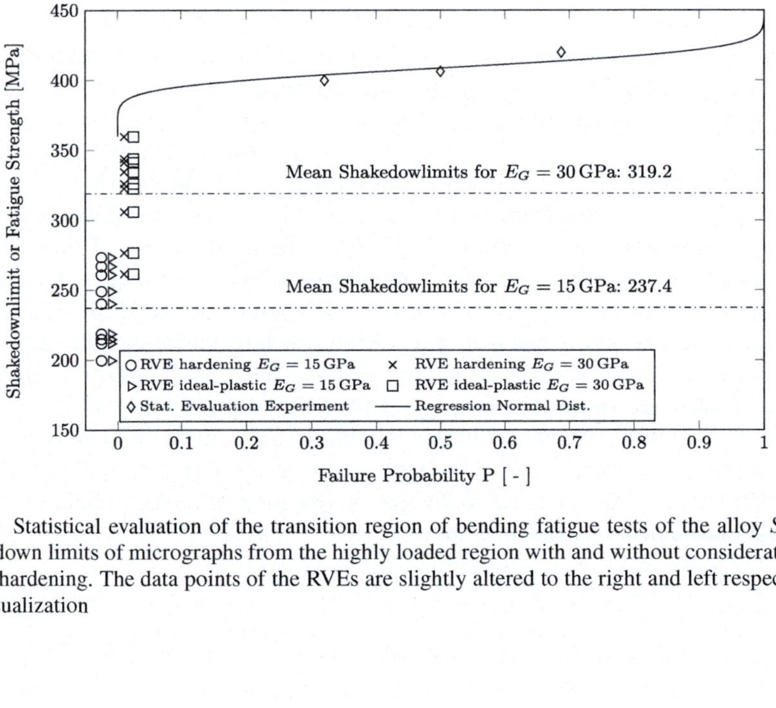

Fig. 9 Statistical evaluation of the transition region of bending fatigue tests of the alloy *S1* and shakedown limits of micrographs from the highly loaded region with and without consideration of strain hardening. The data points of the RVEs are slightly altered to the right and left respectively for visualization

the results, where the data points were alternately positioned slightly to the right or slightly to the left of zero in the x-direction for visualization purposes. These are to be compared with the fatigue strengths of the experiment, i.e. $\sigma_{R=0,P=0.1\%} = 2 \cdot \sigma_{A,P=0.1\%} = 378$ MPa. The calculated shakedown limits for ideal-plastic material behavior are significantly lower than the experimental fatigue strengths. Increasing the elastic modulus of graphite leads to a reduction of the mean relative error between experiment and simulation. However, it should be noted that the scattering of the shakedown limits is high, which is related to the locally varying graphite morphology. If limited kinematic hardening in the ferrite is additionally considered, the shakedown limits do not change in both cases. This phenomenon is well known in the literature as pointed out by Stein, Zhang, and König [43] who have shown that the shakedown limit of the plate with the whole does not change when changing from elastic ideal-plastic material behavior to limited kinematic hardening. Similar observations were observed by Simon and Weichert [42] for a thin-walled tube.

4 Results

4.1 Influence of the RVE Size and the Mesh on the Shakedown Limit

In order to ensure the mesh convergence of the RVE, shakedown simulations were performed using the model parameters for ideal-plastic deformation and an elastic modulus for graphite of $E_G = 15$ GPa on one representative micrograph of each alloy group *S1–S3* and *V*. Figure 10 shows that the shakedown limit converges between 100,000 and 150,000 elements per RVE depending on the alloy group. To determine a sufficient RVE size, five micrographs were randomly cropped out of one micrograph of each alloy group *S1–S3* and *V* with nine different sizes up to a maximum size of 1.1 mm^2, meshed, and the shakedown limit was determined using the above-mentioned model parameters. The results in Fig. 11 show a decreasing scatter of the shakedown limit with increasing RVE size, as indicated by the respective regressions with a power law. The convergence depends on the graphite morphology, with the alloy groups *S1* and *S2* converging faster than *S3* and *V*. The reasons for this are the large graphite sizes and their irregular distribution in the alloy group *S3* as well as the rough interface between graphite and ferrite. With the largest analyzed RVE size, the scatter of the shakedown limit is highest for *S3* with a standard deviation of 15.4 MPa and lowest for S1 with 8.8 MPa, which results in RVE sizes not being representative for this material. Therefore, in addition to a minimum size of the RVE of 1.1mm, the concept of the SRVE (statistically representative volume element) has to be applied. Accordingly, the fatigue strength, i.e. the shakedown limit, results from the statistical evaluation of a high number of shakedown simulations.

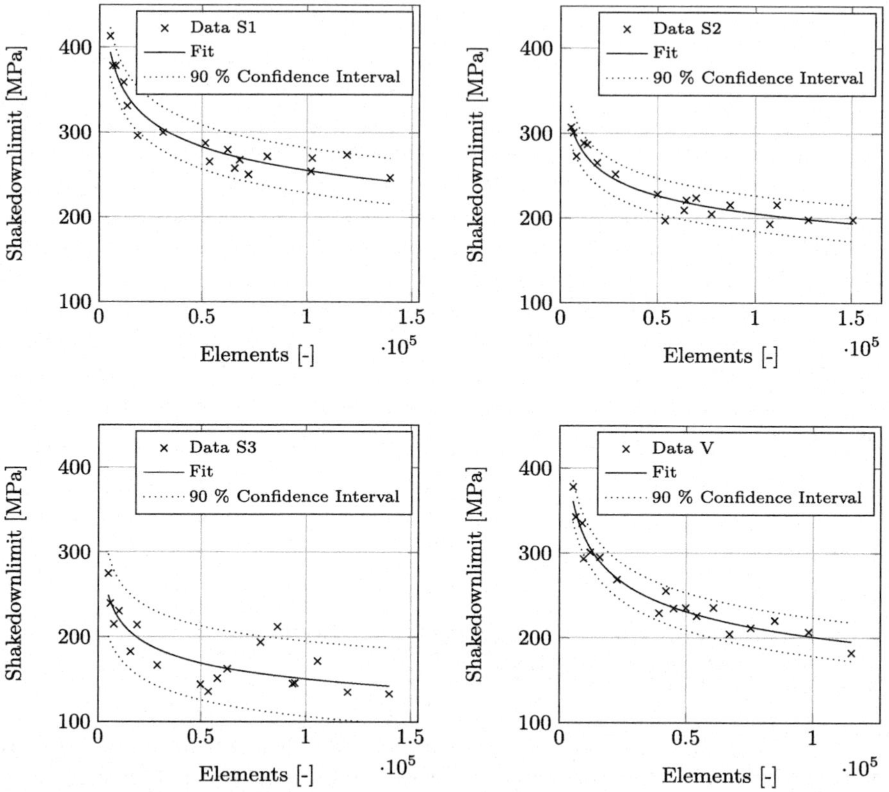

Fig. 10 Influence of the finite element size near the graphite-ferrite interface on the shakedown limit of micrographs of the alloy groups S1 (top left), S2 (top right), S3 (bottom left) and V (bottom right)

4.2 Shakedown Limits of Synthetic RVEs

Table 3 shows the shakedown studies performed on synthetic RVEs. Graphite fraction, nodule count and the nodularity were investigated where the other parameters were kept constant within a small tolerance.

Figure 12 shows the correlation matrix between graphite fraction A_G and nodule count n_G with the shakedown limit Σ_∞, with the distributions of the correlated quantities plotted on the diagonal of the figure. It can be seen that the graphite fraction of a micrograph reduces the simulated shakedown limit—however, the correlation is low with a correlation coefficient of $R = -0.36$. This is due to the fact that the implemented algorithm cannot prevent a simultaneous variation of the nodule count which affects the shakedown limit ($R = -0.4$). The scatter can be explained by the variation of graphite sizes, morphology and distribution. Figure 13 shows the correlation matrix for the study of the nodule count n_G, where the graphite fraction

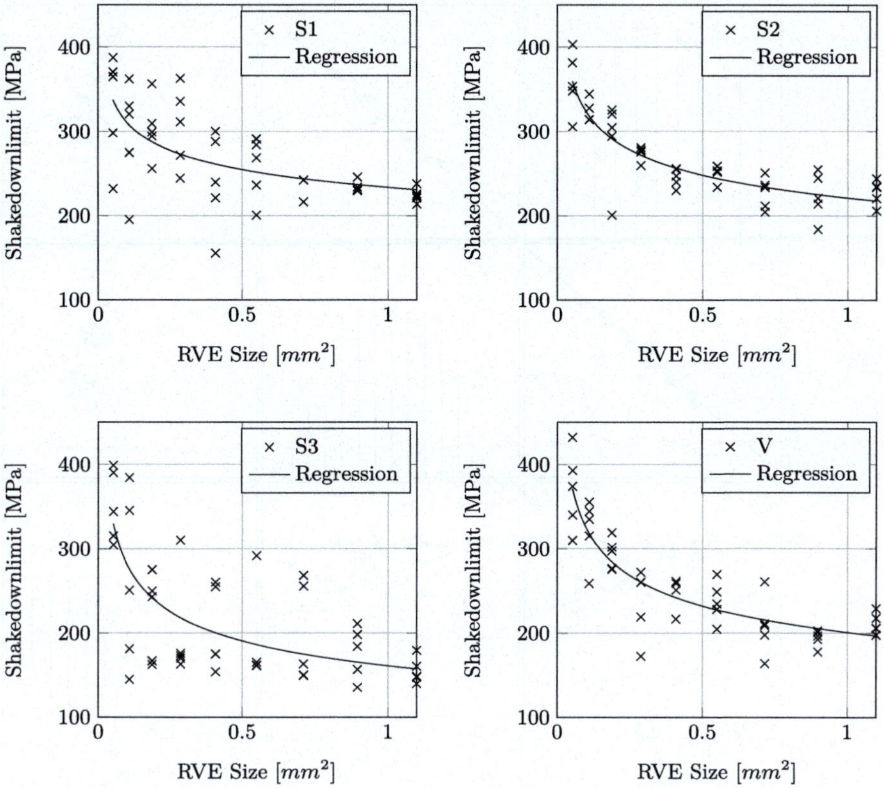

Fig. 11 Influence of the RVE size on the shakedown limit for alloy group *S1* (top left), *S2* (top right), *S3* (bottom left) and *V* (bottom right)

Table 3 Overview of studies performed with synthetic RVEs

Objective	Graphite fraction	Size class	Graphite type
Graphite fraction	Variable	7	VI
Nodule count	Constant	5, 6, 7	VI
Nodularity	Constant	6	III–VI

was kept constant. As the nodule count increases, the shakedown limit ($R = 0.54$) also increases, which is related to lower elastic stresses in the ferritic matrix. At the same time, a dependence of the shakedown limit on the maximum Feret diameter F_{max} of the micrograph becomes noticeable. However, the Feret diameter is linearly dependent on the nodule count n_G of a micrograph. In all studies performed from Table 3 it is noticeable that the maximum Feret diameter of an image correlates with the shakedown limit only if the overall graphite sizes increase. The shakedown limit does not appear to be dominated by individual graphite precipitates.

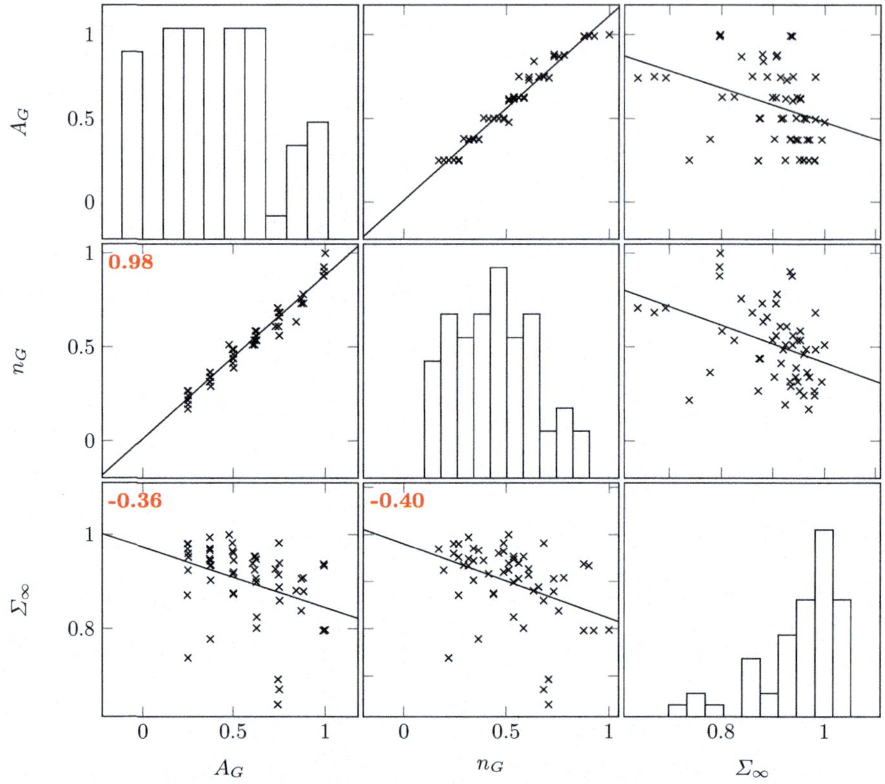

Fig. 12 Correlation matrix between graphite fraction A_G, nodule count n_G and shakedown limit Σ_∞, Pearson correlation coefficients are given in red

Furthermore, the graphite morphology influences the shakedown limit. With the help of the implemented RSA algorithm, synthetic RVEs with varying nodularities were generated using the formula according to the ASTM standard [17]. Figure 14 shows an increase in the shakedown limit with increasing nodularity $N_{\%}$. A slight influence of the maximum Feret diameter is also visible in this simulation study.

4.3 Comparison of Shakedown Limits with Fatigue Experiments

Figure 15 shows the comparison between the simulated shakedown limits and the corresponding fatigue experiments. The shakedown limits are visualized by violin plots. The white dots indicate the respective mean values and the vertical gray lines the quartile distances of the shakedown limits. In contrast to conventional boxplots,

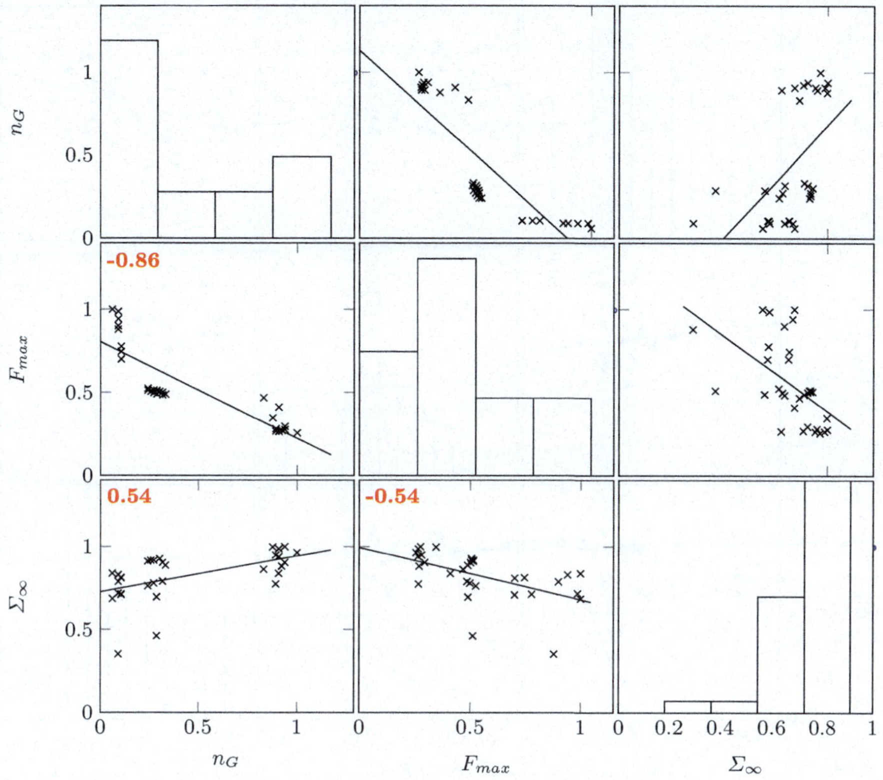

Fig. 13 Correlation matrix between nodule count n_G, maximum Feret diameter F_{max} and shakedown limit Σ_∞, Pearson correlation coefficients are shown in red

the continuous distributions expected from the shakedown limits are also shown in gray and which were determined using kernel density estimation and truncated by whiskers with a quartile spacing of 1.5. For comparison with the pulsating fatigue tests, the fatigue strengths were plotted at $P = 0\%$ and $P = 50\%$ failure probabilities, where the former were determined by regression. As Table 4 shows, good extrapolations are achieved for the alloy groups *S2* and *V*, since the scattering measures T_σ and the error squares R^2 are small. For *S3* and *S1*, on the other hand, only the values for $P = 50\%$ can be sufficiently trusted from a statistical point of view. In all the shakedown analyses performed, it was found that the alloy group *V* has the lowest average shakedown limit, as also seen in the experiment. The alloy group *S3* shows surprisingly high shakedown limits against the background of large graphite precipitates. This is in agreement with the experimental fatigue strength measured for $P = 50\%$, but not for the extrapolations to $P = 0\%$ due to poor experimental regression in the transition region to infinite fatigue life. The absolute values of the shakedown limits for an elastic modulus of graphite of 15GPa are always lower

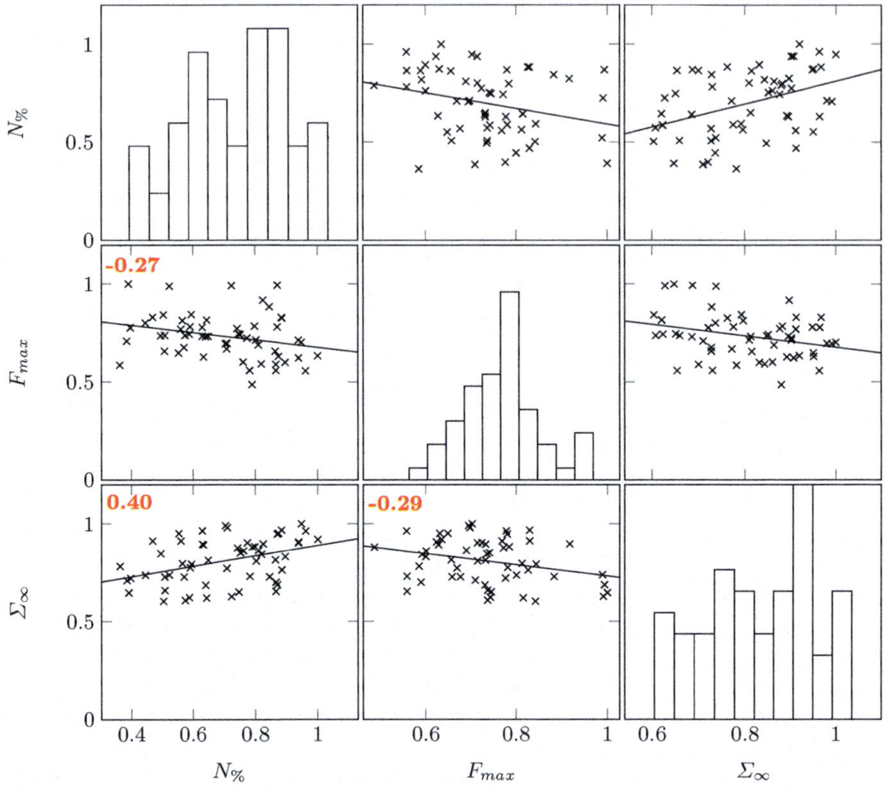

Fig. 14 Correlation matrix between nodularity $N_\%$, maximum Feret diameter F_{max} and shakedown limit Σ_∞, Pearson correlation coefficients are shown in red

than with an elastic modulus of 30GPa for the same mesh. At the same time, the scattering of shakedown limits is high for all alloy groups due to the heterogeneous microstructure. The particular shape of the continuous distribution for alloy group *S2* with a fine mesh, i.e. the high scattering, follows from the presence of degenerated chunky graphite in some of the micrographs. Coarser meshing of about 50,000 wedge elements per micrograph increases the shakedown limit for all alloy groups and smooths the shape of the distributions as the elastic stresses at the graphite-ferrite interface decrease. The calculation time changes non linearly. The coarser FE mesh reduces the computation time to 11% of the computation time of the fine mesh.

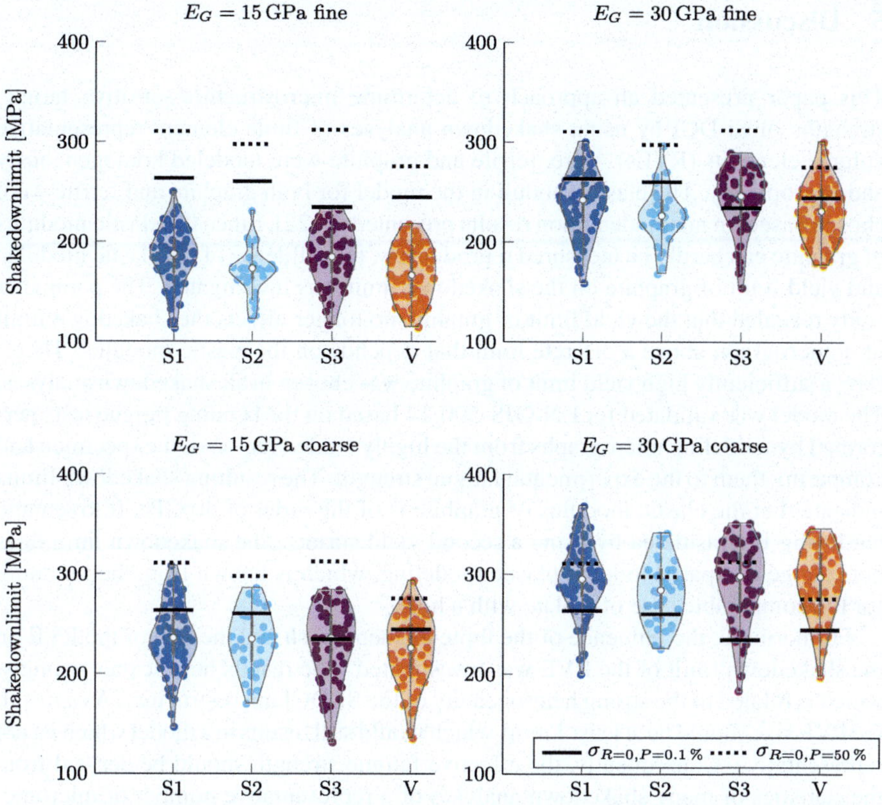

Fig. 15 Results of the shakedown analyses of the real micrographs of the alloy groups *S1–S3* and *V* with different elastic moduli for graphite and two different meshs. The black lines indicate the results of the fatigue experiments

Table 4 Statistical evaluation of the fatigue tests ($R_\sigma = 0$)

	S1	S2	S3	V
$\sigma_{A,P=0.1\%}$	131.79	130.21	117.30	122.33
$\sigma_{A,P=50\%}$	155.47	148.86	156.19	137.80
R^2	0.78	0.99	0.72	0.93
$T_\sigma = 1 : \frac{\sigma_{90}}{\sigma_{10}}$	1:1.20	1:1.17	1:1.40	1:1.14

5 Discussion

This paper presented an approach to determine microstructure-sensitive fatigue strengths of Si-DCI by using shakedown analyses of finite element representative volume elements (RVEs). Here, ferrite and graphite were modeled homogeneously and isotropically. The elastic moduli in the model for both graphite and ferrite were chosen based on nanoindentation results presented in [22]. Since the elastic modulus of graphite can hardly be measured reproducibly, the influence of the elastic modulus and yield limit of graphite on the shakedown limit was investigated. The parameter study revealed that the yield limit of graphite no longer affects the shakedown limit of a micrograph above a discrete limit that depends on the elastic modulus. Therefore, a sufficiently high yield limit of graphite was chosen in all shakedown analyses. The model was validated for EN-GJS-500-14 based on the bending fatigue tests performed by modeling micrographs from the highly loaded region of the specimen and comparing them to the experimental fatigue strength. The resulting shakedown limits indicate that the elastic modulus of graphite is of the order of 30 GPa. If kinematic hardening is considered by using a second yield surface, the shakedown limit does not change compared to ideal-plastic modeling, which is known from the literature for the comparable case of a plate with a hole.

Furthermore, the influence of the finite element mesh and the size of the RVE on the shakedown limit of the RVE was investigated. The results here scatter strongly, which is related to the strong heterogeneity of the Si-DCI microstructure. As a result, the RVE size should be at least 1 mm^2 which would still results in a model which ist not representative. Consequently, the effective fatigue strength should be derived from the statistics of many shakedown analyses of a representative number of micromechanical models. In addition to the nodule density and the graphite fraction, its morphology varies when comparing individual micrographs. For the last reason, the graphite was modeled with approximated B-splines.

In order to investigate the influence of the microstructure on the fatigue strength, shakedown analyses of synthetic RVEs were carried out. In each case, one parameter was kept constant. The correlation analyses reveal that the graphite fraction and the nodule count influence the shakedown limit. With increasing graphite fraction, the shakedown limit decreases slightly. As the nodule count increases while the graphite content remains constant, the shakedown limit increases, which is consistent with experimental observations. In addition, as also observed in the literature, the shakedown limit was found to increase with increasing nodularity. However, the poor correlation coefficients show that the microstructure cannot be described unambiguously with the parameters considered, implying that further influences must exist.

Finally, shakedown analyses of micrographs of the investigated material groups *S1–S3* and *V* were performed, which were taken directly from fatigue specimens and compared with experimental results. Different elastic moduli of the graphite as well as two different meshes were investigated. As a result, good predictions succeed with both fine and somewhat coarser mesh density and with an elastic modulus of graphite of 30 GPa. Vermicular, i.e., graphite nodules deviating from an ideal spheroidal shape, reduce the shakedown limit. As in the experiment, the highest shakedown limits are found with spheroidal graphite and high nodule count. However, the alloy group with comparatively larger graphite nodules also achieved high shakedown limits.

6 Conclusions

The paper presented a modelling approach to predict the fatigue strength of high silicon alloyed nodular cast iron. To this end, finite element models of the microstructure were built and the static shakedown theorem was implemented and used to determine the shakedown limits of the finite element models. The following findings are worth highlighting:

- The fatigue strength of high silicon alloyed nodular cast iron can be estimated from shakedown analyses of 2.5D FE models and is in good agreement with the experimental fatigue strength at low failure probabilities.
- Due to the heterogeneity of the material a representative size of the RVE cannot be found. Consequently, the effective fatigue strength must be obtained from the statistics.

In future work, the limitations of the 2.5D model could be overcome by using 3D-RVEs. However, due to the high experimental effort to obtain the microstructure an approach to synthetically built the models must be developed. Furthermore, more sophisticated material models could be of interest to close the still existing gap between experimental and simulated results. To this end, the direct cyclic method could offer more possibilities to include more physical material models as well as damage.

References

1. ISO/DIS 945-4: Test method for evaluating nodularity in spheroidal graphite cast irons
2. Andriollo, T., Fæster, S., Winther, G.: Probing the structure and mechanical properties of the graphite nodules in ductile cast irons via nano-indentation. Mech. Mater. (2018). https://doi.org/10.1016/j.mechmat.2018.03.010
3. Andriollo, T., Hattel, J.: On the isotropic elastic constants of graphite nodules in ductile cast iron: Analytical and numerical micromechanical investigations. Mech. Mater. **96**, 138–150 (2016). https://doi.org/10.1016/j.mechmat.2016.02.007
4. Andriollo, T., Tiedje, N., Hattel, J.: Distance map based micromechanical analysis of the impact of matrix heterogeneities on the yield stress of nodular cast iron. Mech. Mater. **148**, 103414 (2020). https://doi.org/10.1016/j.mechmat.2020.103414
5. Andriollo, T., Zhang, Y., Fæster, S., Kouznetsova, V.: Analysis of the correlation between micro-mechanical fields and fatigue crack propagation path in nodular cast iron. Acta Mater. **188**, 302–314 (2020). https://doi.org/10.1016/j.actamat.2020.02.026

6. Andriollo, T., Zhang, Y., Fæster, S., Thorborg, J., Hattel, J.: Impact of micro-scale residual stress on in-situ tensile testing of ductile cast iron: digital volume correlation versus model with fully resolved microstructure versus periodic unit cell. J. Mech. Phys. Solids **125**, 714–735 (2019). https://doi.org/10.1016/j.jmps.2019.01.021
7. Auricchio, F., Constantinescu, A., Menna, C., Scalet, G.: A shakedown analysis of high cycle fatigue of shape memory alloys. Int. J. Fatigue **87**, 112–123 (2016). https://doi.org/10.1016/j.ijfatigue.2016.01.017
8. Bargmann, S., Klusemann, B., Markmann, J., Schnabel, J.E., Schneider, K., Soyarslan, C., Wilmers, J.: Generation of 3D representative volume elements for heterogeneous materials: a review. Prog. Mater Sci. **96**, 322–384 (2018). https://doi.org/10.1016/j.pmatsci.2018.02.003
9. Basurto-Hurtado, J., Osornio-Rios, R., Jaen-Cuellar, A., Dominguez-Gonzalez, A., Morales-Hernandez, L.A.: Genetic algorithm-based optimization methodology of Bézier curves to generate a DCI microscale-model. Appl. Sci. **7**(12), 1222 (2017). https://doi.org/10.3390/app7121222
10. Betancur, A., Anflor, C.T.M., Pereira, A., Leiderman, R.: Determination of the effective elastic modulus for nodular cast iron using the boundary element method. Metals **8**(8), 641 (2018). https://doi.org/10.3390/met8080641
11. Bradski, G.: The openCV library. Dr. Dobb's J. Softw. Tools (2000)
12. Charkaluk, E., Constantinescu, A.: Dissipative aspects in high cycle fatigue. Mech. Mater. **41**(5), 483–494 (2009). https://doi.org/10.1016/j.mechmat.2009.01.018
13. Chen, G., Wang, H., Bezold, A., Broeckmann, C., Weichert, D., Zhang, L.: Strengths prediction of particulate reinforced metal matrix composites (PRMMCs) using direct method and artificial neural network. Compos. Struct. **223**, 110951 (2019). https://doi.org/10.1016/j.compstruct.2019.110951
14. Chen, M., Hachemi, A., Weichert, D.: Shakedown and optimization analysis of periodic composites. In: Saxcé, G.D., Oueslati, A., Charkaluk, E., Tritsch, J.B. (eds.) Limit State of Materials and Structures, pp. 45–69. Springer Netherlands, Dordrecht (2013). https://doi.org/10.1007/978-94-007-5425-63
15. DIN Deutsches Institut für Normung e.V.: Schwingfestigkeitsversuch— urchführung und Auswertung von zyklischen Versuchen mit konstanter Lastamplitude für metallische Werkstoffproben und Bauteile
16. DIN Deutsches Institut für Normung e.V.: Gießereiwesen—founding—spheroidal graphite cast irons; German version EN 1563:2018 (2019)
17. E04 Committee.: Test method for determining nodularity and nodule count in ductile iron using image analysis. https://doi.org/10.1520/E2567-16A
18. Endo, M., Matsuo, T.: A practical method for fatigue limit prediction in ductile cast irons. Fatigue Fract. Eng. Mater. Struct. **211**(4), 628 (2019). https://doi.org/10.1111/ffe.13086
19. Fernandino, D.O., Cisilino, A.P., Toro, S., Sanchez, P.J.: Multi-scale analysis of the early damage mechanics of Ferritized ductile iron. Int. J. Fract. **207**(1), 1–26 (2017). https://doi.org/10.1007/s10704-017-0215-1
20. Gaudig, W., Mellert, R., Weber, U., Schmauder, S.: Self-consistent one-particle 3D unit cell model for simulation of the effect of graphite aspect ratio on Young's modulus of cast-iron. Comput. Mater. Sci. **28**(3–4), 654–662 (2003). https://doi.org/10.1016/j.commatsci.2003.08.021
21. Gebhardt, C., Frieß, J., Bezold, A., Broeckmann, C., Bührig-Polaczek, A.: Schwingfestigkeit von hochsiliziumlegiertem Gusseisen mit Kugelgrafit mit variierenden Grafitmorphologien. Giess. Spec. (1) (2019)
22. Gebhardt, C., Sedlatschek, T., Bezold, A., Broeckmann, C.: Full-field inverse identification of Elasto-plastic model parameters for ductile cast iron. Mech. Mater. **162**, 104056 (2021). https://doi.org/10.1016/j.mechmat.2021.104056
23. Gebhardt, C., Trimborn, T., Weber, F., Bezold, A., Broeckmann, C., Herty, M.: Simplified ResNet approach for data driven prediction of microstructure-fatigue relationship. Mech. Mater. 103625 (2020). https://doi.org/10.1016/j.mechmat.2020.103625

24. Gebhardt, C., Zhang, J., Bezold, A., Broeckmann, C.: Microscale fatigue mechanisms in high silicon alloyed nodular cast iron. Int. J. Fatigue 107402 (2022). https://doi.org/10.1016/j.ijfatigue.2022.107402
25. Lloyd, G.: Guideline for the certification of wind turbine (2010)
26. Germann, H., Starke, P., Eifler, D.: Resistivity-based evaluation of the fatigue behavior of cast irons. Metall. and Mater. Trans. A. **43**(8), 2792–2798 (2012). https://doi.org/10.1007/s11661-011-0852-3
27. Gillner, K., Henrich, M., Münstermann, S.: Numerical study of inclusion parameters and their influence on fatigue lifetime. Int. J. Fatigue **111**, 70–80 (2018). https://doi.org/10.1016/j.ijfatigue.2018.01.036
28. Hachemi, A., Weichert, D.: On the problem of interfacial damage in fibre-reinforced composites under variable loads. Mech. Res. Commun. **32**(1), 15–23 (2005). https://doi.org/10.1016/j.mechrescom.2004.06.001
29. Harris, C.R., Millman, K.J., van der Walt, S.J., Gommers, R., Virtanen, P., Cournapeau, D., Wieser, E., Taylor, J., Berg, S., Smith, N.J., Kern, R., Picus, M., Hoyer, S., van Kerkwijk, M.H., Brett, M., Haldane, A., Del Río, J.F., Wiebe, M., Peterson, P., Gérard-Marchant, P., Sheppard, K., Reddy, T., Weckesser, W., Abbasi, H., Gohlke, C., Oliphant, T.E.: Array programming with NumPy. Nature **585**(7825), 357–362 (2020). https://doi.org/10.1038/s41586-020-2649-2
30. Heine, L.M., Bezold, A., Broeckmann, C.: Long crack growth and crack closure in high strength nodular cast iron. Eng. Fract. Mech. **192**, 24–53 (2018). https://doi.org/10.1016/j.engfracmech.2018.02.003
31. Jiang, K., Chen, G., Bezold, A., Broeckmann, C.: Statistics-based numerical study of the fatigue damage evolution in the microstructures of WC-Co hardmetals. Mech. Mater. **164**, 104097 (2022). https://doi.org/10.1016/j.mechmat.2021.104097
32. Kasvayee, A.K., Salomonsson, K., Ghassemali, E., Jarfors, A.E.W.: Microstructural strain distribution in ductile iron; comparison between finite element simulation and digital image correlation measurements. Mater. Sci. Eng., A **655**, 27–35 (2016). https://doi.org/10.1016/j.msea.2015.12.056
33. Klubberg, F., Hempen, M., Schäfer, H.J., Beiss, P.: Praxisgerechte Auswertung von Wöhlerversuchen mit der Software SAFD
34. Magoariec, H., Bourgeois, S., Débordes, O.: Elastic plastic shakedown of 3D periodic heterogeneous media: a direct numerical approach. Int. J. Plast **20**(8–9), 1655–1675 (2004). https://doi.org/10.1016/j.ijplas.2003.11.011
35. Norman, V., Calmunger, M.: On the micro- and macroscopic elastoplastic deformation behaviour of cast iron when subjected to cyclic loading. Int. J. Plast **115**, 200–215 (2019). https://doi.org/10.1016/j.ijplas.2018.11.019
36. Pedegrosa, F., Varoquaux, G., Gramfort, A., Michel, V., Thirion, B., Grisel, O., Blondel, M., Prettenhofer, P., Weiss, R., Dubourg, V., Vanderplas, J., Passos, A., Cournapeau, D., Brucher, M., Perrot, M., Duchesnay, E.: Scikit-learn: machine learning in python. J. Mach. Learn. Res. **12**, 2825–2830 (2011)
37. Piegl, L., Tiller, W.: The NURBS Book. Monographs in Visual Communications, Springer, Berlin and Heidelberg (1995). https://doi.org/10.1007/978-3-642-97385-7
38. Pressestelle: EVK-Schadensstatistik: Schadensstatistik zu Windenergieanlagen 2010–2015
39. Schwabe, F.: Einspieluntersuchungen von Verbundwerkstoffen mit periodischer Mikrostruktur. Dissertation, RWTH Aachen University, Aachen (2000)
40. Simon, J.W.: Shakedown analysis of kinematically hardening structures in n-dimensional loading spaces. In: Spiliopoulos, K.V., Weichert, D. (eds.) Direct Methods for Limit States in Structures and Materials, pp. 57–77. Springer Netherlands, Dordrecht (2014). https://doi.org/10.1007/978-94-007-6827-7_3
41. Simon, J.W., Weichert, D.: Numerical lower bound shakedown analysis of engineering structures. Comput. Methods Appl. Mech. Eng. **200**(41–44), 2828–2839 (2011). https://doi.org/10.1016/j.cma.2011.05.006
42. Simon, J.W., Weichert, D.: Shakedown analysis of engineering structures with limited kinematical hardening. Int. J. Solids Struct. **49**(15–16), 2177–2186 (2012). https://doi.org/10.1016/j.ijsolstr.2012.04.039

43. Stein, E., Zhang, G., König, J.A.: Micromechanical modelling and computation of shakedown with nonlinear kinematic hardening—including examples for 2D-problems. In: Axelrad, D.R., Muschik, W. (eds.) Recent Developments in Micromechanics, pp. 99–114. Springer Berlin Heidelberg, Berlin, Heidelberg (1991). https://doi.org/10.1007/978-3-642-84332-7_8
44. Svensson, I., Sjögren, T.: On modeling and simulation of mechanical properties of cast irons with different morphologies of graphite. Int. J. Metalcast. **3**(4), 67–77 (2009)
45. van der Walt, S., Schönberger, J.L., Nunez-Iglesias, J., Boulogne, F., Warner, J.D., Yager, N., Gouillart, E., Yu, T.: Scikit-image: image processing in Python. PeerJ **2**, e453 (2014). https://doi.org/10.7717/peerj.453
46. Velichko, A., Mücklich, F.: Bildanalytische Formanalyse und Klassifizierung der irregulären Graphitmorphologie in Gusseisen. Pract. Metallogr. **43**(4), 192–207 (2006). https://doi.org/10.3139/147.100297
47. Virtanen, P.: SciPy 1.0: fundamental algorithms for scientific computing in Python. Nat. Methods **17**(3), 261–272 (2020). https://doi.org/10.1038/s41592-019-0686-2
48. Weichert, D., Hachemi, A., Schwabe, F.: Application of shakedown analysis to the plastic design of composites. Arch. Appl. Mech. **69**(9–10), 623–633 (1999). https://doi.org/10.1007/s004190050247
49. You, J.H., Kim, B.Y., Miskiewicz, M.: Shakedown analysis of fibre-reinforced copper matrix composites by direct and incremental approaches. Mech. Mater. **41**(7), 857–867 (2009). https://doi.org/10.1016/j.mechmat.2008.10.007

A Macroscopic Fatigue Criterion for Ductile Porous Materials with Drucker-Prager Matrix

J. Zhang, A. Oueslati, W. Q. Shen, and G. de Saxcé

Abstract In this work, we derive a macroscopic fatigue criterion for ductile porous materials with Drucker-Prager dilatant matrix considering the hollow sphere unit cell subjected to cyclic repeated external loads. The static shakedown theorem is applied basing upon a suitable choice of statically and plastically admissible trial stress fields. The closed-form fatigue criterion in parametric form depends on the friction angle, the porosity, the two stress invariants and the sign of the third invariant of the macroscopic stress tensor. The safety domain is bounded by the proposed fatigue criterion and by the macroscopic yield strength proposed one corresponding to the collapse by development of a mechanism at the first cycle. The established model have been assessed and validated by numerical solutions obtained by finite element computations with a quarter of the hollow sphere subjected to homogeneous strain rate boundary conditions for different values of porosity and frictions angles.

Keywords Fatigue · Shakedown · Porous materials · Drucker-Prager model

1 Introduction

Geological and construction materials are usually widely subjected to cyclic loads such as off-shore structures, foundations and infrastructures of motorway, railway and airports. Their deformation and strength performances can be significantly altered by the high number of loading cycles and load levels [1–3]. As shown on Fig. 1, an elasto-plastic structure subjected to variable repeated actions may fail as a result of incremental collapse (ratchet), or by alternating plasticity (fatigue), comprising

J. Zhang
Key Laboratory of Ministry of Education for Geomechanics and Embankment Engineering, Hohai University, Nanjing, China

A. Oueslati (✉) · W. Q. Shen · G. de Saxcé
Univ. Lille, CNRS, Centrale Lille, UMR 9013 - LaMcube - Laboratoire de mécanique multiphysique et multiéchelle, Lille 59000, France
e-mail: abdelbacet.oueslati@univ-lille.fr

G. Garcea and D. Weichert (eds.), *Direct Methods for Limit State of Materials and Structures*, Lecture Notes in Applied and Computational Mechanics 101,
https://doi.org/10.1007/978-3-031-29122-7_8

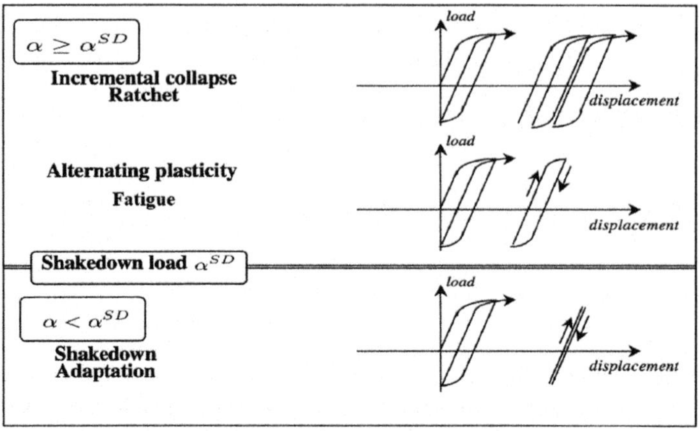

Fig. 1 Three types of asymptotic responses subjected to cyclic loads

equal plastic strains of opposite sign leading eventually to local material failure. It may happen also that the structure endures a finite number of cycles with equivalent elastic behaviors, which is called shakedown. From the point of view of energies, the total dissipation is bounded in time if the structure shakes down [4, 5].

The idea to apply shakedown concepts to determine the fatigue limit load of materials is due to Dang Van [6]. By maximizing the size of the load domain, the shakedown limit load can be obtained until the failure of the structure due to fatigue. Similarly to statical problems [7, 8], the effective strength of porous media subjected to cyclic loadings is also dependent on porosity and local mechanical properties of solid matrix. An appropriate up-scaling method has to be developed for the determination of such a fatigue criterion. One of the logic approaches is to extend the limit analysis methods developed for statical problems without consideration of loading history.

Two reference works in the field are the Melan's statical theorem [9] with the basic idea of time-independent residual stress field and the Koiter's kinematical theorem [10] by the definition of plastic stain increment field. These theorems were developed for linear elastic perfectly plastic solid materials. Their main extensions ware completed for various and more general constitutive laws [11]. It should be pointed out that although the bibliography on the ductile failure of porous materials under monotonic loads is abundant and renewed, there are few papers dealing with the modeling of ductility under cyclic loadings and most of them concern micromechanics-based numerical approaches [13–15].

The aim of the present work is to contribute to the theoretical and numerical studies of the effective shakedown of ductile porous materials under cyclic load by the use of Melan's shakedown theorem. Inspire from the Gurson-like approach within the framework of micromechanics [12], a hollow sphere model with an associated Drucker-Prager matrix is adopted as the Representative Elementary Volume (REV), which has been successfully applied for the limit analysis of porous media [16–18].

Considering the complexity of the fictitious elastic and residual stress fields under general cyclic loads, the case of the hollow sphere subjected to pure hydrostatic loading is firstly studied, in which the exact solution can be provided. Then, the general case involving shear effects with any arbitrary cyclic load fluctuations is concerned with two important steps: the construction of appropriate trial stress and trial residual stress fields and then maximizing the size of the load domain in the spirit of the standard lower shakedown theorem. Following [19], the crucial step is providing a time-independent admissible residual stress field accounting for the influence of the friction angle, which is the most difficult part of the method of solution.

The present study is organized as follows. A brief review of the classical Hill-Mandel lemma for the average-field theory and Melan's shakedown theorem are presented as the starting point in Sect. 2. In Sect. 3, the steady-state behavior of the hollow sphere under repeated pure hydrostatic loading is firstly discussed. Then, the derivation of a macroscopic shakedown of the unit cell under general repeated loads involving shear effects is also studied in this part. In Sect. 4, the established criterion is assessed and validated by comparison with step-by-step FEM numerical simulations. Concluding remarks and some perspectives are finally summarized in the last section.

2 Problem Formulations

The porous material is represented by a thick hollow sphere (REV) Ω made of a pore ω and elastic perfectly plastic solid phase $\Omega_M = \Omega - \omega$. The inner and the outer radii are respectively denoted by a and b. So the porosity is obtained by $f = (a/b)^3 < 1$. Let $\boldsymbol{\sigma}$ be the Cauchy stress tensor and $\boldsymbol{\varepsilon}$ the corresponding strain field in the elasto-plastic body which is decomposed into two parts:

$$\boldsymbol{\varepsilon} = \boldsymbol{\varepsilon}^e + \boldsymbol{\varepsilon}^p \tag{1}$$

with $\boldsymbol{\varepsilon}^p$ being the plastic strain tensor, and the elastic part $\boldsymbol{\varepsilon}^e$ satisfying Hooke's law.

The matrix material obeys to the associated elastic-perfectly plastic Drucker-Prager criterion for which the yield function writes:

$$F(\boldsymbol{\sigma}) = \sigma_{eq}(\boldsymbol{\sigma}) + 3\alpha\sigma_m - \sigma_y \leq 0 \tag{2}$$

where $\sigma_{eq} = \sqrt{\frac{3}{2}\boldsymbol{s} : \boldsymbol{s}}$ is the equivalent stress with $\boldsymbol{s}$ being the deviatoric part, $\sigma_m = \frac{1}{3}tr(\boldsymbol{\sigma})$ the mean stress, σ_y the shear cohesion stress and α the pressure sensitivity factor of solid phase depends on the friction angle ϕ though

$$\tan\phi = 3\alpha \tag{3}$$

The associated flow law ensures that the plastic strain rate obeys to the normality rule:

$$\dot{\boldsymbol{\varepsilon}}^p = \dot{\lambda}\,\frac{\partial F}{\partial \boldsymbol{\sigma}}\,, \tag{4}$$

where $\dot{\lambda} \geq 0$ is the plastic multiplier.

From the classical Hill's lemma [20], the macroscopic stress $\boldsymbol{\Sigma}$ and macroscopic strain $\mathbf{E}$ are obtained as volume averages of their microscopic counterparts $\boldsymbol{\sigma}$ and $\boldsymbol{\varepsilon}$:

$$\boldsymbol{\Sigma} = \frac{1}{\mid \Omega \mid}\int_{\Omega} \boldsymbol{\sigma}\, dV\,, \qquad \mathbf{E} = \frac{1}{\mid \Omega \mid}\int_{\Omega} \boldsymbol{\varepsilon}\, dV\,. \tag{5}$$

The set of kinematical admissible displacement fields classically reads:

$$\mathcal{K}_a = \{\mathbf{u} \quad s.t. \quad \mathbf{u}(\mathbf{x}) = \mathbf{E}\cdot\mathbf{x} \quad on \quad \partial\Omega\}\,. \tag{6}$$

And, the strain field, symmetric part of the displacement gradient, is $\boldsymbol{\varepsilon}(\mathbf{u}) = grad_s\mathbf{u}$.

Similarly, the set of statically admissible stress fields is such as:

$$\mathcal{S}_a = \{\boldsymbol{\sigma} \quad s.t. \quad div\,\boldsymbol{\sigma} = \mathbf{0} \quad in\,\Omega, \quad \boldsymbol{\sigma}\cdot\boldsymbol{n} = \mathbf{0} \quad on\,\partial\omega, \quad \boldsymbol{\sigma} = \mathbf{0} \quad in\,\omega\} \tag{7}$$

where $\boldsymbol{n}$ is the unit outward normal vector of the matrix.

In addition, if $\boldsymbol{\sigma}$ is statically admissible, it is worth to remark that $\boldsymbol{\Sigma}$ depends on it through:

$$\boldsymbol{\Sigma} = \frac{1}{\mid \Omega \mid}\int_{\partial\Omega} (\boldsymbol{\sigma}\cdot\boldsymbol{n})\otimes\boldsymbol{x}\, dS \tag{8}$$

where $\boldsymbol{x}$ is the position vector in the cartesian coordinate system, and dS is an infinitesimal element of the surface area at the external boundary $\partial\Omega$.

For a perfectly elasto-plastic structure Ω under varying loads, the variable loading $\boldsymbol{\Sigma}(t)$ in domain $\mathcal{P}$ is illustrated in Fig. 2. When the load factor α determining the domain dimension reaches a threshold α^{SD} called shakedown limit, the collapses occur by formation of a mechanism (excessive deformation) or by alternating plasticity (fatigue). To eliminate the explicit time dependence of the loadings, König's statement [21] is introduced: *shakedown occurs for any load path in a given convex load domain* $\mathcal{P}$, *if it occurs for a cyclic load path containing all vertices.* This allows us to verify only all the load vertices (Fig. 2b) to predict the safety domain for shakedown instead of the entire loading history (Fig. 2a). Consequently, the cyclic loadings are imposed proportional in each cycle. In fact, Konig's theorem states that, if the load factor α is less than the shakedown factor α^{SD}, then the structures shakes down and the dissipation is bounded in time, otherwise, the asymptotic behavior is alternating plasticity or accomodation leading to failure by accumulation of plastic strains (see Fig. 1).

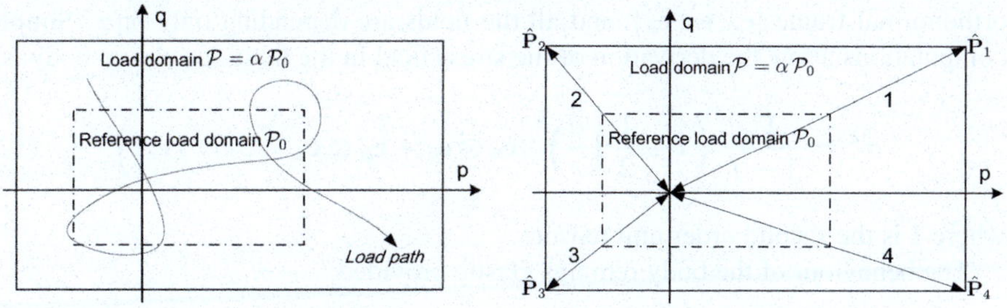

(a) Arbitrary load path (b) Critical cyclic loadings

Fig. 2 Critical cyclic loadings for shakedown analysis and load domain

The residual stresses are computed by subtracting the elastic responses in the fictitious elastic part from the total stresses:

$$\boldsymbol{\rho}(\mathbf{x}, t) = \boldsymbol{\sigma}(\mathbf{x}, t) - \boldsymbol{\sigma}^E(\mathbf{x}, t) \tag{9}$$

In the sense of De Saxcé [22], the admissible time-independent residual stress field $\bar{\boldsymbol{\rho}}$ satisfies the following condition:

$$\bar{\boldsymbol{\rho}} \in \mathcal{S}_a, \quad \& \quad F(\bar{\boldsymbol{\rho}} + \boldsymbol{\sigma}^E) \leq 0 \;\; \forall t \; in \; \Omega \tag{10}$$

3 Macroscopic Fatigue Criterion for Ductile Porous Materials with Drucker-Prager Dilatant Matrix

In this part, the exact solution under pure hydrostatic loading will be firstly studied, and then extended to general loading cases.

3.1 Asymptotic Behavior Under Repeated Pure Hydrostatic Loading

The hollow sphere undergoes a uniform hydrostatic stress q exerted upon its external boundary $\partial\Omega$. Owing to (8), the average mean stress is:

$$\Sigma_m = \frac{1}{3}\,Tr\,(\boldsymbol{\Sigma}) = q \tag{11}$$

Taking into account the central symmetry of the problem, the spherical coordinate (r, θ, φ) are used, r being the radius, θ the inclination angle, φ the azimuth one, with orthonormal frame $\{\mathbf{e}_r, \mathbf{e}_\theta, \mathbf{e}_\phi\}$, and all the fields are depending only on r. Simple computations allow the derivation of the stress field in the fictitious elastic body is:

$$\boldsymbol{\sigma}^E = \frac{\Sigma_m}{1-f}\left(\mathbf{1} + \frac{1}{2}\left(\frac{a}{r}\right)^3 (\mathbf{e}_\theta \otimes \mathbf{e}_\theta + \mathbf{e}_\phi \otimes \mathbf{e}_\phi - 2\,\boldsymbol{e}_r \otimes \boldsymbol{e}_r)\right) \tag{12}$$

where $\mathbf{1}$ is the second order unit tensor.

The behaviour of the body remains elastic provided:

$$\max_{a \le r \le b} \sigma_e = \max_{a \le r \le b}\left(\mid \frac{3}{2}\frac{\Sigma_m}{1-f}\left(\frac{a}{r}\right)^3 \mid +3\alpha \frac{\Sigma_m}{1-f}\right) \le \sigma_0 \tag{13}$$

Hence, the first yielding occurs on the void boundary and the *elastic limit average mean stress* is:

$$\Sigma^E_{m\pm} = \frac{(1-f)}{\pm 3/2 + 3\alpha}\sigma_0 \tag{14}$$

for $\Sigma_m > 0$ or $\Sigma_m < 0$.

Beyond this limit, plastic strains appear and our aim is to determine when the macro-element shakes down using Melan's theorem if the load Σ_m belongs to the domain:

$$\Sigma_{m-} \le \Sigma_m \le \Sigma_{m+} = \Sigma_{m-} + \Delta\Sigma_m \tag{15}$$

where Σ_{m-} and Σ_{m+} are the minimum and maximum macroscopic hydrostatic loads during the cyclic loading process, respectively.

The exact stress field of hollow sphere under pure hydrostatic load beyond the elastic limit is taking the following form [19]:

$$\boldsymbol{\sigma}^* = \frac{A_0}{3\alpha}\left(\left(1 - \left(\frac{a}{r}\right)^{3\gamma}\right)\mathbf{1} + \frac{3}{2}\gamma\left(\frac{a}{r}\right)^{3\gamma} (\boldsymbol{e}_\theta \otimes \boldsymbol{e}_\theta + \boldsymbol{e}_\phi \otimes \boldsymbol{e}_\phi)\right) \tag{16}$$

where $\gamma = 2\alpha/(2\alpha + \epsilon)$ and $\epsilon = \pm 1$ for $A_0 > 0$ or $A_0 < 0$, respectively.

We denote $\boldsymbol{\sigma}^{E(1)}$ the stress field in the corresponding fictitious elastic body, and the residual stress field is defined by:

$$\bar{\boldsymbol{\rho}}^{(1)} = \boldsymbol{\sigma}^* - \boldsymbol{\sigma}^{E(1)} \tag{17}$$

where Σ^E_m takes its extreme values in the inner part $a \le c \le b$ of the body when shakedown occurs:

$$\begin{aligned}\bar{\boldsymbol{\rho}}^{(1)} = A_0 &\left(\left(1 - \left(\frac{a}{r}\right)^{3\gamma}\right)\mathbf{1} + \frac{3}{2}\gamma\left(\frac{a}{r}\right)^{3\gamma} (\boldsymbol{e}_\theta \otimes \boldsymbol{e}_\theta + \boldsymbol{e}_\phi \otimes \boldsymbol{e}_\phi)\right) \\ &- \frac{\Sigma_{m\pm}}{1-f}\left(\mathbf{1} + \frac{1}{2}\left(\frac{a}{r}\right)^3 (\mathbf{e}_\theta \otimes \mathbf{e}_\theta + \mathbf{e}_\phi \otimes \mathbf{e}_\phi - 2\,\boldsymbol{e}_r \otimes \boldsymbol{e}_r)\right)\end{aligned} \tag{18}$$

where A_0 is a constant parameter. It can be verified that γ, A_0 and Σ_m have the same sign ϵ.

Taking the positive sign $\epsilon = 1$ for example, the total stress field under the pure hydrostatic load for $a \leq c \leq b$ is taking the following form:

$$\begin{aligned}\boldsymbol{\sigma}^{(1)} &= A_0 \left(\left(1 - \left(\frac{a}{r}\right)^{3\gamma} \right) \mathbf{1} + \frac{3}{2}\gamma \left(\frac{a}{r}\right)^{3\gamma} (\boldsymbol{e}_\theta \otimes \boldsymbol{e}_\theta + \boldsymbol{e}_\phi \otimes \boldsymbol{e}_\phi) \right) \\ &- \frac{\Sigma_{m+} - \Sigma_m}{1-f} \left(\mathbf{1} + \frac{1}{2} \left(\frac{a}{r}\right)^3 (\mathbf{e}_\theta \otimes \mathbf{e}_\theta + \mathbf{e}_\phi \otimes \mathbf{e}_\phi - 2\, \boldsymbol{e}_r \otimes \boldsymbol{e}_r) \right)\end{aligned} \tag{19}$$

Combining with the yield function (2), the shakedown limit is reached in when the yield function vanishes simultaneously for the extreme values of loading:

$$\left| \frac{3\gamma A_0}{2} \left(\frac{a}{r}\right)^{3\gamma} \right| + 3\alpha \left(\gamma A_0 \left(\frac{a}{r}\right)^{3\gamma} + A_0 \left(1 - \left(\frac{a}{r}\right)^{3\gamma} \right) \right) \leq \sigma_0 \tag{20}$$

$$\left| \frac{3\gamma A_0}{2} \left(\frac{a}{r}\right)^{3\gamma} - \frac{3}{2}\frac{\Delta\Sigma_m}{1-f} \left(\frac{a}{r}\right)^3 \right| + 3\alpha \left(\gamma A_0 \left(\frac{a}{r}\right)^{3\gamma} + A_0 \left(1 - \left(\frac{a}{r}\right)^{3\gamma} \right) - \frac{\Delta\Sigma_m}{1-f} \right) \leq \sigma_0 \tag{21}$$

Because $(a/r)^n$ decreases quickly when r increases, the previous condition is satisfied anywhere in the body if it is fulfilled at $r = a$:

$$\begin{aligned}\left(\frac{3\gamma A_0}{2}\right)^2 + 3\alpha\,(\gamma A_0) \leq \sigma_0 \\ \left(\frac{3\gamma A_0}{2} - \frac{3}{2}\frac{\Delta\Sigma_m}{1-f}\right)^2 + 3\alpha \left(\gamma A_0 - \frac{\Delta\Sigma_m}{1-f}\right) \leq \sigma_0\end{aligned} \tag{22}$$

Combining the two previous equations, the shakedown limit for pure hydrostatic loading reads:

$$\frac{\Delta\Sigma_m}{\sigma_0} \leq \frac{3\,(1-f)}{(3/2 + 3\alpha)\,(3/2 - 3\alpha)} \tag{23}$$

3.2 Macroscopic Shakedown Criterion Under General Cyclic Loadings

For the general case, the choice of a trial stress field is crucial due to the non linearity of Drucker-Prager yield function. It must be rather rich to capture the main physical effects but depends on a few number of parameters in order to obtain a closed-form formula for the macroscopic criterion.

The trial stress field is considered as the sum of the two parts, where a heterogeneous part inspired from the exact field under pure hydrostatic loadings, which has been provided in the previous subsection

$$\boldsymbol{\sigma}^{(1)} = \bar{\boldsymbol{\rho}}^{(1)} + \boldsymbol{\sigma}^{E(1)} \tag{24}$$

where the residual stress field in the inner region $a \leq r \leq c$ is inspired from the exact solution

The other part under the pure deviatoric loadings is expressed in spherical coordinates:

$$\boldsymbol{\sigma}^{(2)} = \bar{\boldsymbol{\rho}}^{(2)} + \boldsymbol{\sigma}^{E(2)} \tag{25}$$

where a statically admissible stress field in the fictitious body, deduced from the Papkovich-Neuber solution [5, 23] for the hollow sphere under the pure deviatoric load, in the following form, in the spherical coordinates:

$$\begin{aligned}\sigma^{E(2)} = -\frac{sign(J_3)\Sigma_e}{3(1-f)} \Bigg\{ & \left[\frac{a^3\left(18a^2+5r^2(-5+\nu)\right)(1+3\cos(2\theta))}{2r^5(-7+5\nu)} - \frac{1+3\cos(2\theta)}{2}\right](\mathbf{e}_r \otimes \mathbf{e}_r) \\ + & \left[\frac{a^3\left(27a^2+5r^2(1-2\nu)-3\left(21a^2+5r^2(-1+2\nu)\right)\cos^2(\theta)\right)}{2r^5(-7+5\nu)} + \frac{-1+3\cos(2\theta)}{2}\right](\mathbf{e}_\theta \otimes \mathbf{e}_\theta) \\ + & \left[\frac{a^3\left(9a^2+25r^2(-1+2\nu)-45\left(a^2+r^2(-1+2\nu)\right)\cos^2(\theta)\right)}{2r^5(-7+5\nu)} + 1\right](\mathbf{e}_\phi \otimes \mathbf{e}_\phi) \\ + & \left[\frac{3a^3\left(12a^2-5r^2(1+\nu)\right)\sin(2\theta)}{2r^5(-7+5\nu)} + \frac{3\sin(2\theta)}{2}\right](\mathbf{e}_r \otimes \mathbf{e}_\theta + \mathbf{e}_\theta \otimes \mathbf{e}_r)\Bigg\}\end{aligned} \tag{26}$$

where ν is Poisson's coefficient, Σ_e the macroscopic equivalent stress and J_3 the third invariant of the macroscopic stress deviator.

Similar to the residual stress field for the pure hydrostatic loading (18), the new deviatoric residual stress field consists of two parts:

$$\bar{\boldsymbol{\rho}}^{(2)} = \bar{\boldsymbol{\rho}}^{(2a)} + \bar{\boldsymbol{\rho}}^{(2b)}\,, \tag{27}$$

where $\bar{\boldsymbol{\rho}}^{(2a)}$ is the extreme value of the elastic stress field in the fictitious elastic body, and the second part $\bar{\boldsymbol{\rho}}^{(2b)}$ is in spherical coordinates:

$$\bar{\boldsymbol{\rho}}^{(2b)} = \rho^{2b}_{rr}\mathbf{e}_r \otimes \mathbf{e}_r + \rho^{2b}_{\theta\theta}\mathbf{e}_\theta \otimes \mathbf{e}_\theta + \rho^{2b}_{\phi\phi}\mathbf{e}_\phi \otimes \mathbf{e}_\phi + \rho^{2b}_{r\theta}(\mathbf{e}_r \otimes \mathbf{e}_\theta + \mathbf{e}_\theta \otimes \mathbf{e}_r) \tag{28}$$

ρ^{2b}_{rr}, $\rho^{2b}_{\theta\theta}$, $\rho^{2b}_{\phi\phi}$ and $\rho^{2b}_{r\theta}$ being the functions of r and θ. We suppose the deviatoric parts $\boldsymbol{s}^{E(2)}$, $\boldsymbol{s}^{(2b)}$ of $\boldsymbol{\sigma}^{E(2)}$, $\bar{\boldsymbol{\rho}}^{(2b)}$ have the following relation:

$$\boldsymbol{s}^{(2b)} = A_1\,K(r)\,\boldsymbol{s}^{E(2)} \quad and \quad \sigma_m(\bar{\boldsymbol{\rho}}^{(2b)}) = A_1\,K(r)\,\sigma_m(\boldsymbol{\sigma}^{E(2)}) \tag{29}$$

where $K(r)$ is a function of r and A_1 the constant to be determined, noticing that the existence of $\bar{\boldsymbol{\rho}}^{(2b)}$ and $K(r)$ was provided in [5].

Consequently, in the matrix Ω_M, the resultant two parameters-based trial stress field in the matrix can be written as:

$$\boldsymbol{\sigma} = \boldsymbol{\sigma}^{(1)} + \boldsymbol{\sigma}^{(2)}, \tag{30}$$

Note that a vanishing stress field is considered in the void ω.

For a variable hydrostatic loading combined with a constant shear loading, we consider the load domain defined by two elementary loads $\boldsymbol{\Sigma}_+$ and $\boldsymbol{\Sigma}_-$, and the axisymmetric macroscopic stress tensor, resulting from (30), takes the form:

$$\boldsymbol{\Sigma}_\pm = \Sigma_{m\pm}\mathbf{1} - sign(J_{3\pm})\frac{\Sigma_{e\pm}}{3}\left(\mathbf{e}_r \otimes \mathbf{e}_r + \mathbf{e}_\theta \otimes \mathbf{e}_\theta - 2\mathbf{e}_\phi \otimes \mathbf{e}_\phi\right) \tag{31}$$

Replacing the stress tensor in the yield function (2) by (30), the shakedown conditions are obtained, considering the collapse by fatigue when the yield function vanishes simultaneously for the extreme values of loading:

$$\begin{aligned}&\left[\left(\frac{3\gamma A_0}{2}\left(\frac{a}{r}\right)^{3\gamma}\right)^2 + (K(r)A_1)^2 P_2(r,\theta) + 3\gamma A_0\left(\frac{a}{r}\right)^{3\gamma} P_1(r,\theta)(K(r)A_1)\right]^{1/2}\\ &+3\alpha\left[\gamma A_0\left(\frac{a}{r}\right)^{3\gamma} + A_0\left(1-\left(\frac{a}{r}\right)^{3\gamma}\right) - \frac{5a^3(\nu+1)\left(3cos^2\theta-1\right)}{r^3(-7+5\nu)}K(r)A_1\right] = \sigma_0\end{aligned} \tag{32}$$

$$\begin{aligned}&\left[\left(\frac{3\gamma A_0}{2}\left(\frac{a}{r}\right)^{3\gamma} - \frac{3}{2}\frac{\Delta\Sigma_m}{1-f}\left(\frac{a}{r}\right)^3\right)^2 + \left(\frac{\Delta(sign(J_3)\Sigma_e)}{3(1-f)} + K(r)A_2\right)^2 P_2(r,\theta)\right.\\ &\left.+\left(3\gamma A_0\left(\frac{a}{r}\right)^{3\gamma} - 3\frac{\Delta\Sigma_m}{1-f}\left(\frac{a}{r}\right)^3\right)P_1(r,\theta)\left(\frac{\Delta(sign(J_3)\Sigma_e)}{3(1-f)} + K(r)A_2\right)\right]^{1/2}\\ &+3\alpha\left[\gamma A_0\left(\frac{a}{r}\right)^{3\gamma} + A_0\left(1-\left(\frac{a}{r}\right)^{3\gamma}\right) - \frac{\Delta\Sigma_m}{1-f}\left(\frac{a}{r}\right)^3\right.\\ &\left.-\frac{5a^3(\nu+1)\left(3cos^2\theta-1\right)}{r^3(-7+5\nu)}\left(\frac{\Delta sign(J_3)\Sigma_e}{3(1-f)} + K(r)A_1\right)\right] = \sigma_0\end{aligned} \tag{33}$$

where

$$P_1(r,\theta) = \frac{3\left(-5\nu + 10\nu(\frac{a}{r})^3 + 18(\frac{a}{r})^5 - 20(\frac{a}{r})^3 + 7\right)}{7-5\nu}\frac{\left(3\cos^2(\theta)-1\right)}{2}$$

and

$$P_2(r,\theta)=\left\{\left[\left(\frac{1125}{2}\nu-\frac{1175}{2}\right)\left(\frac{a}{r}\right)^3+\left(-\frac{1175}{2}\nu-\frac{2205}{2}\right)\left(\frac{a}{r}\right)^5+\left(225\nu^2-1125\nu+\frac{3375}{4}\right)\left(\frac{a}{r}\right)^6\right.\right.$$
$$+\left(1125\nu-\frac{2475}{2}\right)\left(\frac{a}{r}\right)^8+\frac{1215}{4}\left(\frac{a}{r}\right)^{10}\Big]\cos^4(\theta)+\Big[\left(-150\nu^2-165\nu+525\right)\left(\frac{a}{r}\right)^3$$
$$+(675\nu-945)\left(\frac{a}{r}\right)^5+\left(-225\nu^2+900\nu-\frac{1125}{2}\right)\left(\frac{a}{r}\right)^6+(-810\nu+675)\left(\frac{a}{r}\right)^8$$
$$-\frac{135}{2}\left(\frac{a}{r}\right)^{10}\Big]\cos^2(\theta)+\Big[(25\nu^2-70\nu+49)+\left(50\nu^2-\frac{115}{2}\nu-\frac{35}{2}\right)\left(\frac{a}{r}\right)^3+\left(-\frac{135}{2}\nu+\frac{189}{2}\right)\left(\frac{a}{r}\right)^5$$
$$\left.\left.+\left(100\nu^2-175\nu-\frac{475}{4}\right)\left(\frac{a}{r}\right)^6+\left(45\nu-\frac{615}{2}\right)\left(\frac{a}{r}\right)^8+\frac{351}{4}\left(\frac{a}{r}\right)^{10}\right]\right\}\frac{9}{(5\nu-7)^2}$$

Because $(a/r)^n$ decreases quickly when r increases, the previous condition is satisfied anywhere in the body if it is fulfilled at $r=a$. Due to the linear elastic response when shakedown occurs, let us introduce a generalized stress triaxiality

$$\tau=\frac{\gamma A_0}{K(a)\,A_1}=\frac{\gamma A_0-\frac{\Delta\Sigma_m}{1-f}}{\frac{\Delta(sign(J_3)\Sigma_e)}{3(1-f)}+K(a)\,A_1}=\frac{-\frac{\Delta\Sigma_m}{1-f}}{\frac{\Delta(sign(J_3)\Sigma_e)}{3(1-f)}}\tag{34}$$

Replacing γA_0 and $(\gamma A_0-\frac{\Delta\Sigma_m}{1-f})$, (32) and (32) can be respectively written:

$$\left[\frac{9}{4}\tau^2+3\tau P_1(a,\theta)+P_2(a,\theta)-9\alpha^2\left(\tau-\frac{5(\nu+1)\left(3cos^2\theta-1\right)}{(-7+5\nu)}\right)^2\right](K(a)A_1)^2$$
$$+6\alpha\sigma_0\left[\tau-\frac{5(\nu+1)\left(3cos^2\theta-1\right)}{(-7+5\nu)}\right]K(a)A_1-\sigma_0^2=0\tag{35}$$

$$\left[\frac{9}{4}\tau^2+3\tau P_1(a,\theta)+P_2(a,\theta)-9\alpha^2\left(\tau-\frac{5(\nu+1)\left(3cos^2\theta-1\right)}{(-7+5\nu)}\right)^2\right]\left(\frac{\Delta(sign(J_3)\Sigma_e)}{3(1-f)}+K(a)\,A_1\right)^2$$
$$+6\alpha\sigma_0\left[\tau-\frac{5(\nu+1)\left(3cos^2\theta-1\right)}{(-7+5\nu)}\right]\left(\frac{\Delta(sign(J_3)\Sigma_e)}{3(1-f)}+K(a)\,A_1\right)-\sigma_0^2=0\tag{36}$$

Solving (35) and (36) with respect to $K(a)A_1$, we obtain

$$\frac{\Delta(sign(J_3)\Sigma_e)}{3(1-f)\sigma_0}=\frac{2\sqrt{\frac{9}{4}\tau^2+3\tau P_1(a,\theta)+P_2(a,\theta)}}{\frac{9}{4}\tau^2+3\tau P_1(a,\theta)+P_2(a,\theta)-9\alpha^2\left(\tau-\frac{5(\nu+1)\left(3cos^2\theta-1\right)}{(-7+5\nu)}\right)^2}\tag{37}$$

Combining (34), leads to the closed-form macroscopic fatigue criterion:

$$\begin{cases} \dfrac{\Delta\Sigma_m}{\sigma_0} = -\tau(1-f)\dfrac{2\sqrt{\frac{9}{4}\tau^2 + 3\tau P_1(a,\theta) + P_2(a,\theta)}}{\frac{9}{4}\tau^2 + 3\tau P_1(a,\theta) + P_2(a,\theta) - 9\alpha^2\left(\tau - \frac{5(\nu+1)(3cos^2\theta-1)}{(-7+5\nu)}\right)^2} \\ \dfrac{\Delta(sign(J_3)\Sigma_e)}{\sigma_0} = 3(1-f)\dfrac{2\sqrt{\frac{9}{4}\tau^2 + 3\tau P_1(a,\theta) + P_2(a,\theta)}}{\frac{9}{4}\tau^2 + 3\tau P_1(a,\theta) + P_2(a,\theta) - 9\alpha^2\left(\tau - \frac{5(\nu+1)(3cos^2\theta-1)}{(-7+5\nu)}\right)^2} \end{cases} \tag{38}$$

where two events may occur:

- When $J_{3+} > 0$, the condition is satisfied if it is fulfilled at the equator $\theta = \pi/2$ where the left part of the previous shakedown condition (32) and (32) takes its maximum value, where

$$P_1(a, \frac{\pi}{2}) = \frac{3(5\nu + 5)}{2(5\nu - 7)} \qquad P_2(a, \frac{\pi}{2}) = \frac{225(7\nu^2 - 13\nu + 7)}{(5\nu - 7)^2}$$

- When $J_{3+} < 0$, the condition is satisfied if it is fulfilled at the poles $\theta = 0$ and $\theta = \pi$ where the left part of the previous shakedown condition (32) and (32) takes its maximum value, where

$$P_1(a, 0) = \frac{3(5\nu + 5)}{-5\nu + 7} \qquad P_2(a, 0) = \frac{225(\nu^2 + 2\nu + 1)}{(5\nu - 7)^2}$$

The set of Eqs. (38) constitutes the main finding in this study.

4 Numerical Verification of the Established Criterion

4.1 Step-by-Step FEM Numerical Procedure

This part is devoted to validate the accuracy of the established macroscopic fatigue criterion of ductile porous media with Drucker-Prager matrix. As a result, step-by-step incremental elastic-plastic simulations until fatigue collapse by Finite Element Method (FEM) are performed in order to obtain the numerical limit loads.

Because of the geometrical symmetry of hollow sphere model, only a quarter is considered, which is discretized by quadratic axisymmetric elements in Abaqus software (Fig. 3). The velocity field $\boldsymbol{v} = \boldsymbol{D} \cdot \mathbf{x}$ is prescribed on the external surface, achieved by a MPC (Multi-Points Constraints) user subroutine in Abaqus/Standard [24]. The implementation of such MPC program was firstly provided in void interaction and coalescence study of polymeric materials [25], and has been successfully used in the studies of porous media [26]. Hence, the enforced macroscopic triaxiality $T = \Sigma_m/\Sigma_{eq}$ can be maintained during each loading cycle.

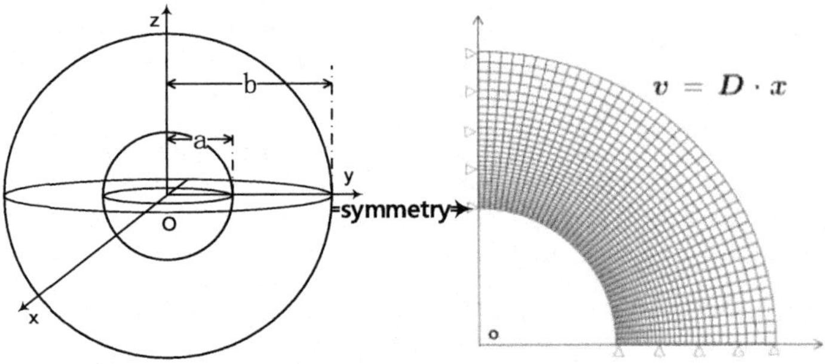

Fig. 3 Hollow sphere model and initial mesh

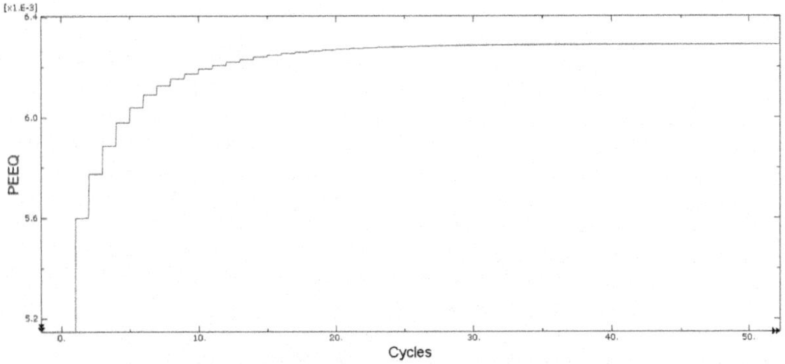

Fig. 4 Accumulated Equivalent Plastic Strain (PEEQ) under pulsating load when shakedown occurs for $f = 0.01$, $\phi = 20°$ and $T = 1.3333$

The computations are performed for different porosities $f \in \{0.001, 0.01\}$, different friction angles $\phi \in \{10°, 20°, 30°\}$ and with $\sigma_0 = 20$ MPa, $E = 14$ GPa and $\nu = 0.2$. Moreover, the following loading cases are considered for instance: alternating load $R = -1$ and an intermediate cyclic loads with $R = 1/5$ and $R = -1/5$, where the stress ratio is defined as $R = \Sigma_-/\Sigma_+$. The state of the long-term behavior is checked by increasing of the amplitude of the imposed deformation on the external boundary until the collapse by fatigue or a mechanism is observed. In practice, the evolution of the equivalent plastic deformation, and the evolution of the components of the plastic strain. In practice, the evolution of the equivalent plastic deformation in the whole structure provides information on the status of the structure as shown in Fig. 4.

4.2 Comparison Between Analytical Solutions and Numerical Results

Figures 5 plotted the macroscopic shakedown domain computed from the established macroscopic fatigue criteria (38) under alternating loading ($R = -1$) for several void volume fractions ($f \in \{0.001, 0.01\}$) and friction angles ($\phi \in \{10°, 20°, 30°\}$). It is worth noting that the collapse by development of a mechanism do not occur in this situation, so the safety domain is only defined by the fatigue criterion. In general, an excellent qualitative agreement can be observed between the analytical prediction and the above step-by-step numerical simulation by FEM for various value of void volume fractions and pressure sensitive parameters.

Interestingly, in the particular case of pure hydrostatic loading, the numerical results fit the exact value $\Delta\Sigma_m^{SD} = \frac{3(1-f)}{(3/2+3\alpha)(3/2-3\alpha)}$. This fact is foreseeable since the trial stress field and the residual stress tensor contain the exact solution for the hollow sphere under hydrostatic load. The location of first occurrence of the fatigue collapse in REV is verified by Fig. 6. The maximum value of the equivalent stress always presents on the void surface ($r = a$) at $\theta = \pi/2$ or 0. The first point where the fatigue occurs is located on the internal boundary $r = a$ at the equator $\theta = \pi/2$ for $J_{3+} > 0$ and at the poles $\theta = 0$ or π for $J_{3+} < 0$, in accord with the analytical solution.

Figure 7 displays the comparison between analytical results and numerical ones of the shakedown limit for the intermediate load corresponding to $R = 1/5$. Two major features must be underlined here. Unlike the alternating load (for which $R = -1$), the safety domain is obtained at the intersection of the domain defined by the new fatigue criterion and the one proposed in Guo et al. [27], corresponding to the collapse by development of a mechanism at the first cycle. In all figures shown hereafter, the analytic safe domain is bounded by solid lines. The second important remark is that the shakedown safe domain is considerably reduced compared to the gauge surface corresponding to the failure under monotonic loading. This ductility reduction is

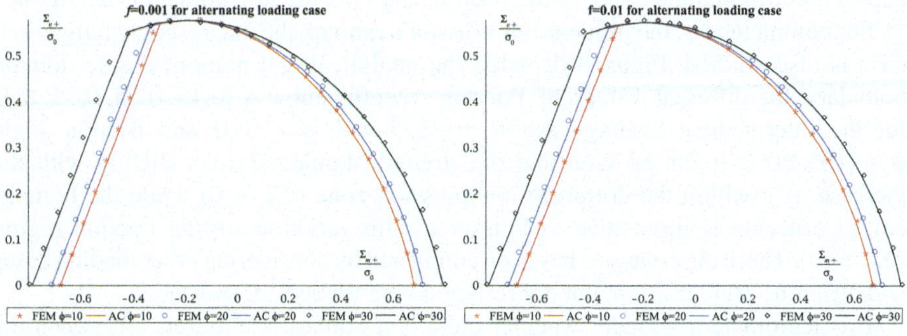

Fig. 5 Comparison between the yield surfaces obtained by the analytic criteria and by numerical simulations under alternating loadings ($R = -1$) for porosity $f = 0.001$ (left) and $f = 0.001$ (right)

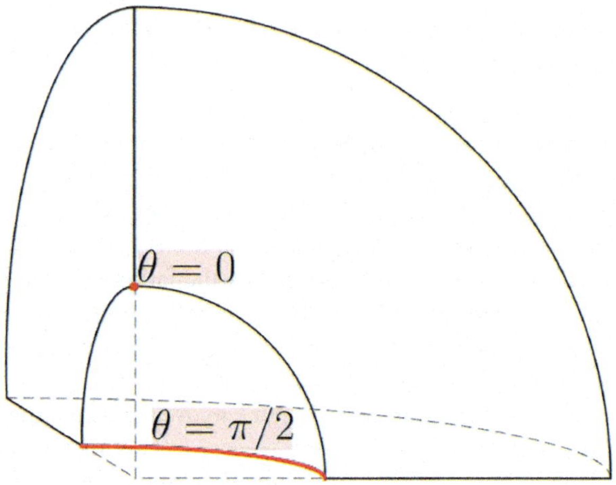

Fig. 6 Location of the occurrence of fatigue collapse

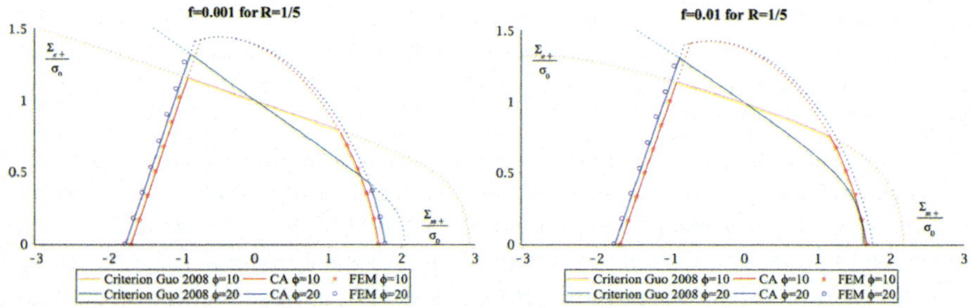

Fig. 7 Interaction curve for the intermediate loads with $R = 1/5$ for porosity $f = 0.001$ (left) and $f = 0.01$ (right). The analytic safe domain is bounded by solid lines

more pronounced in the dominant compression zone ($\Sigma_m < 0$). In addition, these curves confirm that the effect of the friction angle on the safe domain is negligible.

For completeness, the influence of Poisson's ratio on the macroscopic fatigue criteria is also studied. Figure 8 depicted the analytical and numerical safe domain boundary for different values of Poisson's coefficient $\nu \in \{0.15, 0.25, 0.35, 0.4\}$ for the intermediate loading case $R = -1/5$ with $f = 0.01$ and friction angle $\phi \in \{10°, 20°\}$. It can be seen that the strength domain shrinks slightly with the increase of ν within the dominant compression zone ($\Sigma_m < 0$) while the homogenized criterion is insensitive to Poisson's ratio variations in the traction region ($\Sigma_m > 0$). These observations has been confirmed by considering other loading cases and different angle friction, but not reported here for seek of shortness.

It is legitimate to wonder whether there is a noticeable difference between the macroscopic shakedown domains for porous materials with Drucker-Prager model or von Mises criterion, which can be recovered from the proposed model (38) by taking $\phi = 0$. To this end, Fig. 9 depicts the analytic safe domains obtained under

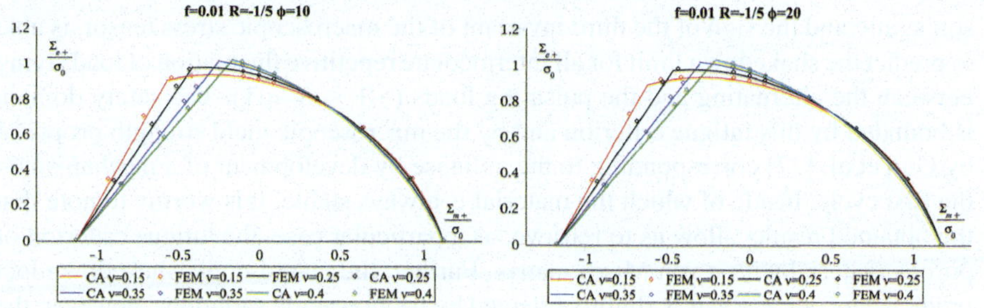

Fig. 8 Effect of Poisson's coefficient on the macroscopic shakedown domain for the intermediate loading case $R = -1/5$

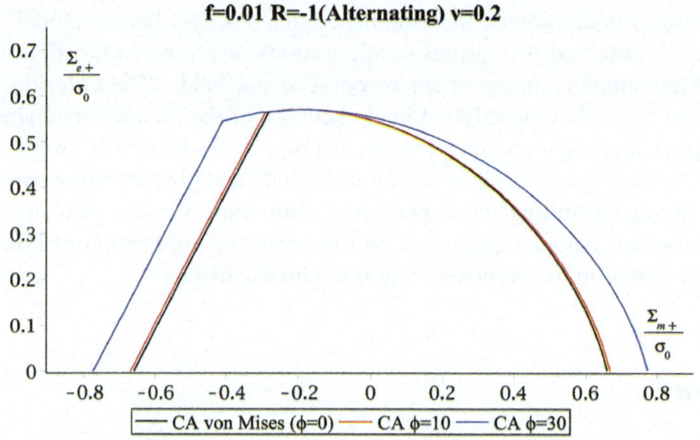

Fig. 9 Comparison of the effective shakedown safety domain with Drucker-Prager dilatant matrix with the one with von Mises model

cyclic loads with Drucker-Prager and von Mises constitutive laws. The difference between the models is noticeable but not considerable as it could be expected because of the presence of the first invariant of the stress in the Drucker-Prager yield function. This remark should be taken with caution because, as we have seen above, the real safe domain is a result of interaction between the homogenized shakedown domain and the limit analysis effective yield.

5 Conclusion

In this study, we applied Melan's shakedown theorem to derive a macroscopic fatigue criterion for ductile porous material with pressure-sensitive dilatant matrix under cyclic repeated loadings, considering the hollow sphere unit cell. The closed-form fatigue criterion in parametric form, depending on the porosity, friction angle, Poisson's ratio and the sign of the third invariant of the macroscopic stress tensor, is able to predict the shakedown limit for all intermediate repetitive fluctuation of loads lying between the alternating and the pulsating loads ($-1 \leq R < 1$). The safety domain is bounded by this fatigue criterion and by the macroscopic yield strength proposed by Guo et al. [27] corresponding to the collapse by development of a mechanism at the first cycle, inside of which the material is always stable. It is worthy to note that the obtained results allow us to retrieve, as a particular case, the fatigue criterion of porous material with a von Mises matrix. Further, the macroscopic shakedown loci predicted by our analysis is slightly effected by friction angle variations, however, the safe domain is considerably reduced compared to the one obtained by limit analysis for monotonic load conditions.

Step-by-step elasto-plastic numerical simulations have been carried out to verify the accuracy of prediction by the established macroscopic fatigue model. The good agreement is observed for various configurations of porosity and frictions angles, owing to the suitable choice of microscopic stress fields. The criterion is strictly conservative to predict the safety domain because of the Melan's statical approach.

For future investigations, inspired from Dang Van's theory of fatigue, we hope apply the shakedown analysis to the monocrystals and polycrystals to propose new fatigue criteria, exhibiting the dependence with respect to the porosity and Poisson's coefficient. Another extension will concern the application of homogenized shakedown criterion to composite material with inclusions.

References

1. Kobayashi, H., Kusumoto, T., Nakazawa, H.: Cyclic J-R-curve and upper limit characteristic of fatigue crack growth in 2-1/2 Cr-Mo steel. In: 11th International Conference on Structural Mechanics in Reactor Technology (SMIRT 11), Tokyo, Japan, paper G27/1 (1991)
2. Meng, Q., Zhang, M., Han, L., et al.: Acoustic emission characteristics of red sandstone specimens under uniaxial cyclic loading and unloading compression. Rock Mech. Rock Eng. **51**(4), 969–988 (2018)
3. Song, Z., Konietzky, H., Fruhwirt, T.: Hysteresis energy-based failure indicators for concrete and brittle rocks under the condition of fatigue loading. Int. J. Fatigue **114**, 298–310 (2018)
4. Zhang, J., Shen, W.Q., Oueslati, A., de Saxcé, G.: Shakedown of porous materials. Int. J. Plast. **2017**(95), 123–141 (2017)
5. Zhang, J., Shen, W.Q., Oueslati, A., de Saxcé, G.: A macroscopic criterion of shakedown limit for ductile porous materials subjected to general cyclic loadings. Mech. Mater. **115**, 76–87 (2017)
6. Dang Van, K.: Sur la résistance à la fatigue des métaux. Ph.D. thesis. University of Paris 6 (1971)
7. Keralavarma, S.M., Benzerga, A.A.: A constitutive model for plastically anisotropic solids with non-spherical voids. J. Mech. Phys. Solids. **58**, 874–901 (2010)
8. El Ghezal, I., Doghri, I., Kondo, D.: Static limit analysis and strength of porous solids with hill orthotropic matrix. Int. J. Solids Struct. **109**, 63–71 (2017)
9. Melan, E.: Theorie statisch unbestimmter Systeme aus ideal plastischem Baustoff. Sitz Ber. Akad. Wiss. Wien IIa, 145-195 (1936)
10. Koiter, W.T.: General Theorems for Elastic-Plastic Solids. In: Sneddon, I.N., Hill, R. (Eds.) Progress in Solid Mechanics, vol. 1. North-Holland, Amsterdam (1960)

11. Weichert, D., Ponter, A.: A historical view on shakedown theory. The History of Theoretical, Material and Computational Mechanics - Mathematics Meets Mechanics and Engineering, pp. 169-193. Springer (2014)
12. Gurson, A.L.: Continuum theory of ductile rupture by void nucleation and growth - part I: yield criteria and flow rules for porous ductile media. J. Eng. Mater. Technol. **99**, 2–15 (1977)
13. Devaux, J., Gologanu, M., Leblond, J.B., Perrin, G.: On continued void growth in ductile metals subjected to cyclic loadings. In: Willis, J. (ed.) Proceedings of the IUTAM Symposium on Nonlinear Analysis of Fracture. Kluwer, Cambridge, GB, pp. 299-310 (1997)
14. Mbiakop, A., Constantinescu, A., Danas, K.: On void shape effects of periodic elasto-plastic materials subjected to cyclic loading. Eur. J. Mech. A Solids **49**, 481e499 (2014)
15. Lacroix, R., Leblond, J.B., Perrin, G.: Numerical study and theoretical modelling of void growth in porous ductile materials subjected to cyclic loadings. Eur. J. Mech./A Solids **55**, 100–109 (2016)
16. Monchiet, V., Cazacu, O., Kondo, D.: Macroscopic yield criteria for plastic anisotropic materials containing spheroidal voids. Int. J. Plast. **24**, 1158–1189 (2008)
17. Cheng, L., Jia, Y., Oueslati, A., de Saxcé, G., Kondo, D.: A bipotential-based limit analysis and homogenization of ductile porous materials with non-associated Drucker-Prager matrix. J. Mech. Phys. Solids **77**, 1–26 (2015)
18. Shen, W., Oueslati, A., de Saxcé, G.: Macroscopic criterion for ductile porous materials based on a statically admissible microscopic stress field. Int. J. Plast. **70**, 60–76 (2015)
19. Cheng, L., Jia, Y., Oueslati, A., de Saxcé, G., Kondo, D.: Plastic limit state of the hollow sphere model with non-associated Drucker-Prager material under isotropic loading. Comput. Mater. Sci. **62**, 210–215 (2012)
20. Hill, R.: Mathematical Theory of plasticity. Oxford University Press, London (1954)
21. König, J.A.: Shakedown of Elastic-Plastic Structures. Elsevier, Amsterdam (1987)
22. Weichert, D., Maier G.: Inelastic Behaviour of Structures Under Variable Repeated Loads: Direct Analysis Methods. Springer Science & Business Media (2002)
23. Soutas-Little, R.W.: Elasticity. Courier Corporation (1999)
24. Hibbitt, Karlsson: Sorensen. ABAQUS/standard, User's Manual (1998)
25. Cheng, L., Guo, T.: Void interaction and coalescence in polymeric materials. Int. J. Solids Struct. **44**(6), 1787–1808 (2007)
26. Cheng, L., de Saxcé, G., Kondo, D.: A stress-based variational model for ductile porous materials. Int. J. Plast. **55**, 133–151 (2014)
27. Guo, T., Faleskog, J., Shih, C.: Continuum modeling of a porous solid with pressure-sensitive dilatant matrix. J. Mech. Phys. Solids **56**, 2188–2212 (2008)

A Direct Method for Cyclic Crystal Plasticity with Application to High-Cycle Fatigue

Insaf Echerradi, Daniel Weisz-Patrault, and Michael Peigney

Abstract The prediction of fatigue in materials and structures is usually based on experimental Wohler curves, relating the number of cycles to failure to the amplitude of the applied cyclic loading. Those curves show some scattering due notably to the variability of the microstructure. Predicting fatigue lifetime can thus be seen as a statistical problem that depends on microstructural descriptors. This paper paves the way to a probabilistic approach for quantitatively linking crystallographic and morphological texture data to fatigue lifetime prediction. In more detail, a simplified mesoscopic model is constructed for calculating the evolution of an elastic-plastic polycrystal with a prescribed texture. That model is limited to high cycle fatigue, corresponding to cyclic loadings of sufficiently low amplitude for plasticity to be mainly confined to few well-separated grains. The model obtained takes details of the texture into account, i.e. the distribution, shape and orientation of the individual grains. It relies on analytical formula and is mesh-free. A comparison with full-field finite element simulation shows that the proposed model leads to satisfactory results in regard to its complexity. In the case of cyclic loading, we show that the model presented leads to a direct method for calculating the asymptotic values of the plastic slips (and cumulated plastic slips) reached in each grain when the number of cycles grows to infinity. We show how that direct approach can be used for upscaling a local failure criterion to the mesoscopic scale and performing probabilistic analysis.

Keywords High cycle fatigue · Mesoscopic model · Crystal plasticity · Incremental energy minimization · Direct method

---

I. Echerradi · M. Peigney (✉)
Lab Navier, Univ Gustave Eiffel, ENPC, CNRS, 77447 Marne la Vallée, France
e-mail: michael.peigney@polytechnique.org

D. Weisz-Patrault
LMS, CNRS, École Polytechnique, Institut Polytechnique de Paris, 91128 Palaiseau, France

G. Garcea and D. Weichert (eds.), *Direct Methods for Limit State of Materials and Structures*, Lecture Notes in Applied and Computational Mechanics 101,
https://doi.org/10.1007/978-3-031-29122-7_9

1 Introduction

It is commonly accepted that fatigue in polycrystals is related to plastic mechanisms occurring at the grain scale. Plastic flow initiates in few critical grains having the least favorable orientations with respect to the applied loading. For small loading amplitudes, plasticity remains confined in those grains but persistent slip bands may appear after a large number of cycles, leading to the possible initiation of a crack. Based on that interpretation, the fatigue life of a polycrystal could in principle be predicted from the elastic-plastic evolution of its constitutive grains. Some work in that direction can notably be found in [1] using full-field simulations. For such problems, direct methods bypassing the incremental simulation over a large number of cycles are of interest. A possible approach is to use shakedown theory. In classical plasticity, Melan and Koiter theorems deliver bounds on the set of loadings for which there is elastic shakedown, i.e. the plastic dissipation is bounded on the time interval $[0, +\infty)$. In more intuitive terms, shakedown means that the medium behaves elastically in the large time limit. Those theorems have been extended to several types of nonlinear behaviors [8, 17–21, 24] and can be used to bound the shakedown domain of a polycrystal with a given microstructure. Since shakedown is beneficial for fatigue, loadings within the shakedown domain are expected to correspond to high-cycle fatigue. A recent illustration of that approach can be found in [15]. However, although shakedown is a requisite for high-cycle fatigue, it does not guarantee an infinite lifetime. Indeed, the number of cycles to failure depends on the asymptotic state of stress reached in the polycrystal, which shakedown theory only provides limited information about. Sections 2–4 of this paper report on a simplified numerical method for a fast evaluation of the plastic slips in elastic-plastic 2D polycrystals. That method is based on incremental variational principles, which have proved to be a fruitful approach in many applications [2, 9, 10, 16, 22, 25, 26]. Starting from elastic-plastic constitutive equations detailed in Sect. 2, the incremental evolution problem is turned into an energy minimization problem over the space of admissible displacement fields and plastic slip fields (Sect. 3). An approximate solution can be obtained by restricting the minimization to a well-chosen finite dimensional subspace, in the spirit of Galerkin's methods. In the present case, we restrict the energy minimization to plastic strain fields that are uniform per grain. For an isotropic elasticity tensor, the incremental problem reduces to a linear complementarity problem. That problem can further be simplified by neglecting the elastic interaction between plastifying grains and assuming that only one plastic slip system is activated in each grain. Such assumptions are expected to be representative of small loadings corresponding to high-cycle fatigue. In that case, the incremental problem can be solved analytically as detailed in Sect. 4. The increment of plastic slip in grain j is obtained as an explicit function of the material parameters, loading parameters and a localization tensor $\boldsymbol{S}^j$ that is entirely determined from the geometry of the grain and the elastic moduli. For an ellipsoidal grain, the tensor $\boldsymbol{S}^j$ corresponds to the Eshelby tensor [5, 14]. The validity of the model presented is briefly discussed in Sect. 5 by comparison with full-field finite element simulations. The approach presented allows one to estimate the step-by-step evolution of plastic slips for a given loading history. In Sect. 6 we focus on cyclic loadings and show that the approach presented leads to a direct method for calculating the asymptotic values taken by the state variables in each grain as the number of cycles grows to infinity. That method is direct in the sense that it gives the asymptotic state without resorting to step-by-step calculation. The direct approach of Sect. 6 is used in Sect. 7 for upscaling a local (microscopic) fatigue criterion to the mesoscopic scale of a polycrystal: starting from a crack initiation criterion at the grain level, we use the asymptotic formulas of the shakedown state to express the fatigue criterion at the mesoscopic level. The mathematical structure of the obtained fatigue criterion is compared with well-established fatigue criteria (Dang Van, Crossland, Sines). In Sect. 8 we show how the model presented can be used in a probabilistic setting.

2 Constitutive Laws

2.1 Single Crystal

First consider a single crystal modeled in the framework of crystalline plasticity at small strains: considering N slip systems $(\boldsymbol{n}_i, \boldsymbol{t}_i)_{1\leq i\leq N}$, the local strain $\boldsymbol{\varepsilon}$ is decomposed as

$$\boldsymbol{\varepsilon} = \boldsymbol{C}^{-1} : \boldsymbol{\sigma} + \sum_{i=1}^{N} \gamma_i \boldsymbol{\tau}_i^0 \tag{1}$$

where

$$\boldsymbol{\tau}_i^0 = \frac{1}{2}(\boldsymbol{n}_i \otimes \boldsymbol{t}_i + \boldsymbol{t}_i \otimes \boldsymbol{n}_i) \tag{2}$$

and $(\gamma_i, \boldsymbol{n}_i, \boldsymbol{t}_i)$ are respectively the plastic slip, the normal to the slip plane and the slip direction for slip system i. In (1), $\boldsymbol{\sigma}$ is the stress and $\boldsymbol{C}$ is the elasticity tensor (assumed to be isotropic in the following). We recall that the strain $\boldsymbol{\varepsilon}$ is related to the displacement $\boldsymbol{u}$ by the relation $\boldsymbol{\varepsilon} = (\nabla \boldsymbol{u} + \nabla^T \boldsymbol{u})/2$. Adopting the framework of generalized standard materials [6], the plasticity flow rule is determined from the free energy Ψ and the dissipation potential Φ, which are respectively chosen as

$$\Psi = \frac{1}{2}\left(\boldsymbol{\varepsilon} - \sum_{i=1}^{N} \gamma_i \boldsymbol{\tau}_i^0\right) : \boldsymbol{C} : \left(\boldsymbol{\varepsilon} - \sum_{i=1}^{N} \gamma_i \boldsymbol{\tau}_i^0\right) + \frac{1}{2}\xi\chi \sum_{i=1}^{N} \gamma_i^2 \tag{3}$$

$$\Phi = \sum_{i=1}^{N} (\tau_c + \xi \eta_i)|\dot{\gamma}_i| \tag{4}$$

where

$$\eta_i = \int_0^t |\dot{\gamma}_i| dt$$

is the accumulated plastic slip, $\tau_c > 0$ is the critical shear stress, $\xi_X \geq 0$ is the linear hardening modulus and $\xi \geq 0$ is the isotropic hardening modulus. The flow rule reads as

$$\boldsymbol{\sigma} : \boldsymbol{\tau}_i^0 - \xi_X \gamma_i \begin{cases} = \tau_c + \xi \eta_i & \text{if } \dot{\gamma}_i > 0 \\ = -\tau_c - \xi \eta_i & \text{if } \dot{\gamma}_i < 0 \\ \in [-\tau_c - \xi \eta_i, \tau_c + \xi \eta_i] & \text{if } \dot{\gamma}_i = 0 \end{cases} \tag{5}$$

2.2 Polycrystal

Now consider a polycrystalline Representative Volume Element (RVE) constituted of M grains occupying the disjoint subdomains $\Omega^1, \ldots, \Omega^M$. The crystalline orientation in grain j is characterized by a rotation $\boldsymbol{R}^j$ relative to the reference crystal, so that the constitutive equations in grain j are obtained by replacing $\boldsymbol{\tau}_i^0$ with $\boldsymbol{\tau}_i^j = {}^T\boldsymbol{R}^j \boldsymbol{\tau}_i^0 \boldsymbol{R}^j$ in (1) and (5), i.e. for $\boldsymbol{x} \in \Omega^j$ we have the relations

$$\boldsymbol{\varepsilon} = \boldsymbol{C}^{-1} : \boldsymbol{\sigma} + \sum_{i=1}^{N} \gamma_i \boldsymbol{\tau}_i^j \tag{6}$$

$$\boldsymbol{\sigma} : \boldsymbol{\tau}_i^j - \xi_X \gamma_i \begin{cases} = \tau_c + \xi \eta_i & \text{if } \dot{\gamma}_i > 0, \\ = -\tau_c - \xi \eta_i & \text{if } \dot{\gamma}_i < 0, \\ \in [-\tau_c - \xi \eta_i, \tau_c + \xi \eta_i] & \text{if } \dot{\gamma}_i = 0. \end{cases} \tag{7}$$

We note than in the more general situation where the elasticity tensor $\boldsymbol{C}$ is not isotropic, the term $\boldsymbol{C}$ in (6) should be replaced with a rotated elasticity tensor depending on $\boldsymbol{R}^j$. The RVE is submitted to traction force $\bar{\boldsymbol{\sigma}}(t) \cdot \boldsymbol{n}$ on its boundary, where $\bar{\boldsymbol{\sigma}}(t)$ can be interpreted as the mesoscopic stress at time t. We consider a proportional loading of the form

$$\bar{\boldsymbol{\sigma}}(t) = f(t) \boldsymbol{\sigma}_0 \tag{8}$$

where $\boldsymbol{\sigma}_0$ is independent of time. Assuming quasistatic evolutions, the stress field needs to satisfy the equilibrium equations

$$\operatorname{div} \boldsymbol{\sigma} = 0 \text{ in } \Omega, \ \boldsymbol{\sigma} \cdot \boldsymbol{n} = \bar{\boldsymbol{\sigma}} \cdot \boldsymbol{n} \text{ on } \partial\Omega \tag{9}$$

where $\Omega = \cup_{j=1}^M \Omega^j$ is the domain occupied by the RVE. To simplify the presentation, the same critical shear stress τ_c and hardening moduli are used for all the grains.

3 Incremental Energy Minimization

For solving the evolution problem defined by Eqs. (6)–(9), a common approach is to resort to time-discretization: the evolution is calculated in a time marching approach, using an incremental problem for estimating the state variable $(\boldsymbol{\varepsilon}, \boldsymbol{\sigma}, \gamma_i, \eta_i)$ at time $t^0 + \delta t$ ($\delta t > 0$), assuming their values $(\boldsymbol{\varepsilon}^0, \boldsymbol{\sigma}^0, \gamma_i^0, \eta_i^0)$ to be known. Adopting the backward Euler scheme, the corresponding incremental problem reads as

$$
\begin{aligned}
&\operatorname{div} \boldsymbol{\sigma} = 0 \text{ in } \Omega, \ \boldsymbol{\sigma} \cdot \boldsymbol{n} = \bar{\boldsymbol{\sigma}} \cdot \boldsymbol{n} \text{ on } \partial\Omega \\
&\boldsymbol{\varepsilon} = \boldsymbol{C}^{-1} : \boldsymbol{\sigma} + \sum_{i=1}^{N} \gamma_i \boldsymbol{\tau}_i^j \text{ in } \Omega^j \\
&\boldsymbol{\sigma} : \boldsymbol{\tau}_i^j - \xi_X \gamma_i \begin{cases} = \tau_c + \xi \eta_i & \text{if } \gamma_i > \gamma_i^0 \\ = -\tau_c - \xi \eta_i & \text{if } \gamma_i < \gamma_i^0 \\ \in [-\tau_c - \xi \eta_i \tau_c + \xi \eta_i] & \text{if } \gamma_i = \gamma_i^0 \end{cases} \text{ in } \Omega^j,
\end{aligned}
\tag{10}
$$

with $\eta_i = \eta_i^0 + |\gamma_i - \gamma_i^0|$. A variational formulation is attached to the incremental problem (10). Setting $\boldsymbol{\gamma} = (\gamma_1, \ldots, \gamma_N)$, it can be verified indeed that the displacement field $\boldsymbol{u}$ and plastic slips field $\boldsymbol{\gamma}$ in (10) are solution to the minimization problem

$$
\inf_{\boldsymbol{u}, \boldsymbol{\gamma}} F \tag{11}
$$

where

$$
F = \sum_{j=1}^{M} \int_{\Omega^j} \Psi^j d\Omega + \sum_{i=1}^{N} \int_{\Omega} (\tau_c + \xi \eta_i^0 + \frac{1}{2} \xi \, |\gamma_i - \gamma_i^0|) |\gamma_i - \gamma_i^0| d\Omega - \int_{\partial\Omega} (\bar{\boldsymbol{\sigma}} \cdot \boldsymbol{n}) \cdot \boldsymbol{u} dS
$$

and Ψ^j is the free energy in grain j, defined by replacing $\boldsymbol{\tau}_j^0$ with $\boldsymbol{\tau}_j^i$ in (3). Observing that $\int_{\partial\Omega} (\bar{\boldsymbol{\sigma}}.\boldsymbol{n}).\boldsymbol{u} \mathrm{d}S = \int_{\Omega} \bar{\boldsymbol{\sigma}} : \boldsymbol{\varepsilon} d\Omega$, we can rewrite F as

$$
\begin{aligned}
F = &\sum_{j=1}^{M} \int_{\Omega^j} \frac{1}{2} \left(\boldsymbol{\varepsilon} - \sum_{i=1}^{N} \gamma_i \boldsymbol{\tau}_i^j \right) : \boldsymbol{C} : (\boldsymbol{\varepsilon} - \sum_{i=1}^{N} \gamma_i \boldsymbol{\tau}_i^j) d\Omega + \frac{1}{2} \xi_X \int_{\Omega} \sum_{i=1}^{N} \gamma_i^2 d\Omega \\
&+ \sum_{i=1}^{n} \int_{\Omega} \left(\tau_c + \xi \eta_i^0 + \frac{1}{2} \xi \, |\gamma_i - \gamma_i^0| \right) |\gamma_i - \gamma_i^0| d\Omega - \int_{\Omega} \bar{\boldsymbol{\sigma}} : \boldsymbol{\varepsilon} d\Omega
\end{aligned}
$$

In general, problem (10) or (11) needs to be solved numerically, using e.g. the finite element method. Following [23], we note that problem (11) is mathematically equivalent to a linear complementarity problem. Dedicated numerical algorithms are available for such problems [3]. In the following we introduce a series of simplifying assumptions that allow us to solve (10) in a semi-analytical form. Those assumptions are expected to be relevant for small loadings corresponding to the regime of high-cycle fatigue.

4 Simplified Problem

4.1 Piecewise-Constant Plastic Slips

We restrict the minimization in (11) to plastic slips that are uniform in each grain. That assumption notably allows one to make some progress on the minimization of (11) with respect to $\boldsymbol{u}$. For a given $\boldsymbol{\gamma}$ (uniform per grain), observe indeed that

$$\inf_{\boldsymbol{u}} F = \frac{1}{2}\xi\chi \int_\Omega \sum_{i=1}^N \gamma_i^2 d\Omega + \sum_{i=1}^N \int_\Omega \left(\tau_c + \xi\eta_i^0 + \frac{1}{2}\xi \, |\gamma_i - \gamma_i^0|\right) |\gamma_i - \gamma_i^0| d\Omega$$
$$+\frac{1}{2}\sum_{j=1}^M |\Omega^j| \boldsymbol{\tau}^j : \boldsymbol{C} : \boldsymbol{\tau}^j + W$$

where

$$W = \inf_{\boldsymbol{u}} \frac{1}{2}\int_\Omega \boldsymbol{\varepsilon} : \boldsymbol{C} : \boldsymbol{\varepsilon} d\Omega - \sum_{j=1}^M \int_{\Omega^j} \boldsymbol{\varepsilon} : \boldsymbol{C} : \boldsymbol{\tau}^j d\Omega - \int_\Omega \bar{\boldsymbol{\sigma}} : \boldsymbol{\varepsilon} d\Omega \tag{12}$$

and $\boldsymbol{\tau}^j = \sum_{i=1}^N \gamma_i \boldsymbol{\tau}_i^j$. Let χ^j be the characteristic function of Ω^j, i.e. $\chi^j(\boldsymbol{x}) = 1$ if $\boldsymbol{x} \in \Omega^j$ and $\chi^j(\boldsymbol{x}) = 0$ otherwise. Setting $\boldsymbol{\tau}(\boldsymbol{x}) = \boldsymbol{C}^{-1} : \bar{\boldsymbol{\sigma}} + \sum_j \chi^j(\boldsymbol{x})\boldsymbol{\tau}^j$, we obtain

$$W = \inf_{\boldsymbol{u}} \frac{1}{2}\int_\Omega \boldsymbol{\varepsilon} : \boldsymbol{C} : \boldsymbol{\varepsilon} d\Omega - \int_\Omega \boldsymbol{\varepsilon} : \boldsymbol{C} : \boldsymbol{\tau} d\Omega. \tag{13}$$

The solution $\boldsymbol{u}^*$ to the minimization problem (13) satisfies the stationarity condition

$$\int_\Omega \boldsymbol{\varepsilon}^* : \boldsymbol{C} : \boldsymbol{\varepsilon} d\Omega = \int_\Omega \boldsymbol{\tau} : \boldsymbol{C} : \boldsymbol{\varepsilon} d\Omega \tag{14}$$

for any compatible strain field $\boldsymbol{\varepsilon}$. Using (14) with $\boldsymbol{\varepsilon} = \boldsymbol{\varepsilon}^*$ gives $\int_\Omega \boldsymbol{\varepsilon}^* : \boldsymbol{C} : \boldsymbol{\varepsilon}^* d\Omega = \int_\Omega \boldsymbol{\tau} : \boldsymbol{C} : \boldsymbol{\varepsilon}^* \mathrm{d}\Omega$ so that

$$W = -\frac{1}{2}\int_\Omega \boldsymbol{\tau} : \boldsymbol{C} : \boldsymbol{\varepsilon}^* d\Omega. \tag{15}$$

Choosing $\boldsymbol{\varepsilon} = \int_\Omega (\boldsymbol{\varepsilon}^* - \boldsymbol{\tau})\mathrm{d}\Omega$ in (14) gives $\int_\Omega \boldsymbol{\varepsilon}^* d\Omega = \int_\Omega \boldsymbol{\tau} d\Omega = |\Omega| \boldsymbol{C}^{-1} : \bar{\boldsymbol{\sigma}} + \sum_j |\Omega^j| \boldsymbol{\tau}^j$ so that

$$\begin{aligned}\int_{\Omega} \boldsymbol{\tau} : \boldsymbol{C} : \boldsymbol{\varepsilon}^* d\Omega &= \bar{\boldsymbol{\sigma}} : \int_{\Omega} \boldsymbol{\varepsilon}^* d\Omega + \sum_j \boldsymbol{\tau}^j : \boldsymbol{C} : \left(\int_{\Omega^j} \boldsymbol{\varepsilon}^* d\Omega \right) \\ &= |\Omega| \bar{\boldsymbol{\sigma}} : \boldsymbol{C}^{-1} : \bar{\boldsymbol{\sigma}} + \sum_j |\Omega^j| \bar{\boldsymbol{\sigma}} : \boldsymbol{\tau}^j + \sum_j \boldsymbol{\tau}^j : \boldsymbol{C} : \left(\int_{\Omega^j} \boldsymbol{\varepsilon}^* d\Omega \right).\end{aligned} \tag{16}$$

Note that $\boldsymbol{u}^*$ is a solution to the linear elasticity problem

$$\begin{aligned}&\boldsymbol{\sigma}^* = \boldsymbol{C} : (\boldsymbol{\varepsilon}^* - \boldsymbol{\tau}^j) \text{ in } \Omega^j, \\ &\boldsymbol{\varepsilon}^* = (\nabla \boldsymbol{u}^* + \nabla^T \boldsymbol{u}^*)/2, \\ &\operatorname{div} \boldsymbol{\sigma}^* = 0 \text{ in } \Omega, \\ &\boldsymbol{\sigma}^* \cdot \boldsymbol{n} = \bar{\boldsymbol{\sigma}} \cdot \boldsymbol{n} \text{ on } \partial\Omega.\end{aligned} \tag{17}$$

The superposition principle implies that

$$\boldsymbol{\varepsilon}^* = \boldsymbol{C}^{-1} : \bar{\boldsymbol{\sigma}} + \sum_j \boldsymbol{\varepsilon}^j \tag{18}$$

where $\boldsymbol{\varepsilon}^j$ is the strain field corresponding to the solution of the *single inclusion problem* (defined for any given $j = 1, \ldots, M$)

$$\begin{aligned}&\boldsymbol{\sigma} = \boldsymbol{C} : (\boldsymbol{\varepsilon}^j - \boldsymbol{\tau}^j) \text{ in } \Omega^j, \\ &\boldsymbol{\sigma} = \boldsymbol{C} : \boldsymbol{\varepsilon}^j \text{ in } \Omega - \Omega^j, \\ &\boldsymbol{\varepsilon}^j = (\nabla \boldsymbol{u}^j + \nabla^T \boldsymbol{u}^j)/2, \\ &\operatorname{div} \boldsymbol{\sigma} = 0 \text{ in } \Omega, \\ &\boldsymbol{\sigma} \cdot \boldsymbol{n} = 0 \text{ on } \partial\Omega.\end{aligned} \tag{19}$$

Problem (19) is linear in $\boldsymbol{\tau}^j$ so that $\boldsymbol{\varepsilon}^j(\boldsymbol{x})$ can be written as

$$\boldsymbol{\varepsilon}^j(\boldsymbol{x}) = \boldsymbol{S}^j(\boldsymbol{x}) : \boldsymbol{\tau}^j \tag{20}$$

where the fourth-order tensor $\boldsymbol{S}^j$ does not depend on $\boldsymbol{\tau}^j$ and is entirely determined from Ω, Ω^j and $\boldsymbol{C}$. It follows that

$$\int_{\Omega^j} \boldsymbol{\varepsilon}^* \mathrm{d}\Omega = |\Omega^j| \left(\boldsymbol{C}^{-1} : \bar{\boldsymbol{\sigma}} + \sum_k \boldsymbol{S}^{jk} : \boldsymbol{\tau}^k \right) \tag{21}$$

where

$$\boldsymbol{S}^{jk} = \frac{1}{|\Omega^j|} \int_{\Omega^j} \boldsymbol{S}^k(\boldsymbol{x}) \mathrm{d}\Omega. \tag{22}$$

For $j \neq k$, the tensor $\boldsymbol{S}^{jk}$ in (22) captures the elastic interaction between grains j and k. The norm of that tensor is expected to decrease with the distance between grains j and k. Substituting (21) in (16) we finally arrive at

$$W = -\frac{1}{2}|\Omega|\bar{\sigma} : C^{-1} : \bar{\sigma} - \sum_j |\Omega^j|\bar{\sigma} : \tau^j - \sum_{j,k} \frac{1}{2}|\Omega^j|\tau^j : C : S^{jk} : \tau^k. \quad (23)$$

Dropping the constant term $|\Omega|\bar{\sigma} : C^{-1} : \bar{\sigma}/2$, minimization problem (11) thus reduces to

$$\begin{aligned}\inf_\gamma & -\sum_{j=1}^{M} |\Omega^j|\bar{\sigma} : \tau^j - \sum_{j,k=1}^{M} \frac{1}{2}|\Omega^j|\tau^j : C : S^{jk} : \tau^k + \frac{1}{2}\sum_{j=1}^{M} |\Omega^j|\tau^j : C : \tau^j \\ & + \frac{1}{2}\xi_X \int_\Omega \sum_{i=1}^{N} \gamma_i^2 d\Omega + \sum_{i=1}^{N} \int_\Omega \left(\tau_c + \xi\eta_i^0 + \frac{1}{2}\xi\ |\gamma_i - \gamma_i^0|\right) |\gamma_i - \gamma_i^0| d\Omega.\end{aligned} \quad (24)$$

4.2 Critical Grains

Consider an initial state ($t = 0$) in which $\gamma = 0$. In the general case of a nonproportional loading, several slip systems may be successively activated in each grain depending on the direction of $\bar{\sigma}(t)$. However, for a proportional loading as considered in (8), only one slip system is expected to be activated in each grain (at least for small loading level $f(t)$). That slip system $I(j)$ corresponds to the maximum resolved shear stress in each grain, i.e.

$$I(j) = \text{argmax}_{1 \le i \le N}\ |\sigma^0 : \tau_i^j|.$$

Assuming that only slip system $I(j)$ is activated in grain j and denoting by γ^j (resp. η^j) the corresponding plastic slip (resp. cumulated plastic slip), (24) becomes

$$\begin{aligned}\inf_\gamma & -\sum_j |\Omega^j|\bar{\sigma} : \tau^j - \sum_{j,k} \frac{1}{2}|\Omega^j|\tau^j : C : S^{jk} : \tau^k + \frac{1}{2}\sum_j |\Omega^j|\tau^j : C : \tau^j \\ & + \frac{1}{2}\xi_X \sum_{j=1}^{M} |\Omega^j|(\gamma^j)^2 + \sum_{j=1}^{M} |\Omega^j| \left(\tau_c^0 + \frac{1}{2}\xi\ |\gamma^j - \gamma^{j,0}|\right) |\gamma^j - \gamma^{j,0}|\end{aligned} \quad (25)$$

where $\gamma^{j,0}$ and $\eta^{j,0}$ are the plastic slip and the accumulated plastic slip in grain j at time t^0. The updated critical shear stress τ_c^0 in (25) is defined by $\tau_c^0 = \tau_c + \xi\eta^{j,0}$. All the grains remains elastic if the loading level $f(t)$ in (8) is such that $|f(t)\sigma^0 : \tau_{I(j)}^j| \le \tau_c$ for all j, i.e. the elastic limit is

$$\frac{\tau_c}{\sup_j |\sigma^0 : \tau_{I(j)}^j|}.$$

For loading levels slightly above that elastic limit, plastic flow is expected to be limited to few critical grains characterized by the highest value of the resolved shear stress $|\boldsymbol{\sigma}^0 : \boldsymbol{\tau}^j_{I(j)}|$. Without loss of generality, we can assume that those critical grains are grains $1, \ldots, m$. We assume that the critical grains are far away from one another, so that their elastic interaction is negligible. In that case, (24) reduces to

$$\inf_{\gamma} -\sum_{j=1}^{m} |\Omega^j| \bar{\boldsymbol{\sigma}} : \boldsymbol{\tau}^j - \sum_{j=1}^{m} \frac{1}{2} |\Omega^j| \boldsymbol{\tau}^j : \boldsymbol{C} : \boldsymbol{S}^{jj} : \boldsymbol{\tau}^j + \frac{1}{2} \sum_{j=1}^{m} |\Omega^j| \boldsymbol{\tau}^j : \boldsymbol{C} : \boldsymbol{\tau}^j$$
$$+ \frac{1}{2} \xi_X \sum_{j=1}^{n} |\Omega^j| (\gamma^j)^2 + \sum_{j=1}^{m} |\Omega^j| \left(\tau_c^0 + \frac{1}{2} \xi \, |\gamma^j - \gamma^{j,0}| \right) |\gamma^j - \gamma^{j,0}| \tag{26}$$

where we recall that $\boldsymbol{\tau}^j = \gamma^j \boldsymbol{\tau}^j_{I(j)}$. Setting

$$\alpha^j = \boldsymbol{\tau}^j_{I(j)} : \boldsymbol{C} : (\boldsymbol{I} - \boldsymbol{S}^{jj}) : \boldsymbol{\tau}^j_{I(j)}, \beta^j = \bar{\boldsymbol{\sigma}} : \boldsymbol{\tau}^j_{I(j)} + \xi \gamma^{j,0}, \tag{27}$$

and

$$\tilde{\alpha}^j = \alpha^j + \xi + \xi_X,$$

the local optimality condition in (26) reads as

$$\beta^j - \tilde{\alpha}^j \gamma^j \begin{cases} \in [-\tau_c^0, \tau_c^0] & \text{if } \gamma^j = \gamma^{j,0} \\ = \tau_c^0 & \text{if } \gamma^j > \gamma^{j,0} \\ = -\tau_c^0 & \text{if } \gamma^j < \gamma^{j,0} \end{cases}$$

which yields

$$\gamma^j = \begin{cases} \gamma^{j,0} & \text{if } |\beta^j - \tilde{\alpha}^j \gamma^{j,0}| \le \tau_c^0, \\ \dfrac{\beta^j - \tau_c^0}{\tilde{\alpha}^j} & \text{if } \beta^j - \tilde{\alpha}^j \gamma^{j,0} > \tau_c^0, \\ \dfrac{\beta^j + \tau_c^0}{\tilde{\alpha}^j} & \text{if } \beta^j - \tilde{\alpha}^j \gamma^{j,0} < \tau_c^0. \end{cases} \tag{28}$$

for $j = 1, \ldots, m$. The aforementioned assumptions thus allow one to obtain a closed form solution to the incremental problem of elastic-plastic evolutions in a polycrystal. We can observe in (28) that the plastic slips $\gamma^1, \ldots, \gamma^m$ are determined independently from one another, i.e. there is no coupling between the plastifying grains. This results from the fact that the elastic interaction between grains has been neglected. Apart from loading and material parameters, the plastic slip in (28) depends on the shape and the crystalline orientation of the grain through the scalars α^j and β^j in (27). In more detail, the tensor $\boldsymbol{S}^{jj}$ is determined from the shape of grain j whereas the crystalline orientation plays a role through the selection of the active slip system $I(j)$ and the corresponding strain $\boldsymbol{\tau}^j_{I(j)}$.

5 Illustration

5.1 Mesoscopic Model Construction

A 2D polycrystalline RVE of 500 grains is generated randomly using *Neper* -a generation and meshing software package based on the Voronoi-Laguerre tessellation. The obtained tessellation is shown in Fig. 1a. The implementation of the formula (28) requires the evaluation of the fourth-order tensor $\boldsymbol{S}^j$ for each grain. Note that problem (19) defining the tensor $\boldsymbol{S}^j$ is formally similar to Eshelby's inclusion problem. However, the irregularity of the grain geometries ruins any hope of solving (19) analytically. Rather than evaluating $\boldsymbol{S}^j$ numerically (which could be done using finite element calculations), we choose to approximate each grain into an ellipsoid, as shown in Fig. 1b. That strategy allows Eshelby's explicit solution [5, 14] to be used for $\boldsymbol{S}^j$. The approximation of each grain into an ellipsoid can be performed using the maximum volume inscribed ellipsoid optimization algorithm in *Matlab* [27] as well as a modeling system for convex optimizations (CVX *Matlab* package).

Regarding slip systems, the face-centered cubic lattive (FCC) is one of the most common crystalline structures in metals (Fig. 2a). In that case, there are 12 slip systems defined according to the Thompson tetrahedron. In the 2D setting considered in this paper, we use 6 slip systems defined according to an equilateral triangle shown in Fig. 2b. The inclination angle Θ_j shown in Fig. 2b is the Euler angle describing the crystallographic orientation of grain j. The vectors $(\boldsymbol{n}_i, \boldsymbol{t}_i)$ defining the 6 slip systems in the reference single crystal take the form

$$\begin{cases} \boldsymbol{n}_i = \begin{pmatrix} -\sin(\phi_i) \\ \cos(\phi_i) \end{pmatrix}, \text{ where } \phi_i \in \{0°, 60°, 120°\} \\ \\ \boldsymbol{t}_i = \begin{pmatrix} \cos(m_i\phi_i) \\ \sin(m_i\phi_i) \end{pmatrix}, \text{ where } m_i \in \{-1, 1\} \end{cases} \tag{29}$$

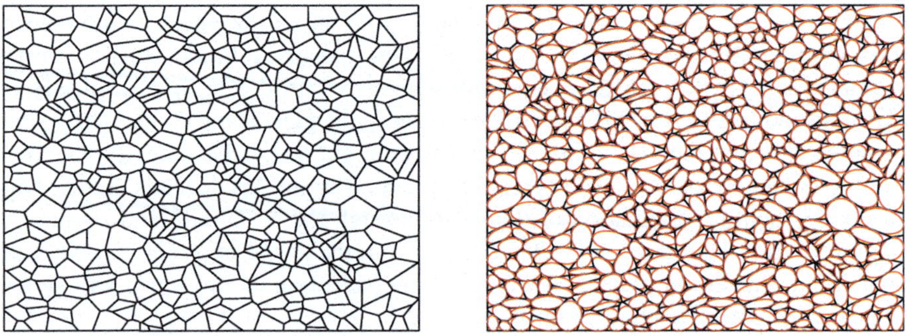

Fig. 1 **a** A 500-grain tessellation generated by *Neper* and **b** its approximated geometry in *Matlab*

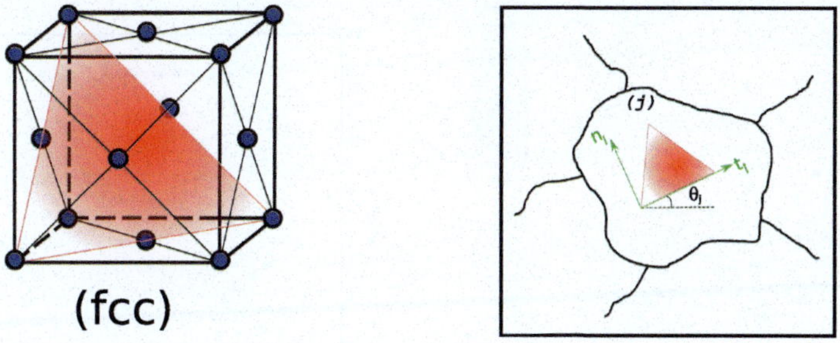

Fig. 2 **a** The FCC crystalline structure. **b** Slip system in grain j of a 2D polycrystal

At this point, all the ingredients are in place for using the simplified model presented in Sect. 4: In a first stage, the active slip system and the best fitting ellipsoid are calculated in each grain, which allows one to evaluate the parameters (α^j, β^j) in (27). Subsequently, the plastic slip in each grain is updated at each time step via formula (28).

5.2 *Finite Element Model Construction*

In order to validate the solution procedure presented in Sect. 4, Finite Element (FE) simulations have been performed in *Freefem++* [7], importing the tessellation shown in Fig. 1a into a 1×1 square domain that was finely meshed. That unit square was embedded into a coarsely meshed 2×2 square serving as an elastic matrix for reducing border effects (Fig. 4a). Material and loading parameters are reported in Table 1. The tensor $\boldsymbol{\sigma}^0$ is set to $(\boldsymbol{u}_1 \otimes \boldsymbol{u}_2 + \boldsymbol{u}_2 \otimes \boldsymbol{u}_1)/2$ (pure shear). The plastic slip at each Gauss point is calculating using Lemke's algorithm [3] for solving the linear complementarity problem arising from (11). In Fig. 3a are shown the plastic slips as calculated by the finite element method after the first half-cycle. An average plastic slip per grain has been extracted from the FE simulations and compared with the results obtained from the formula (28). The relative error between those two values has been calculated for each grain and is shown in Fig. 3b as a function of the average plastic slip computed in *Freefem++*. The average relative error is 4.46% (4.28% when restricted to the critical grains with the highest plastic slips as shown in Fig. 3b) with a maximum of approximately 13% (7% when restricted to critical grains). A finer

Table 1 Material and loading parameters

E (GPa)	ν	f^a (MPa)	f^m (MPa)	ζ/μ (MPa^{-1})	τ_c^0 (MPa)
210	0.3	101	0	1615.38	100

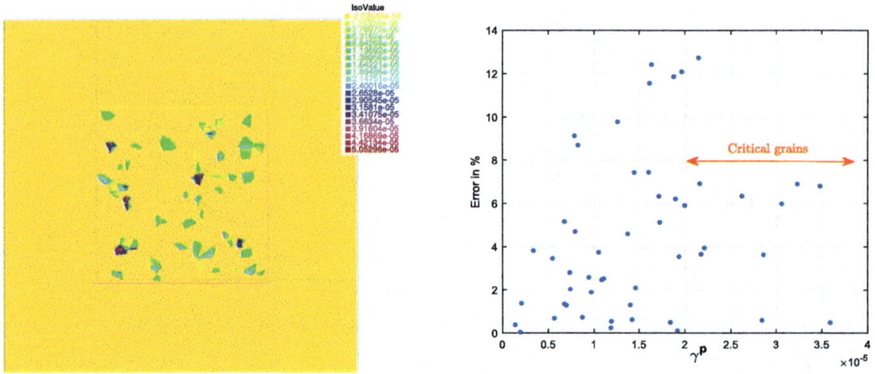

(a) FE results for the plastic slip field in the polycrystal

(b) Model error as a function of the plastic slip (FE value)

Fig. 3 Validation of the mesoscopic model

meshing density reduces the average error on critical grains to 3.79%. That error is partly due to the nonuniformity of the plastic slip within each grain, the elastic interaction between plastifying grains and the irregularity of the grain geometries, all of which are not taken into account in the simplified method.

6 Direct Approach for Cyclic Loadings

The simplified method presented in Sect. 4 allows one to estimate the step-by-step evolution of plastic slips for a given loading history. In this Section, we consider a cyclic loading and are interested in the asymptotic state reached by the polycrystal as the number of cycles grows to infinity. As detailed in the following, we show that asymptotic values of the plastic slips (and cumulated plastic slips) in each grain can be obtained directly, i.e. without carrying out step-by-step calculations. Since there is no coupling between grains in (28), we first note that critical grains can studied independently from one another. Hence, from now on we simplify the notations by dropping the superscript j in (28) and consider a single critical grain. The active slip system is denoted by $(\boldsymbol{n}, \boldsymbol{t})$. We have

$$\gamma = \begin{cases} \gamma^0 & \text{if } |\beta - \tilde{\alpha}\gamma^0| \leq \tau_c^0 \\ \dfrac{\beta - \tau_c^0}{\tilde{\alpha}} & \text{if } \beta - \tilde{\alpha}\gamma^0 > \tau_c^0 \\ \dfrac{\beta + \tau_c^0}{\tilde{\alpha}} & \text{if } \beta - \tilde{\alpha}\gamma^0 < \tau_c^0 \end{cases} \tag{30}$$

where

$$\alpha = \boldsymbol{\tau} : \boldsymbol{C} : (\boldsymbol{I} - \boldsymbol{S}) : \boldsymbol{\tau}, \quad \beta = \bar{\boldsymbol{\sigma}} : \boldsymbol{\tau} + \xi \gamma^0, \tag{31}$$

and

$$\boldsymbol{\tau} = \frac{1}{2}(\boldsymbol{t} \otimes \boldsymbol{n} + \boldsymbol{n} \otimes \boldsymbol{t}), \quad \tilde{\alpha} = \alpha + \xi + \xi_X.$$

We consider a cyclic loading of the form (8) where $f(t)$ alternates between a minimum value f_{min} and a maximum value f_{max}. To fix ideas, we assume that $f(0) = f(T) = f_{min}$ and $f(T/2) = f_{max}$ where T is the period of the loading. The function f is monotonically increasing (resp. decreasing) on the time interval $[0, T/2]$ (resp. $[T/2, T]$) and the plastic slip γ is expected to have a similar behavior. This motivates calculating the evolution of γ by solving the incremental problem (28) on each of the half-cycles $[nT, nT + T/2]$ and $[nT + T/2, (n+1)T]$. We denote by γ_n (resp. $\gamma_{n+\frac{1}{2}}$) the value of γ at time nT (resp $nT + T/2$). Similarly we denote by $\tau_{c,n}$ (resp. $\tau_{c,n+\frac{1}{2}}$) the value of the updated critical shear stress τ_c at time nT (resp $nT + T/2$). We obtain

$$\gamma_{n+\frac{1}{2}} = \begin{cases} \gamma_n & \text{if } |\beta_{n+\frac{1}{2}} - \tilde{\alpha}\gamma_n| \le \tau_{c,n} \\ \dfrac{\beta_{n+\frac{1}{2}} - \tau_{c,n}}{\tilde{\alpha}} & \text{if } \beta_{n+\frac{1}{2}} - \tilde{\alpha}\gamma_n > \tau_{c,n} \end{cases} \tag{32}$$

where $\beta_{n+\frac{1}{2}} = f_{max}\boldsymbol{\sigma}^0 : \boldsymbol{\tau} + \xi\gamma_n$, and

$$\gamma_{n+1} = \begin{cases} \gamma_{n+\frac{1}{2}} & \text{if } |\beta_{n+1} - \tilde{\alpha}\gamma_{n+\frac{1}{2}}| \le \tau_{c,n+\frac{1}{2}} \\ \dfrac{\beta_{n+1} + \tau_{c,n+\frac{1}{2}}}{\tilde{\alpha}} & \text{if } \beta_{n+1} - \tilde{\alpha}\gamma_{n+\frac{1}{2}} < \tau_{c,n+\frac{1}{2}} \end{cases} \tag{33}$$

where $\beta_{n+1} = f_{min}\boldsymbol{\sigma}^0 : \boldsymbol{\tau} + \xi\gamma_{n+\frac{1}{2}}$. Formulas (32) and (33) give a recurrence relation for calculating the sequence of plastic slips $\gamma_1, \gamma_2, \ldots$. Let us consider the case where shakedown occurs, i.e. the plastic slip converge towards a time-independent value γ_∞ as time tends to infinity. Taking the limit $n \to \infty$ in (32)-(33) and assuming plastic flow on each half-cycle, we obtain the relations

$$\begin{aligned} \gamma_\infty &= \frac{f_{max}\boldsymbol{\sigma}^0 : \boldsymbol{\tau} + \xi\gamma_\infty - \tau_{c,\infty}}{\tilde{\alpha}} \\ \gamma_\infty &= \frac{f_{min}\boldsymbol{\sigma}^0 : \boldsymbol{\tau} + \xi\gamma_\infty + \tau_{c,\infty}}{\tilde{\alpha}} \end{aligned} \tag{34}$$

where $\tau_{c,\infty} = \tau_c + \xi\eta_\infty$ and $\eta_\infty = \sum_{n=0}^{\infty}(|\gamma_{n+1} - \gamma_{n+\frac{1}{2}}| + |\gamma_{n+\frac{1}{2}} - \gamma_n|)$ is the asymptotic value of the cumulated plastic slip. It follows that

$$\gamma_\infty = \boldsymbol{\sigma}^0 : \boldsymbol{\tau}\frac{f_{max} + f_{min}}{2(\alpha + \xi_X)}, \quad \tau_{c,\infty} = \boldsymbol{\sigma}^0 : \boldsymbol{\tau}\frac{f_{max} - f_{min}}{2} \tag{35}$$

from which we obtain the value of η_∞ as

$$\eta_\infty = \frac{1}{\xi}\left(\frac{f_{max} - f_{min}}{2}\boldsymbol{\sigma}^0 : \boldsymbol{\tau} - \tau_c\right). \tag{36}$$

The obtained expressions for γ_∞ and η_∞ can be rewritten in a more compact fashion as

$$\gamma_\infty = \frac{\sigma_{nt}^m}{\alpha + \xi_X}, \quad \eta_\infty = \frac{\sigma_{nt}^a - \tau_c}{\xi} \tag{37}$$

where

$$\sigma_{nt}^m = \frac{f_{max} + f_{min}}{2}\boldsymbol{\sigma}^0 : \boldsymbol{\tau}, \sigma_{nt}^a = \frac{f_{max} - f_{min}}{2}\boldsymbol{\sigma}^0 : \boldsymbol{\tau} \tag{38}$$

are respectively the mean value and the amplitude of the resolved macroscopic stress $f(t)\boldsymbol{\sigma}^0 : \boldsymbol{\tau}$. Formula (37) shows that there is a simple relation between the loading parameters and the plastic state variables in the shakedown state. However, there are some restrictions on the loading parameters for that formula to apply. We have indeed

$$|\gamma_\infty| = |\sum_{n=0}^{\infty}(\gamma_{n+1} - \gamma_{n+\frac{1}{2}}) + (\gamma_{n+\frac{1}{2}} - \gamma_n)| \leq \sum_{n=0}^{\infty}|\gamma_{n+1} - \gamma_{n+\frac{1}{2}}| + |\gamma_{n+\frac{1}{2}} - \gamma_n| = \eta_\infty$$

which implies that (37) holds only if

$$\frac{|\sigma_{nt}^m|}{\alpha + \xi_X} \leq \frac{\sigma_{nt}^a - \tau_c}{\xi}. \tag{39}$$

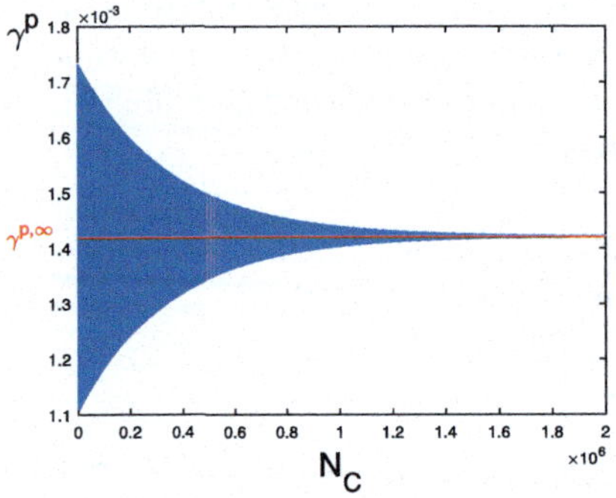

(a) Plastic slip as a function of the number of cycles

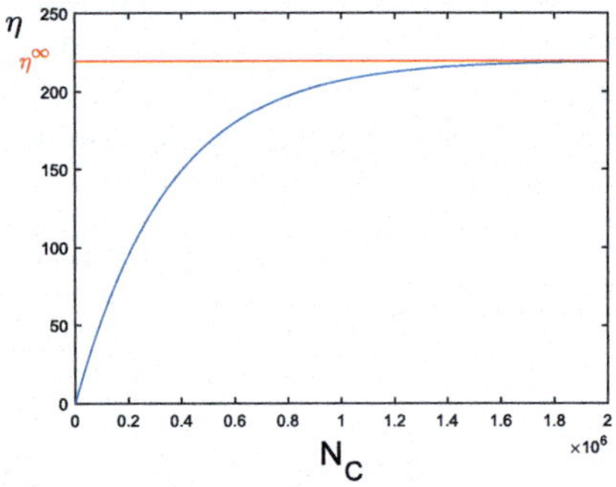

(b) Cumulated plastic slip as a function of the number of cycles

Fig. 4 Asymptotic state in a critical grain

Table 2 Material and loading parameters

E (GPa)	ν	f^a (MPa)	f^m (MPa)	ζ (MPa^{-1})	τ_c^0 (MPa)
210	0.3	101	50	5.10^{-2}	90

When the loading parameters $(\sigma_{nt}^m, \sigma_{nt}^a)$ do not satisfy that condition, shakedown is found to occur after a finite number of cycles. There is no direct formula for estimating the shakedown state in that case.

Expressions (37) for the asymptotic state are verified through a comparison with a step-by-step simulation of a 100-grain randomly generated tessellation submitted to a cyclic shear. In Fig. 4a is shown the plastic slip γ_{Nc} as a function of N_c, as calculated from (30). The cumulated slip η_{Nc} is shown in Fig. 4b. Those values correspond to the grain that plastifies first in the tessellation. The material and loading parameters used are reported in Table 2. The values γ_∞ and η_∞ given by (37) are shown as red lines in Fig. 4. As expected, (γ_{Nc}, η_{Nc}) converge towards γ_∞ and η_∞ as $N_c \to \infty$. For $N_c > 1.1\ 10^6$, the relative difference between γ_{Nc} and γ_∞ is smaller than 1%.

Formulas (37) gives a direct approach for estimating the asymptotic state in each grain of the polycrystal. This is valuable for fatigue analysis and probabilistic studies, as illustrated in the next Sections.

7 Mesoscopic Fatigue Criterion

Experimental observations show that fatigue cracks usually appear in persistent slip bands inside critical grains. Motivated by such observations, we consider that a fatigue crack may initiate in a given critical grain if $X \geq b$ where

$$X = \sup_t \sigma_{nt}(t) + a\sigma_{nn}(t) \tag{40}$$

and (a, b) are (positive) material parameters. In (40), $\sigma_{nt}(t)$ and $\sigma_{nn}(t)$ are defined as

$$\sigma_{nt}(t) = \boldsymbol{n} \cdot \boldsymbol{\sigma}(t) \cdot \boldsymbol{t}, \quad \sigma_{nn}(t) = \boldsymbol{n} \cdot \boldsymbol{\sigma}(t) \cdot \boldsymbol{n}$$

where $\boldsymbol{n}$ is the normal to the active slip plane, $\boldsymbol{t}$ is the direction of the slip and $\boldsymbol{\sigma}(t)$ is the local stress in the shakedown state. Correspondingly, fatigue life is infinite if

$$X \leq b. \tag{41}$$

A condition similar to (41) is considered in Dang Van's fatigue criterion [4] and is also related to damage indicators proposed by [13]. In the original construction of Dang Van's fatigue criterion, the purely elastic stress in critical grains is approximated using Lin-Taylor's approximation and the stabilized plastic strain is estimated using a heuristic ansatz. The approach presented allows one to improve on those assumptions. In more detail, when shakedown is reached in a critical grain, the local stress takes the form

$$\boldsymbol{\sigma}(t) = \bar{\boldsymbol{\sigma}}(t) - \gamma_\infty \boldsymbol{C} : (\boldsymbol{I} - \boldsymbol{S}) : \boldsymbol{\tau}.$$

Recalling that $\boldsymbol{C}$ is isotropic (with Lame coefficients (λ, μ)) and noting that $\boldsymbol{n} \cdot \boldsymbol{t} = 0$, we have

$$\sigma_{nt}(t) = f(t)\sigma^0_{nt} - \gamma_\infty \alpha, \quad \sigma_{nn}(t) = f(t)\sigma^0_{nn} - \gamma_\infty \alpha' \tag{42}$$

where α is defined as in (31) and

$$-\alpha' = \lambda \operatorname{tr}(\boldsymbol{S} : \boldsymbol{\tau}) + 2\mu \boldsymbol{n} \otimes \boldsymbol{n} : \boldsymbol{S} : \boldsymbol{\tau}. \tag{43}$$

In (42), we have set $\sigma^0_{nt} = \boldsymbol{n} \cdot \boldsymbol{\sigma}^0 \cdot \boldsymbol{t}$ and $\sigma^0_{nn} = \boldsymbol{n} \cdot \boldsymbol{\sigma}^0 \cdot \boldsymbol{n}$. Using expression (37) for γ_∞ yields

$$\sigma_{nt}(t) + a\sigma_{nn}(t) = f(t)(\sigma^0_{nt} + a\sigma^0_{nn}) - (\alpha + a\alpha')\frac{\sigma^m_{nt}}{\alpha + \xi_X}.$$

The value of $f(t)$ reaching the maximum of the expression above is either f_{max} or f_{min}, depending on the sign of $\sigma^0_{nt} + a\sigma^0_{nn}$. Setting

$$\sigma^m_{nn} = \frac{f_{max} + f_{min}}{2}\sigma^0_{nn}, \; \sigma^a_{nn} = \frac{f_{max} - f_{min}}{2}\sigma^0_{nn}, \tag{44}$$

we arrive at $X = \sup\{X_+, X_-\}$ where

$$X_\pm = \sigma^m_{nt} + a\sigma^m_{nn} \pm (\sigma^a_{nt} + a\sigma^a_{nn}) - (\alpha + a\alpha')\frac{\sigma^m_{nt}}{\alpha + \xi_X}. \tag{45}$$

In the case $\xi_X = 0$ (which we consider in the following), the expression above simplifies as

$$X_\pm = a\sigma^m_{nn} \pm (\sigma^a_{nt} + a\sigma^a_{nn}) - a\frac{\alpha'}{\alpha}\sigma^m_{nt}. \tag{46}$$

7.1 Influence of the Grain Shape

The quantity $X = \sup\{X_+, X_-\}$ drives the fatigue life of critical grains. For a given loading, X depends on the shape of critical grains through the ratio α'/α where α and α' are defined in (31) and (43) respectively. That ratio can be interpreted as a geometric amplification factor of the mean resolved shear stress σ^m_{nt}. Evaluating α'/α requires evaluating the tensor $\boldsymbol{S}$ introduced in (20), which can only be done numerically in general. A notable exception is the case of ellipsoidal grains, for which $\boldsymbol{S}$ coincides with the Eshelby tensor. Let $r \leq 1$ be the aspect ratio of the ellipsoidal

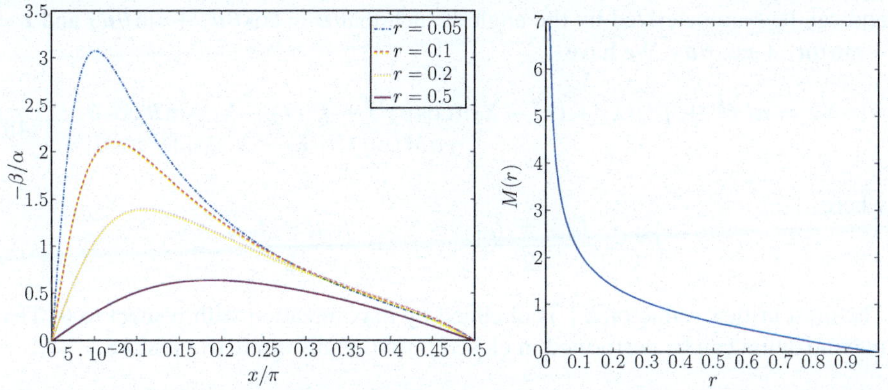

Fig. 5 Geometric amplification factor capturing the influence of the grain shape (orientation x, aspect ratio r) on the fatigue indicator

grain and let x be the angle between the long axis of the grain and the direction $\boldsymbol{n}$ of the slip plane. Some calculations based on the expression of the Eshelby tensor leads to

$$\frac{-\alpha'}{\alpha} = \frac{2(1-r)(1+r+(1-r)\cos 2x)\sin 2x}{(1+r)^2-(1-r)^2\cos 4x}. \tag{47}$$

Interestingly, the ratio $-\alpha'/\alpha$ is found to be independent of the elastic moduli. It is plotted in Fig. 5(left) as a function of the angle x, for several values of the aspect ratio r. In the limit case $r = 1$ (circular inclusion), α'/α is equal to 0. In the limit case $r = 0$ (flat grains), we obtain $-\alpha'/\alpha = \text{cotan}\, x$ which may take any value in $\mathbb{R}$ depending on the angle x. We have in particular $-\alpha'/\alpha = +\infty$ when $x = 0$, i.e. when the normal to slip plane is parallel to the (flat) grain. More generally, for a given aspect ratio r, it can easily be seen that $-\alpha'/\alpha$ is $\pi-$periodic and covers an interval of the form $[-M(r), M(r)]$ as x varies between 0 and π. The function $M(r)$ is plotted in Fig. 5(right) and can be interpreted as a maximum amplification factor of σ^m_{nt} in the fatigue life indicator X.

7.2 *Influence of the Crystalline Orientation*

For a given aspect ratio r, we consider the worst situation where the angle x gives the largest value of $|\alpha'/\alpha|$ in which case

$$X_\pm = a\sigma^m_{nn} \pm (\sigma^a_{nt} + a\sigma^a_{nn}) + aM(r)|\sigma^m_{nt}|.$$

Let $(\boldsymbol{u}_1, \boldsymbol{u}_2)$ be eigenvectors of $\boldsymbol{\sigma}^0$, so that $\boldsymbol{\sigma}^0 = \lambda_1 \boldsymbol{u}_1 \otimes \boldsymbol{u}_1 + \lambda_2 \boldsymbol{u}_2 \otimes \boldsymbol{u}_2$ with $\lambda_1 \leq \lambda_2$. The orientation θ of the active slip system depends on the crystalline orientation and can be parameterized by the angle θ such that $\boldsymbol{n} = \cos\theta \boldsymbol{u}_1 + \sin\theta \boldsymbol{u}_2$ and $\boldsymbol{t} = -\sin\theta \boldsymbol{u}_1 + \cos\theta \boldsymbol{u}_2$. We have

$$X_+ = a(f^m + f^a)(\lambda_2 + (\lambda_1 - \lambda_2)(\cos\theta)^2) + f^a(\lambda_2 - \lambda_1)\sin\theta\cos\theta \\ +aM(r)|f^m|(\lambda_2 - \lambda_1)|\sin\theta\cos\theta| \tag{48}$$

where

$$f^m = \frac{f_{max} + f_{min}}{2}, \quad f^a = \frac{f_{max} - f_{min}}{2}.$$

The most critical value of X_+ is obtained by maximization with respect to θ. That maximization can be performed in closed form (details omitted) yielding

$$\sup_\theta X_+ = a(f^m + f^a)\frac{\lambda_1 + \lambda_2}{2} + \frac{1}{2}(\lambda_2 - \lambda_1)\sqrt{(f^a + aM(r)|f^m|)^2 + a^2(f^m + f^a)^2}.$$

A similar calculation gives

$$\sup_\theta X_- = a(f^m - f^a)\frac{\lambda_1 + \lambda_2}{2} + \frac{1}{2}(\lambda_2 - \lambda_1)\sqrt{(f^a + aM(r)|f^m|)^2 + a^2(f^m - f^a)^2}.$$

Hence, if

$$a(f^m \pm f^a)\frac{\lambda_1 + \lambda_2}{2} + \frac{1}{2}(\lambda_2 - \lambda_1)\sqrt{(f^a + aM(r)|f^m|)^2 + a^2(f^m \pm f^a)^2} \leq b \tag{49}$$

then a critical grain of aspect ratio r satisfies the infinite lifetime condition, whatever its geometric or crystalline orientation. Now consider a polycrystal made of ellipsoidal grains with varying aspect ratios in a given interval $[r_0, r_1]$. Observing that the left-hand side of (49) is increasing with $M(r)$ and noting from Fig. 5 (right) that M is decreasing with r, we obtain that the condition

$$a(f^m \pm f^a)\frac{\lambda_1 + \lambda_2}{2} + \frac{1}{2}(\lambda_2 - \lambda_1)\sqrt{(f^a + aM(r_0)|f^m|)^2 + a^2(f^m \pm f^a)^2} \leq b \tag{50}$$

ensures that all critical grains satisfy the infinite lifetime condition, whatever their orientation (both geometric and crystalline) and whatever the exact value of their aspect ratios. Condition (50) can thus be interpreted as an infinite lifetime condition for the polycrystal. It is expected to be representative of polycrystals that contains many grains covering almost all possible orientations. In that case, there is indeed a high probability that there exists a critical grain close to the worst orientation, thus achieving the value $\max X$ in (50). For a RVE with few grains or a with special texture, condition (50) is expected to be conservative. More details on such issues are provided in Sect. 8.

In condition (50), the only parameters related to the shape of the grains is the scalar $M(r_0)$. The most favorable situation corresponds to circular inclusions ($r_0 = 1$), in which case $M(r_0) = 0$ and (50) reduces to

$$a(f^m \pm f^a)\frac{\lambda_1+\lambda_2}{2} + \frac{1}{2}(\lambda_2-\lambda_1)\sqrt{(f^a)^2 + a^2(f^m \pm f^a)^2} \leq b. \quad (51)$$

Equation (51) is expressed only in terms of macroscopic loading parameters. It can be interpreted as an upscaled version of the local expression (41) used in individual grains. It is interesting to compare (50) with 2D versions of widely used macroscopic fatigue criteria such as Crossland's, Sines' or Dang Van's criterion. For the loading considered, the infinite lifetime condition obtained from a 2D version of Dang Van's criterion takes the form

$$a(f^m \pm f^a)\frac{\lambda_1+\lambda_2}{2} + \frac{1}{2}(\lambda_2-\lambda_1)f^a \leq b. \quad (52)$$

There is a close resemblance between (52) and (50), especially in the case of circular grains. In particular, the term in $\frac{\lambda_1+\lambda_2}{2}$ is the same. Condition (51) can be observed to be more conservative than (52) because the multiplying factor of $\frac{\lambda_2-\lambda_1}{2}$ is larger. The infinite lifetime condition obtained from Crossland's and Sines' criteria in 2D read respectively as

$$a\left(f^m\frac{\lambda_1+\lambda_2}{2} + f^a\left|\frac{\lambda_1+\lambda_2}{2}\right|\right) + \frac{1}{2}(\lambda_2-\lambda_1)f^a \leq b \quad (53)$$

$$af^m\frac{\lambda_1+\lambda_2}{2} + \frac{1}{2}(\lambda_2-\lambda_1)f^a \leq b \quad (54)$$

which differ from (52) by the term in $(\lambda_1+\lambda_2)/2$ capturing the influence of the hydrostatic stress. In Dang Van's, Crossland's and Sines' criterion, the term in $(\lambda_2 - \lambda_1)/2$ (capturing the influence of the deviatoric stress) is the same.

8 Probabilistic Aspects

8.1 *Survival Function for a Critical Grain of Random Orientation*

The direct expressions (37) for the asymptotic state are well suited to a probabilistic analysis of fatigue. This Section elaborates on this point for the simplest case of a zero-mean loading (i.e. $f^m = 0$) and circular grains. In that case, we have from (48)

$$X_+ = -X_- = af^a(\lambda_2 + (\lambda_1-\lambda_2)(\cos\theta)^2) + f^a(\lambda_2-\lambda_1)\sin\theta\cos\theta \quad (55)$$

where we recall that θ is the orientation of the active slip system. The angle θ depends on the crystalline orientation Θ in the critical grain, as shown in Fig. 6 for the case of the 6 slip systems listed in (29). In that case, it can be calculated that

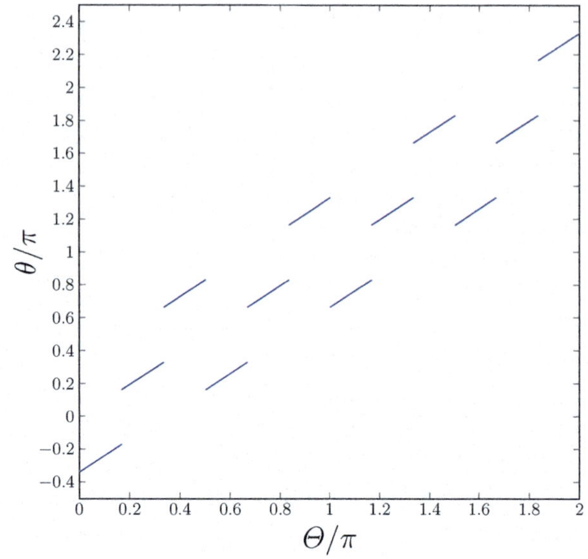

Fig. 6 Angle θ of the active slip system as a function of the crystalline orientation Θ

$\theta = \Theta - \pi/3$ if $0 \leq \Theta \leq \pi/6$ (modulo $\pi/2$), $\theta = \Theta$ if $\pi/6 \leq \Theta \leq \pi/3$ (modulo $\pi/2$), $\theta = \Theta + \pi/3$ if $\pi/3 \leq \Theta \leq \pi/2$ (modulo $\pi/2$). Setting $\psi = \arctan(1/a)$, some straightforward manipulations lead to

$$\frac{X_+}{f^a} = a\frac{\lambda_1 + \lambda_2}{2} + \sqrt{1+a^2}\frac{\lambda_1 - \lambda_2}{2} y \tag{56}$$

where

$$y = \cos(2\theta + \psi). \tag{57}$$

Provided that $\pi/6 \leq \psi \leq \pi/3$ (which corresponds to a in the range $[1/\sqrt{3}, \sqrt{3}]$), it can be verified that y takes values in $[-y_2, -y_1] \cup [y_1, y_2]$ where

$$y_1 = \cos\left(\psi - \frac{2\pi}{3}\right), \quad y_2 = \cos\left(\psi - \frac{\pi}{3}\right).$$

If now Θ is seen as a random variable with a prescribed probability density, then relations (56) and (57) allow one to calculate the probability densities of y and $X = \max(X_+, X_-)$. In the particular case where Θ is uniformly distributed on $[0, 2\pi]$, it can be calculated that the variable y has a probability density p given by

$$p(y) = \begin{cases} \dfrac{3}{2\pi}\dfrac{1}{\sqrt{1-y^2}} & \text{if } |y| \in [y_1, y_2], \\ 0 & \text{otherwise.} \end{cases} \tag{58}$$

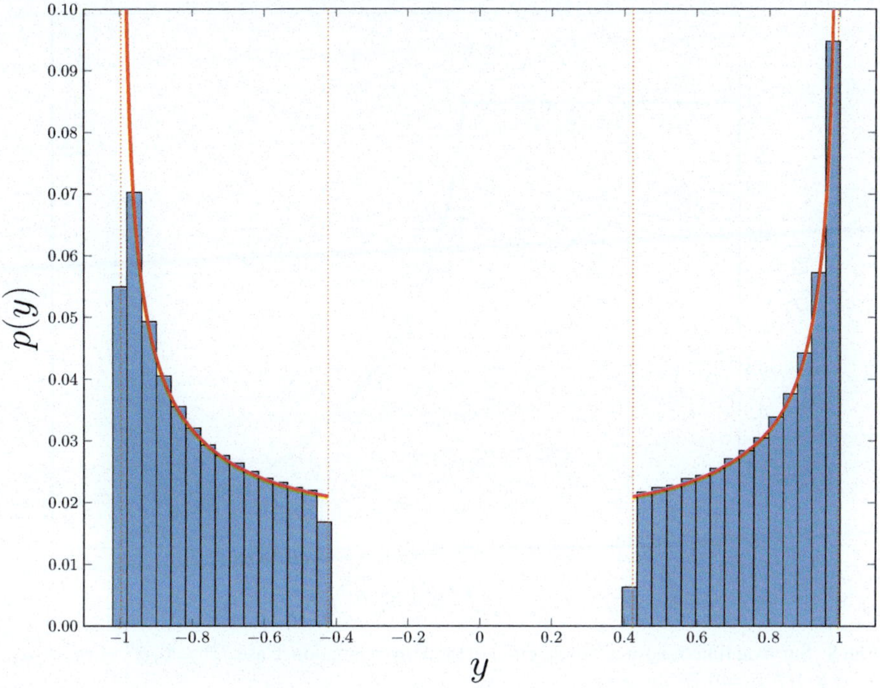

Fig. 7 Probability density p of $y = \cos 2\theta + \psi$ (red curve). The vertical dashed lines show the support of p. The histogram shows the probability density approximated from a sample of 1000000 randomly generated crystalline orientations

The probability density p in (58) is shown in Fig. 7 (red curve). To validate expression (58), the values of y corresponding to 1000000 randomly generated crystalline orientations have been calculated. The results, shown as a normalized histogram in Fig. 7, are in good agreement with (58).

The probability density of X can be obtained from (58) and relation (55). Of particular interest is the survival function for X, defined as the probability that $X \leq b$ when f^a is prescribed. That quantity is denoted by $S(f^a)$ in the following. It can be interpreted as the probability of infinite lifetime for a critical grain of random orientation when submitted to a loading amplitude f^a. Equation (56) shows that the condition $X \leq b$ is equivalent to

$$\frac{1}{B}\left(A - \frac{b}{f_a}\right) \leq y \leq \frac{1}{B}\left(A + \frac{b}{f_a}\right)$$

where

$$A = a\frac{\lambda_1 + \lambda_2}{2}, \quad B = \sqrt{1 + a^2}\frac{\lambda_2 - \lambda_1}{2}. \tag{59}$$

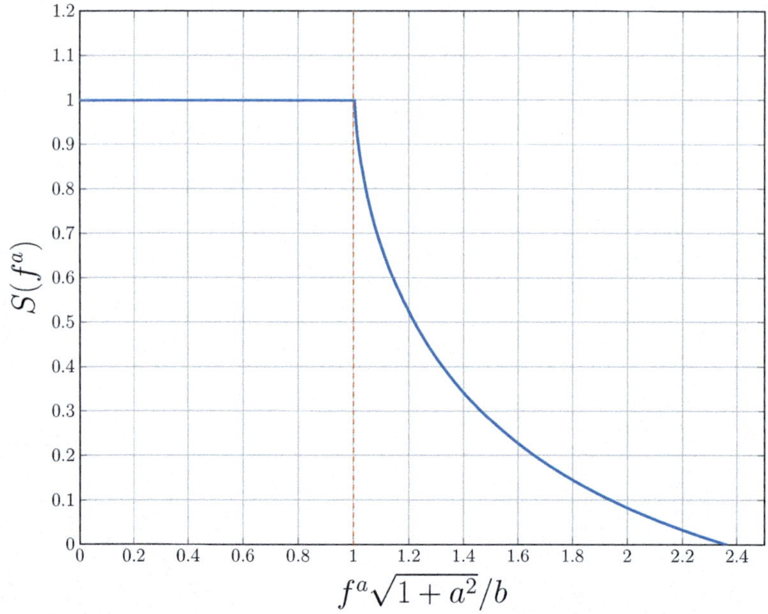

Fig. 8 Survival function for a critical grain of random orientation. Case $f^m = 0, \boldsymbol{\sigma}^0 = (\boldsymbol{u}_1 \otimes \boldsymbol{u}_2 + \boldsymbol{u}_2 \otimes \boldsymbol{u}_1)/2$, $a = 0.7$

It follows that

$$S(f^a) = \int_{\frac{A-b/f^a}{B}}^{\frac{A+b/f^a}{B}} p(y)dy. \tag{60}$$

Using expression (58) for p, there is no substantial difficulty in evaluating the integral in (60). The result depends on the loading mode $\boldsymbol{\sigma}^0$ in (8) through the parameters (A, B) in (59). Both those parameters depend indeed on the eigenvalues (λ_1, λ_2) of $\boldsymbol{\sigma}^0$. Let us consider the case of pure shear, i.e. $\boldsymbol{\sigma}^0 = (\boldsymbol{u}_1 \otimes \boldsymbol{u}_2 + \boldsymbol{u}_2 \otimes \boldsymbol{u}_1)/2$. In that case we have $A = 0$, $B = \sqrt{1+a^2}$ and we obtain

$$S(f^a) = \begin{cases} 0 & \text{if } f_a \geq \dfrac{b}{y_1\sqrt{1+a^2}} \\ \dfrac{1}{2} + \dfrac{3}{\pi}\left(\arcsin\left(\dfrac{b}{f^a\sqrt{1+a^2}}\right) - \psi\right) & \text{if } \dfrac{b}{y_2\sqrt{1+a^2}} \leq f^a \leq \dfrac{b}{y_1\sqrt{1+a^2}} \\ 1 & \text{if } f_a \leq \dfrac{b}{y_2\sqrt{1+a^2}} \end{cases} \tag{61}$$

The survival function given by (61) is plotted in Fig. 8 for the case $a = 0.7$. For any given k in $[0, 1]$, we denote by f_k^a the maximum loading amplitude f^a such that $S(f^a) = k$. It can be verified from (61) that the median survival loading $f_{50\%}^a$ (i.e. the loading f^a corresponding to a survival probability of 50%) is equal to b. The value

$f^a_{100\%}$ (corresponding to a guaranteed infinite lifetime) is equal to $b/y_2\sqrt{1+a^2}$, with $y_2 \simeq 0.996$ for the case $a = 0.7$ depicted in Fig. 8. It is interesting to compare that value with the maximum value f^a_{meso} corresponding to an infinite lifetime according to the mesoscopic fatigue criterion (50). For the loading under consideration, we obtain from (50) that $f^a_{meso} = b/\sqrt{1+a^2}$ which is slightly smaller that $f^a_{100\%}$. The reason is that Eq. (50) was derived by allowing the orientation θ of the active slip system to take any value in $[0, 2\pi]$, which is actually not the case as shown in Fig. 6. However, the relative difference between $f^a_{100\%}$ and f^a_{meso} is really small in the case $a = 0.7$ (this can be verified to remain true for any value of a in $[0, 1]$).

8.2 *Survival Function for a RVE with M Critical Grains*

Consider now a RVE containing M critical grains and submitted to a given macroscopic loading $(f^m, f^a, \boldsymbol{\sigma}^0)$. The RVE survives (i.e. has an infinite time) is all its critical grains do. If the crystalline orientations in the critical grains are statically independent variables, the survival function S_{meso} of the RVE is

$$S_{meso} = S^M \tag{62}$$

In (62) it is assumed implicitly that all the critical grains have the same survival function S, which is notably the case if $f^m = 0$ and the grains are circular. For a shear loading with zero-mean value ($\boldsymbol{\sigma}^0 = (\boldsymbol{u}_1 \otimes \boldsymbol{u}_2 + \boldsymbol{u}_2 \otimes \boldsymbol{u}_1)/2,\ f^m = 0$), it follows from (61) and (62) that the loading amplitude $f^a_{50\%}$ corresponding to a 50% probability of survical for the RVE is

$$\frac{b}{\sqrt{1+a^2}} \frac{1}{\sin(\psi + \frac{\pi}{6}(2^{1-\frac{1}{M}} - 1))}. \tag{63}$$

The value $f^a_{50\%}$ given by (63) is plotted in Fig. 9 as a function of the number M of critical grains. It can be verified that $f^a_{50\%}$ converges towards $b/y_2\sqrt{1+a^2} \simeq f^a_{meso}$ as $M \to \infty$. This means that the mesoscopic criterion (50) is a good fatigue indicator for RVE with many critical grains. In fact, 20 critical grains is enough for the relative difference between $f^a_{50\%}$ and f^a_{meso} to be smaller than 1%. By contrast, for a RVE with a small number (<10) of critical grains, $f^a_{50\%}$ may be significantly larger than f^a_{meso}. Such a situation may occur for structures with large stress gradients. In that case, the stress field varies rapidly in space, which imposes the RVE to be small, consequently reducing the potential number of critical grains. This in accordance with the experimental observation that fatigue limits generally increase in the presence of local stress gradients [11, 12].

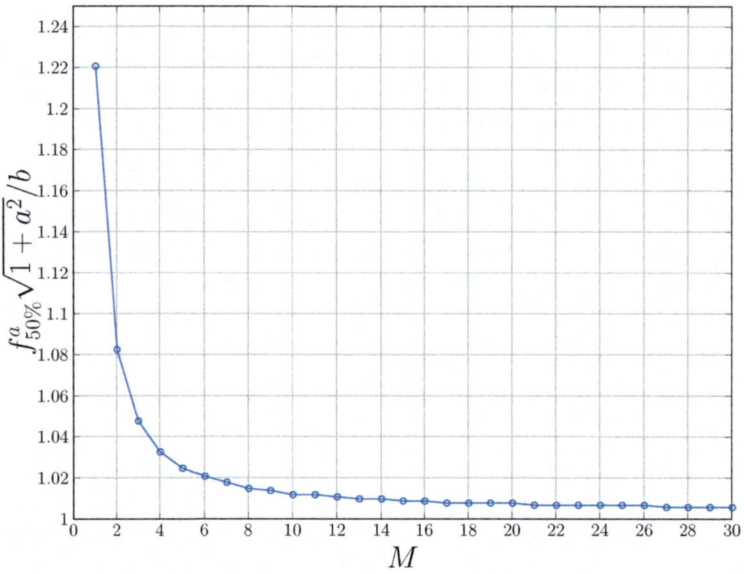

Fig. 9 Median loading amplitude $f^a_{50\%}$ for a RVE with M critical grains. Case $f^m = 0$, $\boldsymbol{\sigma}^0 = (\boldsymbol{u}_1 \otimes \boldsymbol{u}_2 + \boldsymbol{u}_2 \otimes \boldsymbol{u}_1)/2$, $a = 0.7$

9 Concluding Remarks

In this paper we have presented a simplified method for calculating the elastic-plastic evolutions of polycrystals submitted to loadings of relatively low amplitude. The comparison with full-field FE simulations shows that the method presented is relatively accurate, at least in situations where plastifying grains are relatively far away from each other. The main advantage of the proposed approach lies in its numerically efficiency as it relies on analytical formula. In the case of cyclic loadings, the proposed approach leads to a direct method for estimating the asymptotic state. The proposed approach could be extended in several ways. For instance, more advanced hardening laws could be considered. The accuracy of the method could also be improved by taking the elastic interaction between critical grains into account, which is the topic of ongoing work. An other important issue is the extension to the 3D setting. In that regard, we note that the simplified model of Sect. 4 and the analytical formula of Sect. 6 for the shakedown state can be directly extended to 3D. However, the construction of the mesoscopic fatigue criterion in Sect. 7 as well as the probabilistic analysis in Sect. 8 should be revisited and involve more complex calculations.

References

1. Bertolino, G., Constantinescu, A., Ferjani, M., Treiber, P.: A multiscale approach of fatigue and shakedown for notched structures. Theor. Appl. Fracture Mech. **48**(2), 140–151 (2007)
2. Bluthé, J., Weisz-Patrault, D., Ehrlacher, A.: Energetic approach for a sliding inclusion accounting for plastic dissipation at the interface, application to phase nucleation. Int. J. Solids Struct. **121**, 163–173 (2017), https://doi.org/10.1016/j.ijsolstr.2017.05.023
3. Cottle, R.W., Pang, J.S., Stone, R.E.: The Linear Complementarity Problem. SIAM (2009)
4. Dang Van, K.: Introduction to fatigue analysis in mechanical design by the multiscale approach. In: High-Cycle Metal Fatigue, pp. 57–88. Springer (1999)
5. Eshelby, J.D.: The determination of the elastic field of an ellipsoidal inclusion, and related problems. In: Proceedings of the Royal Society of London A: Mathematical, Physical and Engineering Sciences, vol. 241, pp. 376–396. The Royal Society (1957)
6. Halphen, B., Nguyen, Q.S.: Sur les matériaux standard généralisés. J. de mécanique **14**(1), 39–63 (1975)
7. Hecht, F.: New development in freefem++. J. Numer. Math. **20**(3–4), 251–265 (2012)
8. Klarbring, A., Barber, J., Spagnoli, A., Terzano, M.: Shakedown of discrete systems involving plasticity and friction. Eur. J. Mech. A **64**, 160–164 (2017)
9. Miehe, C.: Strain-driven homogenization of inelastic microstructures and composites based on an incremental variational formulation. Int. J. Numer. Methods Eng. **55**(11), 1285–1322 (2002)
10. Miehe, C., Lambrecht, M., Gürses, E.: Analysis of material instabilities in inelastic solids by incremental energy minimization and relaxation methods: evolving deformation microstructures in finite plasticity. J. Mech. Phys. Solids **52**(12), 2725–2769 (2004)
11. Morel, F., Palin-Luc, T.: A non-local theory applied to high cycle multiaxial fatigue. Fatigue & Fract. Eng. Mater. Struct. **25**(7), 649–665 (2002)
12. Morel, F., Palin-Luc, T., Froustey, C.: Comparative study and link between mesoscopic and energetic approaches in high cycle multiaxial fatigue. Int. J. Fatigue **23**(4), 317–327 (2001)
13. Mróz, Z., Seweryn, A.: Damage description with related crack initiation and propagation conditions. J. de Physique IV Proc. **06**(C6), 529–538 (1996)
14. Mura, T.: Micromechanics of Defects in Solids. Springer Science & Business Media (1987)
15. Nguyen, P.H., Le, C.V., Ho, P.L.: Numerical evaluation of macroscopic fatigue criterion of anisotropic materials using computational homogenisation and conic programming. Eur. J. Mech.-A/Solids 104654 (2022)
16. Peigney, M., Seguin, J.: An incremental variational approach to coupled thermo-mechanical problems in an elastic solids. Application to shape-memory alloys. Int. J. Solids Struct. **50**(24), 4043–4054 (2013)
17. Peigney, M.: Shakedown theorems and asymptotic behaviour of solids in non-smooth mechanics. Eur. J. Mech. A **29**, 784–793 (2010)
18. Peigney, M.: On shakedown of shape memory alloys structures. Ann. Solid Struct. Mech. **6**, 17–28 (2014)
19. Peigney, M.: Shakedown of elastic-perfectly plastic materials with temperature-dependent elastic moduli. J. Mech. Phys. Solids **71**, 112–131 (2014)
20. Peigney, M.: Cyclic steady states in diffusion-induced plasticity with applications to lithium-ion batteries. J. Mech. Phys. Solids **111**, 530–556 (2018)
21. Peigney, M.: Static and kinematic shakedown theorems in diffusion-induced plasticity. J. Theor. Appl. Mech. **58** (2020)
22. Peigney, M., Scalet, G., Auricchio, F.: A time integration algorithm for a 3d constitutive model for SMAS including permanent inelasticity and degradation effects. Int. J. Numer. Methods Eng. **115**(9), 1053–1082 (2018)
23. Peigney, M., Seguin, J., Hervé-Luanco, E.: Numerical simulation of shape memory alloys structures using interior-point methods. Int. J. Solids Struct. **48**(20), 2791–2799 (2011)
24. Pham, D.C.: Consistent limited kinematic hardening plasticity theory and path-independent shakedown theorems. Int. J. Mech. Sci. **130**, 11–18 (2017)

25. Sakout, S., Weisz-Patrault, D., Ehrlacher, A.: Energetic upscaling strategy for grain growth. I: Fast mesoscopic model based on dissipation. Acta Materialia **196**, 261–279 (2020)
26. Scalet, G., Peigney, M.: A robust and efficient radial return algorithm based on incremental energy minimization for the 3D Souza-Auricchio model for shape memory alloys. Eur. J. Mech. A/Solids **61**, 364–382 (2017)
27. Vandenberghe, L., Mutapcic, A.: Maximum volume inscribed ellipsoid in a polyhedron. Boyd and Vandenberghe "Convex Optimization" (2006)

Masonry Domes Under Complex Loading Conditions: A Shell-Based Static Limit Analysis Approach

Nicola A. Nodargi and Paolo Bisegna

Abstract A shell-based static limit analysis approach is proposed for the structural assessment of masonry domes subject to complex loading conditions, such as involving horizontal forces or differential settlements of the supports. The problem formulation resorts to the statics of shells and to the shell stress resultants on the dome mid-surface to characterize the equilibrated and statically admissible stress states in the dome. By exploiting differential or integral shell equilibrium conditions, alongside classical Heyman's assumptions, collapse and minimum-thrust analysis problems are formulated. Finite difference or finite volume discretization strategies are discussed to arrive at their discrete counterparts, which are efficiently solved as second-order cone programming problems. Numerical simulations are presented, dealing with the pseudo-static seismic analysis of spherical and catenary domes subject to uniformly or linearly distributed horizontal accelerations along the height of the dome, and with the minimum-thrust analysis of spherical domes under non-standard distributions of support differential settlements. The obtained results prove the computational merit of the proposed framework.

Keywords Historical monuments · Limit analysis · Vulnerability assessment

1 Introduction

The seismic assessment of historical masonry structures is a part of the societal challenge to preserve the architectural heritage. Attention is here focused on masonry domes, resorting to static limit analysis as structural analysis framework and to Heyman's assumptions of infinite compressive strength, vanishing tensile strength, and no-sliding condition for the constitutive description of masonry [22].

N. A. Nodargi (✉) · P. Bisegna
Department of Civil Engineering and Computer Science, University of Rome Tor Vergata, Rome, Italy
e-mail: nodargi@ing.uniroma2.it

G. Garcea and D. Weichert (eds.), *Direct Methods for Limit State of Materials and Structures*, Lecture Notes in Applied and Computational Mechanics 101,
https://doi.org/10.1007/978-3-031-29122-7_10

A vast literature, tracing back to several historical contributions (e.g., see [15, 24, 26, 33, 45, 46]) exists on the structural behavior of masonry domes under their self-weight. From a computational standpoint, in the last decades two main approaches have emerged. On the one hand, lunar-slices formulations aim at improving the classical sliced equilibrium model, which regards the dome as a collection of independent lunar slices and thus reduces the problem to the thrust line analysis of an arch with variable width [21], by the introduction of suitable distributions of compressive hoop stresses (e.g., see [1, 25, 44]). On the other hand, membrane formulations search for a compressed thrust membrane contained within the thickness of the dome in equilibrium with its self-weight. A continuous description of the unknown thrust membrane and of the relevant membrane forces is adopted in the thrust surface analysis method [2–4, 16], whereas a discrete description of the unknown thrust membrane as a network of compressed truss elements is the rationale for the thrust network analysis method (e.g., see [6–9, 17, 18, 30, 43]).

By contrast, the problem of masonry domes subjected to complex loading conditions, e.g. related to the presence of horizontal forces (as required in pseudo-static seismic analyses) or differential settlements of the supports, has received less attention [13].

In [49], experimental results on the collapse capacity of block masonry domes subject to horizontal forces proportional to their self-weight have been derived by testing small-scale models on a tilting table. A simple lunar-slice formulation considering the equilibrium of the two opposite lunar slices of the dome in the tilting direction has been therein proposed for an interpretation of the experimental evidences. Concerning membrane formulations, proportional horizontal forces have been introduced in the thrust surface analysis method in [12], by considering a suitably rotated configuration of the dome in which the verticality of external loads is recovered. In [29], an extension of the thrust network analysis method has been proposed for computing the structural capacity under horizontal forces. That requires determining those horizontal forces which imply the "deepest" and "shallowest" configurations of the thrust network to coincide.

In the context of kinematic limit analysis, an adaptive approach has been proposed in [20]. It is based on the assumption that the failure of a dome is driven by the opening of few curved flexural hinges, which produce a mechanism of rigid bodies. The actual collapse mechanism is determined through the solution of a nonlinear optimization problem, which searches for the optimal position of the curved flexural hinges within a NURBS parameterization of the dome. Furthermore, an application of block-based methods to masonry domes under horizontal forces has been carried out in [10], based on a point contact model simplifying the failure conditions to be imposed at block interfaces.

Recently, a novel shell-based approach to the static limit analysis of masonry domes has been proposed. It exploits the classical statics of shells and the shell stress resultants on the dome mid-surface to characterize the set of equilibrated and statically admissible stress states in the dome. Thanks to the non-customary choice to explicitly include the bending moments in the formulation, the resulting framework has been proven to be more simple and more general compared to competing approaches when considering axisymmetric masonry domes under their self-weight [37, 38, 40]. In particular, it allows recovering lunar-slices formulations, thrust surface analysis method, and thrust network analysis method as special cases. More importantly, in conjunction with suitable discretization of the shell stress resultants on the dome mid-surface, such a framework has been demonstrated to enable the treatment of horizontal forces, thus leading to the pseudo-static seismic assessment of masonry domes [36, 39].

In the present work, an extension of such a shell-based static limit analysis approach for masonry domes is proposed for addressing general, complex loading conditions, e.g. due to the presence of horizontal forces or to differential settlements of the supports. Normal-force and bending-moment tensors, and shear-force vector on the dome mid-surface are introduced as the problem unknowns. Exploiting differential or integral shell equilibrium conditions, alongside classical Heyman's assumptions, collapse and minimum-thrust analysis problems are formulated. Two alternative discretization strategies, respectively based on finite difference or finite volume methods, are adopted for deriving the discrete counterpart of those optimization problems, to be solved in the form of second-order cone programming problems by standard and effective optimization tools. Numerical simulations are addressed for assessing the computational performances of the methodology. The pseudo-static seismic analysis of spherical and catenary domes under uniformly or linearly distributed horizontal accelerations along the height of the dome is presented, with parametric analyses enlightening on the role played by the dome geometry. Furthermore, the minimum-thrust analysis of spherical domes under non-standard distributions of support differential settlements is discussed. The obtained results, which provide an estimate of the structural safety of masonry domes on the safe side because of the underlying static limit analysis approach, highlight the potential of the proposed approach.

The paper is organized as follows. Section 2 deals with problem formulation, whereas the problem discretization is addressed in Sect. 3. Section 4 is devoted to numerical applications and conclusions are outlined in Sect. 5.

2 Shell-based Static Limit Analysis

A masonry dome of revolution is considered. It is described by its mid-surface and its thickness h measured along the normal direction to the mid-surface (Fig. 1a). In a Cartesian reference frame $(O; x, y, z)$ with coordinate axes respectively parallel to the unit vectors $\boldsymbol{i}$, $\boldsymbol{j}$ and $\boldsymbol{k}$, the typical point on the dome mid-surface is parameterized as:

$$\boldsymbol{x}(t, \vartheta) = r(t)\,\boldsymbol{e}_r(\vartheta) + z(t)\,\boldsymbol{k}. \tag{1}$$

Here, t and ϑ denote a meridional parameter and the longitude angle, and respectively span meridian and parallel curves on the dome mid-surface. Moreover, r and z respectively denote the distance from the revolution axis along the radial direc-

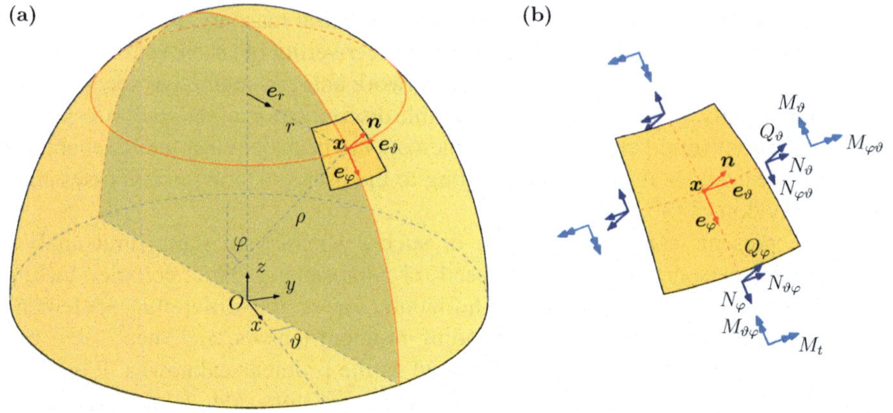

Fig. 1 Formulation: **a** three-dimensional view of the mid-surface of an axially symmetric masonry dome, and **b** shell stress resultants acting on the mid-surface of the dome

tion $\boldsymbol{e}_r = \cos\vartheta\, \boldsymbol{i} + \sin\vartheta\, \boldsymbol{j}$, and the elevation, of the point $\boldsymbol{x}$. The meridian curves are characterized by a tangential angle φ and by a radius of curvature ρ given by:

$$\tan\varphi = -\frac{z_{/t}}{r_{/t}}, \quad \rho = -\frac{(r_{/t}^2 + z_{/t}^2)^{3/2}}{z_{/tt} r_{/t} - z_{/t} r_{/tt}}, \tag{2}$$

the slash symbol standing for differentiation with respect to the indicated variable. Henceforth, the tangential angle φ will be used as meridional parameter in place of t, exploiting relationship $(2)_1$. The following physical basis vectors are introduced on the dome mid-surface:

$$\boldsymbol{e}_\varphi = \cos\varphi\, \boldsymbol{e}_r - \sin\varphi\, \boldsymbol{k}, \quad \boldsymbol{e}_\vartheta = -\sin\vartheta\, \boldsymbol{i} + \cos\vartheta\, \boldsymbol{j}, \quad \boldsymbol{n} = \sin\varphi\, \boldsymbol{e}_r + \cos\varphi\, \boldsymbol{k}, \tag{3}$$

such that $\boldsymbol{e}_\varphi$ and $\boldsymbol{e}_\vartheta$ are unit vectors respectively tangent to meridian and parallel curves, whereas $\boldsymbol{n}$ is the exterior normal unit vector to the mid-surface. For future use, the metric tensor ${}^s\boldsymbol{I}$, the Weingarten tensor ${}^s\boldsymbol{L}$ and the alternating tensor ${}^s\boldsymbol{W}$ on the dome mid-surface are respectively introduced:

$$\begin{aligned} {}^s\boldsymbol{I} &= {}^s\nabla\boldsymbol{x} = \boldsymbol{e}_\varphi \otimes \boldsymbol{e}_\varphi + \boldsymbol{e}_\vartheta \otimes \boldsymbol{e}_\vartheta, \\ {}^s\boldsymbol{L} &= -{}^s\nabla\boldsymbol{n} = -\rho^{-1}\boldsymbol{e}_\varphi \otimes \boldsymbol{e}_\varphi - r^{-1}\sin\varphi\, \boldsymbol{e}_\vartheta \otimes \boldsymbol{e}_\vartheta, \\ {}^s\boldsymbol{W} &= -\boldsymbol{e}_\varphi \otimes \boldsymbol{e}_\vartheta + \boldsymbol{e}_\vartheta \otimes \boldsymbol{e}_\varphi, \end{aligned} \tag{4}$$

with $\otimes$ denoting tensor product and ${}^s\nabla$ as surface gradient operator (e.g., see [48]).

A shell-based approach is resorted to for formulating the static limit analysis problem for the dome under external loads. That requires:

i. reducing external loads on the dome to statically equivalent surface distributions of forces $\boldsymbol{q}$ and couples $\boldsymbol{m}$ applied to its mid-surface;

ii. describing the stress state in the dome by the normal-force tensor $\boldsymbol{N}$, shear-force vector $\boldsymbol{Q}$, and bending-moment tensor $\boldsymbol{M}$ introduced as tangent fields on its mid-surface as customary in classical statics of shells (e.g., see [19, 23]).

In particular, surface distributions of forces and couples corresponding to the dome self-weight are given by [34–36, 39]:

$$\boldsymbol{q}^0 = -q\boldsymbol{k}, \quad \boldsymbol{m}^0 = m \sin\varphi\, \boldsymbol{e}_\varphi,$$
$$q = \left(1 + \frac{h^2}{12\rho}\frac{\sin\varphi}{r}\right)\gamma h, \quad m = \left(\frac{1}{\rho} + \frac{\sin\varphi}{r}\right)\frac{\gamma h^3}{12}, \tag{5}$$

where γ stands for the specific weight of the constituting masonry material. On the other hand, the following representation within the physical basis of vectors $(\boldsymbol{e}_\varphi, \boldsymbol{e}_\vartheta, \boldsymbol{n})$ is assumed for the tangent fields $\boldsymbol{N}$, $\boldsymbol{Q}$, and $\boldsymbol{M}$ on the dome mid-surface (Fig. 1(b)):

$$\begin{aligned}
\boldsymbol{N} &= N_\varphi\, \boldsymbol{e}_\varphi \otimes \boldsymbol{e}_\varphi + N_{\vartheta\varphi}\, \boldsymbol{e}_\vartheta \otimes \boldsymbol{e}_\varphi + N_{\varphi\vartheta}\, \boldsymbol{e}_\varphi \otimes \boldsymbol{e}_\vartheta + N_\vartheta\, \boldsymbol{e}_\vartheta \otimes \boldsymbol{e}_\vartheta, \\
\boldsymbol{Q} &= Q_\varphi\, \boldsymbol{e}_\varphi + Q_\vartheta\, \boldsymbol{e}_\vartheta, \\
\boldsymbol{M} &= M_\varphi\, \boldsymbol{e}_\varphi \otimes \boldsymbol{e}_\varphi + M_{\varphi\vartheta}\left(\boldsymbol{e}_\varphi \otimes \boldsymbol{e}_\vartheta + \boldsymbol{e}_\vartheta \otimes \boldsymbol{e}_\varphi\right) + M_\vartheta\, \boldsymbol{e}_\vartheta \otimes \boldsymbol{e}_\vartheta.
\end{aligned} \tag{6}$$

It is noticed that non-symmetric normal-force and symmetric bending-moment tensors are considered. The latter assumption, which is not strictly required in classical statics of shells, may be motivated by a consistent derivation of the shell stress resultants from a 3D stress state via a thickness integration involving the Cauchy stress tensor (e.g., see [32, 36]).

Within such a framework, the equilibrium and static admissibility conditions to be enforced on the stress state in the dome are discussed.

2.1 Equilibrium

The differential shell equilibrium equations of the dome mid-surface read (e.g., see [32]):

$$\begin{aligned}
\boldsymbol{0} &= {}^s\boldsymbol{I}\,{}^s\mathrm{div}\,\boldsymbol{N} - {}^s\boldsymbol{L}\,\boldsymbol{Q} + \boldsymbol{p}, \\
0 &= {}^s\mathrm{div}\,\boldsymbol{Q} + {}^s\boldsymbol{L}\cdot\boldsymbol{N} + q, \\
\boldsymbol{0} &= {}^s\boldsymbol{I}\,{}^s\mathrm{div}\,\boldsymbol{M} - \boldsymbol{Q} + \boldsymbol{m}, \\
0 &= {}^s\boldsymbol{W}\cdot\left(\boldsymbol{N} - \boldsymbol{M}\,{}^s\boldsymbol{L}\right).
\end{aligned} \tag{7}$$

where $\cdot$ and ${}^s\mathrm{div}$ respectively denote scalar product and surface divergence operator (e.g., see [48]). Equations $(7)_{1,2}$ imply translational equilibrium within the tangent plane and along the normal direction to the dome mid-surface, provided the decomposition $\boldsymbol{q} = \boldsymbol{p} + q\boldsymbol{n}$ of the external surface forces $\boldsymbol{q}$ is introduced. Analogously, equations $(7)_{3,4}$ imply rotational equilibrium about the tangent plane and about the normal direction, under the assumption that the external surface couples $\boldsymbol{m}$ obey the condition $\boldsymbol{m} \cdot \boldsymbol{n} = 0$.

Exploiting relationships (4) and (6), Eqs. (7) can be expanded component-wise. When multiplied by $r\rho$, they yield (e.g., see [19, 23], with a slight change of notation due to switching the indices of the off-diagonal components of the stress tensors and with the assumption $M_{\vartheta\varphi} = M_{\varphi\vartheta}$):

$$
\begin{aligned}
0 &= \left(rN_\varphi\right)_{/\varphi} + \rho N_{\varphi\vartheta/\vartheta} - \rho\cos\varphi N_\vartheta + rQ_\varphi + r\rho q_\varphi,\\
0 &= \left(rN_{\vartheta\varphi}\right)_{/\varphi} + \rho N_{\vartheta/\vartheta} + \rho\cos\varphi N_{\varphi\vartheta} + \rho\sin\varphi\, Q_\vartheta + r\rho q_\vartheta,\\
0 &= \left(rQ_\varphi\right)_{/\varphi} + \rho Q_{\vartheta/\vartheta} - rN_\varphi - \rho\sin\varphi N_\vartheta + r\rho q_n,\\
0 &= \left(rM_\varphi\right)_{/\varphi} + \rho M_{\varphi\vartheta/\vartheta} - \rho\cos\varphi M_\vartheta - r\rho Q_\varphi + r\rho m_\varphi,\\
0 &= \left(rM_{\varphi\vartheta}\right)_{/\varphi} + \rho M_{\vartheta/\vartheta} + \rho\cos\varphi M_{\varphi\vartheta} - r\rho Q_\vartheta + r\rho m_\vartheta,\\
0 &= r\rho\left(N_{\vartheta\varphi} - N_{\varphi\vartheta}\right) + \left(r - \rho\sin\varphi\right)M_{\varphi\vartheta}.
\end{aligned}
\tag{8}
$$

The differential equilibrium equations (7) or (8) need to be complemented by the boundary conditions on the free part of the mid-surface boundary. If that is e.g. assumed to be unloaded, the boundary conditions result to be:

$$
\boldsymbol{N}\boldsymbol{\nu} = \boldsymbol{0}, \quad \boldsymbol{Q}\cdot\boldsymbol{\nu} = 0, \quad {}^s\boldsymbol{W}\boldsymbol{M}\boldsymbol{\nu} = \boldsymbol{0}, \tag{9}
$$

where, for $\boldsymbol{\tau}$ the unit tangent vector to the boundary and $\times$ denoting cross product, $\boldsymbol{\nu} = \boldsymbol{\tau} \times \boldsymbol{n}$. For the sake of simplicity, it is assumed that the supporting structures of the dome are sufficiently resistant to withstand the transmitted actions. Accordingly, no boundary conditions need to be enforced on the supported boundary of the dome and equations analogous to (9) allow the post-computation of the support constraint reactions.

By integrating the differential shell equilibrium equations (7) on a typical part $\mathscr{P}$ of the dome mid-surface, and invoking the surface divergence theorem (e.g., see [48]), an integral shell equilibrium characterization is obtained:

$$\mathbf{0} = \int_{\partial \mathscr{P}} (\boldsymbol{N} + \boldsymbol{n} \otimes \boldsymbol{Q})\, \boldsymbol{\nu}\, \mathrm{d}l + \int_{\mathscr{P}} \boldsymbol{q}\, \mathrm{d}a,$$

$$\mathbf{0} = \int_{\partial \mathscr{P}} \left[(\boldsymbol{x} - O) \times (\boldsymbol{N} + \boldsymbol{n} \otimes \boldsymbol{Q})\, \boldsymbol{\nu} + {}^{s}\boldsymbol{W}\boldsymbol{M}\boldsymbol{\nu}\right] \mathrm{d}l + \int_{\mathscr{P}} \left[(\boldsymbol{x} - O) \times \boldsymbol{q} + {}^{s}\boldsymbol{W}\boldsymbol{m}\right] \mathrm{d}a. \tag{10}$$

Those equations, respectively consisting in translational and rotational equilibrium conditions, involve the tractions, i.e. internal forces and couples per unit length, associated to the shell stress tensors on the boundary $\partial \mathscr{P}$ of the part $\mathscr{P}$. Within such a formulation, boundary conditions can be possibly considered on the free part of the mid-surface boundary prescribing the boundary integrals of the relevant tractions.

2.2 Static Admissibility

The classical Heyman's assumptions of infinite compressive and vanishing tensile strengths are adopted to characterize the admissible stress states in the dome [22]. Therefore, the shell stress resultants are required to obey the following unilateral conditions (e.g., see [27, 36, 39]):

$$\operatorname{sym} \boldsymbol{N} \preceq \mathbf{0}, \quad \operatorname{sym}(\boldsymbol{M} - \boldsymbol{N}h/2) \succeq \mathbf{0}, \quad \operatorname{sym}(\boldsymbol{M} + \boldsymbol{N}h/2) \preceq \mathbf{0}, \tag{11}$$

where sym denotes the symmetric part operator and the notation $\boldsymbol{S} \succeq \mathbf{0}$ [resp., $\boldsymbol{S} \preceq \mathbf{0}$] translates the symmetric tensor $\boldsymbol{S}$ to be positive [resp., negative] semidefinite. Condition $(11)_1$ implies that, for any tangent unit vector $\boldsymbol{\nu}$ to the dome mid-surface, the normal force $\boldsymbol{N}\boldsymbol{\nu} \cdot \boldsymbol{\nu}$ is compressive. Likewise, conditions $(11)_{2,3}$ imply the bending moments $\boldsymbol{M}\boldsymbol{\nu} \cdot \boldsymbol{\nu}$ to be bounded by $\pm h/2\, \boldsymbol{N}\boldsymbol{\nu} \cdot \boldsymbol{\nu}$, i.e. the center of pressure to be contained within the dome thickness for any tangent unit vector $\boldsymbol{\nu}$ to the dome mid-surface. In passing, it is noticed that condition $(11)_1$ is linearly dependent on the remaining two, and can be thus dropped off.

A sufficiently large friction coefficient is assumed to be available, such that Heyman's no-sliding conditions are tacitly satisfied. The influence of finite friction has been discussed for general masonry structures e.g. in [5, 14, 28, 41, 47] and for masonry domes in [36, 39].

2.3 Collapse and Minimum-Thrust Analysis Problems

Collapse and minimum-thrust analysis problems for a masonry dome subjected to general loading conditions are here formulated in the context of static limit analysis.

To that end, let the set of equilibrated and statically admissible stress states in the dome be introduced:

$$\mathcal{H} = \left\{ (\boldsymbol{N},\ \boldsymbol{Q},\ \boldsymbol{M}) \,|\, \text{conditions (7) or (10), and conditions } (11)_{2,3},\ \text{hold} \right\}. \quad (12)$$

For a collapse analysis, it is assumed that the external forces $\boldsymbol{q}$ and couples $\boldsymbol{m}$ applied to the dome mid-surface are the sum of dead and live load contributions, the latter obtained by amplification of a basic live load distribution through a load multiplier λ:

$$\boldsymbol{q} = \boldsymbol{q}^{\mathrm{d}} + \lambda \boldsymbol{q}^{\mathrm{l}}, \quad \boldsymbol{m} = \boldsymbol{m}^{\mathrm{d}} + \lambda \boldsymbol{m}^{\mathrm{l}}. \quad (13)$$

The static limit analysis theorem states that the collapse multiplier λ_{c} of the basic live loads, i.e. the collapse capacity of the dome, is achieved by maximizing λ within the set $\mathcal{H}$ (e.g., see [11, 22]):

$$\max_{\{\lambda,\ \boldsymbol{N},\ \boldsymbol{Q},\ \boldsymbol{M}\}} \quad \lambda, \qquad \text{subject to} \quad (\boldsymbol{N},\ \boldsymbol{Q},\ \boldsymbol{M}) \in \mathcal{H}. \quad (14)$$

In passing, it is noticed that a pseudo-static seismic analysis matches such a formulation, as detailed in Sect. 4.1.

By contrast, a minimum-thrust analysis problem aims at providing quantitative assessment of the structural safety of the dome under completely prescribed loads (also possibly reducing to the self-weight). A given settlement distribution $\boldsymbol{\delta}$ is considered on the supported boundary $\mathcal{C}$ of the dome mid-surface, which produces a settlement mechanism of the dome resisted by the reactions $\boldsymbol{r}$ of the settled constraints. The static theorem of the minimum thrust guarantees that a minimum thrust state in the dome is achieved by minimizing the opposite of the work done by those reactions (coinciding with the complementary energy of the dome due to the support settlement):

$$\mathcal{W} = -\int_{\mathcal{C}} \boldsymbol{r} \cdot \boldsymbol{\delta} \, \mathrm{d}l, \quad (15)$$

within the set $\mathcal{H}$, i.e. by addressing the following optimization problem (e.g., see [11, 22]):

$$\min_{\{\boldsymbol{N},\ \boldsymbol{Q},\ \boldsymbol{M}\}} \quad \mathcal{W}, \qquad \text{subject to} \quad (\boldsymbol{N},\ \boldsymbol{Q},\ \boldsymbol{M}) \in \mathcal{H}. \quad (16)$$

The solution of either problems (14) or (16) requires a suitable discretization of the set $\mathcal{H}$ of equilibrated and statically admissible stress states. In the next section, two alternative discretization methods will be discussed for achieving efficient computational strategies.

3 Problem Discretization

The proposed shell-based static limit analysis formulation offers a general framework for developing computational strategies. Since amounting to a discretization of the set $\mathcal{H}$ of equilibrated and statically admissible stress states in the dome, those strategies require a discretization of the shell stress tensors $\boldsymbol{N}$, $\boldsymbol{Q}$, and $\boldsymbol{M}$ on the dome mid-surface, and the imposition of discretized equilibrium and static admissibility conditions descending from their continuous counterpart, i.e. conditions (7) or (10), and $(11)_{2,3}$, respectively. In the following, two alternative strategies, based on finite difference and finite volume methods are discussed.

3.1 *Finite Difference Method*

For a finite difference discretization of the shell-based static limit analysis formulation, a main and an auxiliary rectangular grids are considered in the parameter space (φ, ϑ) (Fig. 2a). They are characterized by grid spacing $\Delta\varphi$ and $\Delta\vartheta$, and shifted by half the grid spacing with respect to each other in both directions. The nodes of the main grid are labeled by indices (i, j), respectively running along φ- and ϑ-direction, whereas the nodes of the auxiliary grid, which coincide by construction with the centers of the finite difference cells of the main grid, are labeled by indices $(i + 1/2, j + 1/2)$.

The unknown physical components of the shell stress tensors are located at the nodes of the main grid. Accordingly, the vector $\boldsymbol{X}$ is introduced, which column-stacks the nodal unknowns at each node:

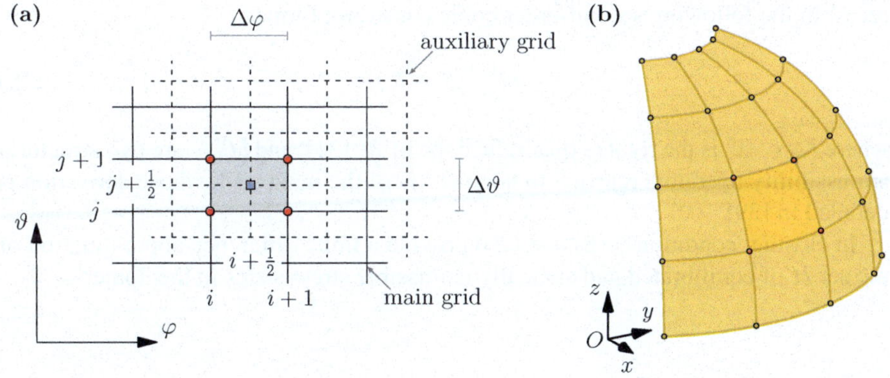

Fig. 2 Problem discretization: **a** main and auxiliary staggered rectangular grids in the parameter space (φ, ϑ) for finite difference discretization, and **b** mesh on the dome mid-surface for finite volume discretization

$$X^{(i,j)} = \left(N_{\varphi}^{(i,j)};\, N_{\vartheta\varphi}^{(i,j)};\, N_{\varphi\vartheta}^{(i,j)};\, N_{\vartheta}^{(i,j)};\;\; Q_{\varphi}^{(i,j)};\, Q_{\vartheta}^{(i,j)};\, M_{\varphi}^{(i,j)};\, M_{\vartheta\varphi}^{(i,j)};\, M_{\vartheta}^{(i,j)}\right). \tag{17}$$

The differential equilibrium equations (7), in their component-wise form (8), are enforced in the finite difference sense at the nodes of the auxiliary grid. To that end, let the typical physical component S of a shell stress tensor be considered. A finite difference approximation of S and of its partial derivatives $S_{/\varphi}$ and $S_{/\vartheta}$ can be derived at the nodes of the auxiliary grid in terms of the unknown values at the nodes of the main grid:

$$\begin{aligned}
S^{(i+1/2,\,j+1/2)} &= \frac{1}{4}\left[S^{(i,j)} + S^{(i+1,j)} + S^{(i+1,j+1)} + S^{(i,j+1)}\right],\\
S_{/\varphi}^{(i+1/2,\,j+1/2)} &= \frac{1}{2\Delta\varphi}\left[-S^{(i,j)} + S^{(i+1,j)} + S^{(i+1,j+1)} - S^{(i,j+1)}\right],\\
S_{/\vartheta}^{(i+1/2,\,j+1/2)} &= \frac{1}{2\Delta\vartheta}\left[-S^{(i,j)} - S^{(i+1,j)} + S^{(i+1,j+1)} + S^{(i,j+1)}\right].
\end{aligned} \tag{18}$$

Accordingly, after some algebra, the finite difference approximation of Eqs. (8) can be compactly written as [39]:

$$\boldsymbol{E}\boldsymbol{X} + \boldsymbol{f} = \boldsymbol{0}, \tag{19}$$

with $\boldsymbol{E}$ and $\boldsymbol{f}$ respectively as the structural equilibrium matrix and the vector of nodal forces. Due to their algebraic nature, possible boundary conditions (9) on the free boundary of the dome mid-surface may be straightforwardly enforced at the nodes of the main grid, resulting in linear conditions in the unknown $\boldsymbol{X}$.

The static admissibility conditions $(11)_{2,3}$ on the shell stress tensors are also enforced at the nodes of the main grid. For computational convenience, they can be recast in the following second-order conic constraint format:

$$\boldsymbol{\mathcal{U}}_{\pm}^{(i,j)}\boldsymbol{X} \in \mathcal{K}_{\mathrm{r}}, \tag{20}$$

where $\mathcal{K}_{\mathrm{r}} \subset \mathbb{R}^3$ is the rotated quadratic cone in $\mathbb{R}^3$ [31], and $\boldsymbol{\mathcal{U}}_{\pm}^{(i,j)}$ are two structural admissibility operators relevant to node $(i,\ j)$ of the main grid, whose derivation is detailed in [39].

In closing, conditions (18) and (20) provide a finite difference approximation of the set $\mathcal{H}$ of equilibrated and statically admissible stress states in the dome.

3.2 Finite Volume Method

For a finite volume discretization of the shell-based static limit analysis formulation, the dome mid-surface is discretized into finite volumes (or cells) that are the image of a rectangular mesh in the parameter domain (φ, ϑ) (Fig. 2b).

The nodal values of the physical components of the shell stress tensors are assumed as problem unknowns. They are collected in the vector $\boldsymbol{X}$, which column-stacks the nodal unknowns at each node:

$$\boldsymbol{X}^i = (N^i_\varphi;\, N^i_{\vartheta\varphi};\, N^i_{\varphi\vartheta};\, N^i_\vartheta;\;\; Q^i_\varphi;\, Q^i_\vartheta;\;\; M^i_\varphi;\, M^i_{\varphi\vartheta};\, M^i_\vartheta). \tag{21}$$

The integral equilibrium equations (10) are enforced for any cell of the mesh. Upon noticing that the shell stress tensors $\boldsymbol{N}$, $\boldsymbol{Q}$, and $\boldsymbol{M}$ are only involved on the cell boundaries, a piecewise-linear Lagrangian interpolation of the physical components of the shell stress tensors is thereon adopted in terms of the nodal values (21). Let the typical physical component S of a shell stress tensor be considered. Its approximation on the edge γ joining nodes i and j is given by:

$$S|_\gamma(\sigma) \approx L_1(\sigma)\, S^i + L_2(\sigma)\, S^j, \quad L_1(\sigma) = 1 - \frac{\sigma}{l}, \quad L_2(\sigma) = \frac{\sigma}{l}, \tag{22}$$

where l and $\sigma \in [0, l]$ are the length of, and a curvilinear abscissa along, the edge γ, respectively, and L_1 and L_2 are the linear Lagrange functions on $[0, l]$. From their interpolated physical components, the shell stress tensors at any point of γ are finally reconstructed by representations (6), where the exact physical basis vectors at that point are substituted. Therefore, after some algebra, the finite volume approximation of Eqs. (10) can be expressed as [36]:

$$\boldsymbol{E}\boldsymbol{X} + \boldsymbol{f} = \boldsymbol{0}, \tag{23}$$

which is formally identical to the finite difference equilibrium conditions, Eq. (19). Also in this case, possible boundary conditions (9) on the free boundary of the dome mid-surface may be straightforwardly enforced at the free cell boundaries, resulting in linear conditions in the unknown $\boldsymbol{X}$.

The static admissibility conditions $(11)_{2,3}$ on the shell stress tensors are imposed at the nodes of the mesh and amount to the following requirements:

$$\boldsymbol{\mathcal{U}}^i_\pm \boldsymbol{X} \in \mathcal{K}_\mathrm{r}, \tag{24}$$

which have analogous interpretation to the ones pertaining to the finite difference method, Eq. (20), and are detailed in [36].

In summary, conditions (23) and (24) yield a finite volume approximation of the set $\mathcal{H}$ of equilibrated and statically admissible stress states in the dome.

3.3 Discrete Collapse and Minimum-Thrust Analysis Problems

Substitution of the finite difference or finite volume approximations of the set $\mathcal{H}$ of equilibrated and statically admissible stress states in the dome within the collapse analysis problem (14) or the minimum-thrust analysis problem (16) yields their discrete counterpart. The former discrete problem reads:

$$\begin{aligned} \max_{\{\lambda, \boldsymbol{X}\}} \quad & \lambda, \\ \text{s.t.} \quad & \boldsymbol{E}\boldsymbol{X} + \boldsymbol{f}^{\mathrm{d}} + \lambda \boldsymbol{f}^{\mathrm{l}} = \boldsymbol{0}, \quad \boldsymbol{B}\boldsymbol{X} = \boldsymbol{0}, \\ & \mathcal{U}_{\pm}^{\bullet}\boldsymbol{X} \in \mathcal{K}_{\mathrm{r}}, \quad \text{at all nodes,} \end{aligned} \tag{25}$$

where the decomposition (13) of the external loads is reflected in the decomposition of the vector $\boldsymbol{f}$ of nodal forces, $\boldsymbol{B}$ is a suitable matrix enforcing possibile boundary conditions on the free boundary of the dome mid-surface, and the symbol $\bullet$ respectively stands for (i, j) or i for finite difference or finite volume discretization, Eqs. (20) and (24). In a similar fashion, the discrete minimum-thrust analysis problem is given by:

$$\begin{aligned} \max_{\boldsymbol{X}} \quad & -\boldsymbol{\delta}^T \boldsymbol{W}\boldsymbol{X}, \\ \text{s.t.} \quad & \boldsymbol{E}\boldsymbol{X} + \boldsymbol{f} = \boldsymbol{0}, \quad \boldsymbol{B}\boldsymbol{X} = \boldsymbol{0}, \\ & \mathcal{U}_{\pm}^{\bullet}\boldsymbol{X} \in \mathcal{K}_{\mathrm{r}}, \quad \text{at all nodes,} \end{aligned} \tag{26}$$

where $\boldsymbol{\delta}$ is the prescribed vector collecting the nodal values of the constraint settlements, and $\boldsymbol{W}$ is the operator associated to the bilinear form in Eq. (15). Those in Eqs. (25) and (26) are convex optimization problems, known in mathematical programming as second-order cone programming problems. For their solution standard and effective optimization softwares are available (e.g., see [31]).

4 Numerical Results

In this section, numerical simulations are presented to assess the computational performances of the proposed shell-based static limit analysis approach. Pseudo-static seismic analyses of spherical and catenary domes with parameterized geometry and minimum-thrust analyses of spherical domes under differential settlements are respectively discussed in Sects. 4.1 and 4.2.

All numerical analyses have been performed by means of an in-house MATLAB® code, and the computations have been done on a single machine with dual Intel® Xeon® CPU Gold 6226R @ 2.89 GHz and 256 GB RAM. The optimization problems (25) and (26) have been solved by Mosek® optimization software (version 10.0) [31].

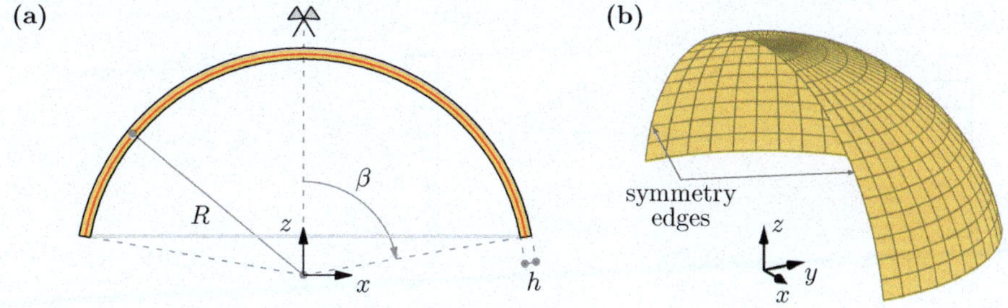

Fig. 3 Pseudo-static seismic analysis: **a** section geometry, with highlighted mid-surface and geometric parameters, and **b** typical finite volume mesh on half of the mid-surface, with highlighted problem symmetry for proportional horizontal forces along x-direction, of a spherical dome

4.1 Pseudo-static Seismic Analysis

Pseudo-static seismic analyses fit the format of collapse analysis problems, Eqs. (13)–(14), by assuming the self-weight loads $(\boldsymbol{q}^0, \boldsymbol{m}^0)$ in Eq. (5) as dead loads $(\boldsymbol{q}^\mathrm{d}, \boldsymbol{m}^\mathrm{d})$, and suitable distributions of seismic loads as basic live loads $(\boldsymbol{q}^1, \boldsymbol{m}^1)$. Concerning the latter, horizontal accelerations that are uniformly or linearly distributed along the height of the dome are considered, in accordance to the Italian norms for constructions NTC 2018 [42]. As a consequence, the basic live loads boil down to:

$$\boldsymbol{q}^1 = \frac{W}{S}\,\ell q \boldsymbol{\imath}, \quad \boldsymbol{m}^1 = \frac{W}{S}\,\ell m\, {}^s\boldsymbol{\imath}, \tag{27}$$

where $\boldsymbol{\imath}$ and ${}^s\boldsymbol{\imath}$ respectively denote the seismic acceleration direction and its projection on the tangent plane to the dome mid-surface, the choices $\ell = 1$ or $\ell = z$ respectively correspond to uniform or linear seismic acceleration distributions, and W, i.e. the weight of the dome, and S are respectively computed as the integral of q and of ℓq over the dome mid-surface. Taking advantage of the axially-symmetric geometry of spherical and catenary domes investigated in the following, in all the simulations it is assumed without loss of generality that the seismic acceleration direction $\boldsymbol{\imath}$ coincides with the direction $\boldsymbol{i}$. As the descending problem is symmetric with respect to the xz-plane, only half of the domes is modeled with suitable boundary conditions imposed on the symmetry edges.

Spherical Domes. A spherical dome is considered, characterized by mid-surface radius R, normalized thickness h/R, and half-embrace angle β (Fig. 3a). For finite difference simulations, the main and auxiliary staggered rectangular grids are obtained by subdividing φ- and ϑ-domains respectively into m and $2m$ intervals. For finite volume simulations, the image of the main rectangular grid onto the dome mid-surface is adopted for a discretization (Fig. 3b). Evidences from of a convergence analysis not reported here show that, in engineering terms, both methods are at convergence for a 32×64 mesh. Relevant numerical results are henceforth discussed.

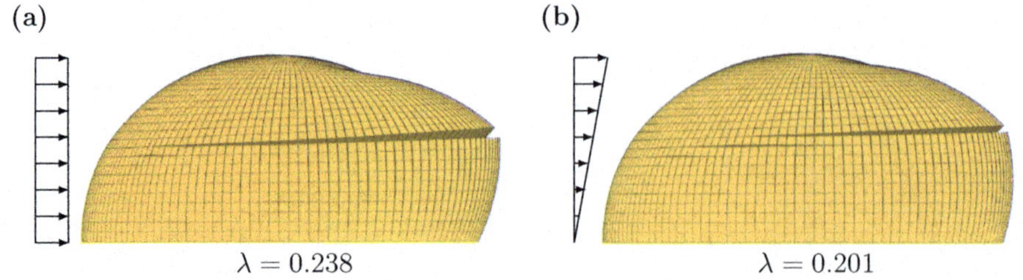

Fig. 4 Pseudo-static seismic analysis: incipient collapse mechanism of a hemispherical dome with normalized thickness $h/R = 0.07$ under **a** uniformly or **b** linearly distributed pseudo-static seismic loads along the height (lateral view). Results obtained from finite difference discretization strategy are shown

The case of hemispherical dome with normalized thickness $h/R = 0.07$ is first investigated to explore how its collapse capacity depends on the seismic loads distribution. By recalling that the normalized minimum thickness of a hemispherical dome, as estimated under Heyman's assumptions, is 0.04284 [37], the geometric safety factor [22] of the dome under investigation results to be 1.634. The proposed finite difference discretization predicts a collapse load multiplier $\lambda = 0.238$ [resp., $\lambda = 0.201$] for uniformly [resp., linearly] distributed pseudo-static seismic loads. Correspondingly, the predictions from the proposed finite volume scheme are $\lambda = 0.241$ [resp., $\lambda = 0.203$], thus proving a satisfactory agreement between the two discretization approaches. Figure 4 shows the incipient collapse mechanisms derived by the finite difference discretization in the two cases of uniform and linear seismic load distributions. It is remarked that those mechanisms are computed by applying the second-order cone programming duality to the results from the collapse analysis problem (25) (e.g., see [36, 39]). From a qualitative point of view, both mechanisms exhibit the formation of three curved flexural hinges along parallel curves in the half of the dome in the positive direction of the horizontal forces. Two of them, developing at the extrados of the dome, are located at its springing and in the vicinity of its crown, whereas one develops at the intrados of the dome in the haunch region. Despite the qualitative similarities of the two mechanisms, the distribution of seismic loads affects the position of the uppermost curved hinge, which shifts upward when linearly distributed seismic loads are concerned and is responsible for a slight reduction in the collapse multiplier compared to the case of uniformly distributed seismic loads.

A parametric analysis is carried out to systematically explore the seismic capacity of spherical domes in relationship with their geometrical features and the seismic loads distribution. Results are shown in Fig. 5, where the collapse multiplier λ is plotted versus the normalized thickness h/R of the dome for the half-embrace angles $\beta = \{70°, 80°, 90°\}$. In addition to the good agreement between predictions from finite difference and finite volume approaches, it is noticed that the collapse capacity of spherical domes increases with h/R (in particular, it is vanishing in the minimum thickness configuration) and decreases with β. The slight reduction of the

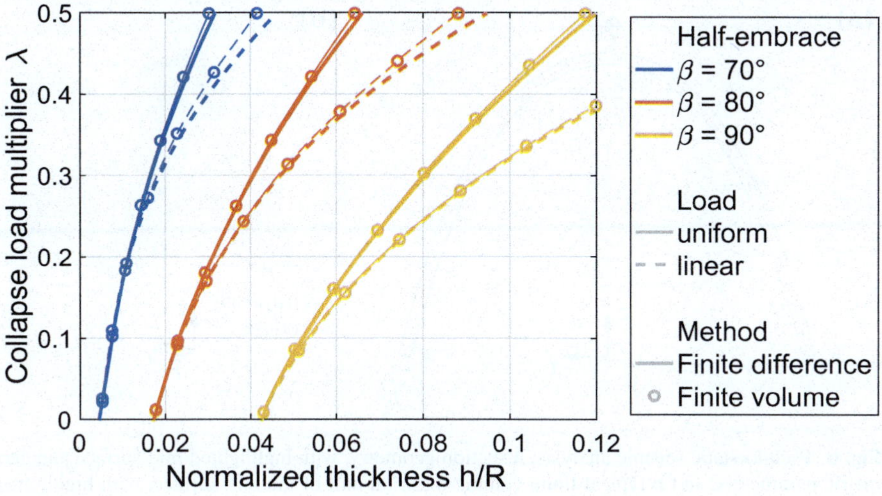

Fig. 5 Pseudo-static seismic analysis: collapse load multiplier λ of spherical domes versus normalized thickness h/R, for selected values of the embrace angle β and considering uniformly or linearly distributed seismic loads along the height of the dome. Results obtained from finite difference and finite volume discretization strategies are compared

collapse multiplier for linearly distributed seismic loads compared to uniform ones, which becomes more pronounced for increasing h/R, is confirmed.

In closing, the present results highlight the computational merit of the proposed shell-based static limit analysis approach and its versatility to alternative discretization strategies. From an applicative viewpoint, when Heyman's assumptions on masonry material and in particular the no-sliding hypothesis are satisfied, masonry domes exhibit a moderate resistance to seismic loads (for the influence of finite shear strength, see [36, 39]).

Catenary Domes. A catenary dome is considered, characterized by a midspan l, a rise-to-midspan ratio f/l, and a normalized thickness h/l (Fig. 6). The dome parameterization matches that in Eq. (1) by assuming the following radial distance and elevation functions:

$$r(t) = a \operatorname{arcsinh}(t/a), \quad z(t) = (f + a) - a\sqrt{1 + (t/a)^2}, \tag{28}$$

where the parameter a is determined by imposing that the curve (r, z) passes through the point $(l, 0)$. Finite difference and finite volume discretization analogous to the ones introduced for spherical domes are adopted (the typical finite volume mesh is shown in Fig. 6b). Also in this case, results relevant to a 32×64 mesh, which practically guarantees convergence, are discussed.

Initially, a catenary dome characterized by rise-to-midspan ratio $f/l = 1.5$ and normalized thickness $h/l = 0.07$ is investigated. As the minimum thickness of cate-

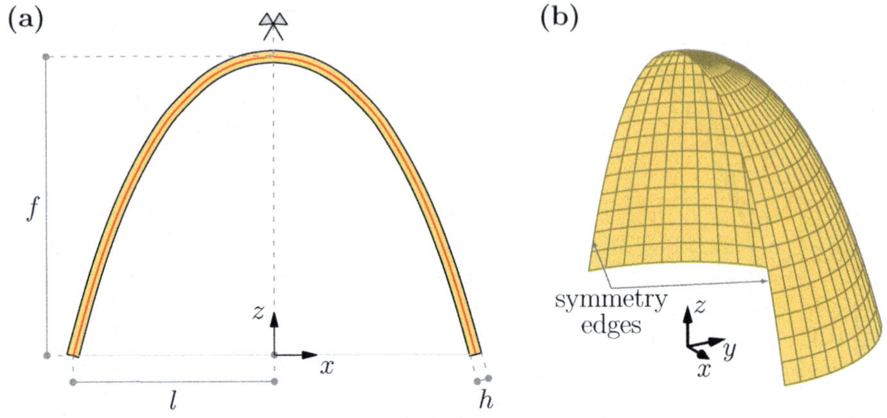

Fig. 6 Pseudo-static seismic analysis: **a** section geometry, with highlighted mid-surface and geometric parameters, and **b** typical finite volume mesh on half of the mid-surface, with highlighted problem symmetry for proportional horizontal forces along x-direction, of a catenary dome

nary domes is vanishing, its geometric safety factor is theoretically unbounded. The collapse load multiplier estimated by the present finite difference discretization is $\lambda = 0.413$ [resp., $\lambda = 0.296$] for uniformly [resp., linearly] distributed pseudo-static seismic loads. Correspondingly, the finite volume scheme estimates $\lambda = 0.444$ [resp., $\lambda = 0.304$]. The incipient collapse mechanisms derived by the finite volume discretization for uniform and linear seismic load distributions are depicted in Fig. 7. As for spherical domes, both mechanisms are qualitatively caused by the opening of three parallel hinges, two at the extrados and one at the intrados, in the half of the dome in the positive direction of the horizontal forces. In the case of linear seismic load distribution, those hinges are not as much clearly identified, and relative rotations appear to be diffused between a family of parallel curves. It is also confirmed that the reduction in collapse capacity due to linearly distributed seismic loads is related to a higher position of the uppermost parallel hinge.

A parametric analysis on the collapse load multiplier is then performed for catenary domes with parameterized geometry. Results relevant to the finite difference method are shown in Fig. 8, where the collapse load multiplier λ is plotted as a function of the normalized thickness h/l for the values $f/l = \{0.75, 1, 1.25, 1.5, 2\}$ of the rise-to-midspan ratio. As expected, the collapse capacity of catenary domes increases with h/l and decreases with f/l. In addition, it markedly reduces when switching form uniform to linear seismic loads distributions. Although a completely fair comparison is hindered by intrinsic geometric differences, it is noticed that the seismic resistance of catenary domes largely outperforms the one of spherical domes. On the one hand, especially when considering small rise-to-midspan ratios, i.e. shallow geometries, accounting for the influence of finite shear strength may be advisable. On the other hand, the present evidences extend the well-known result that catenary geometries exhibit excellent structural performances under vertical loads to the

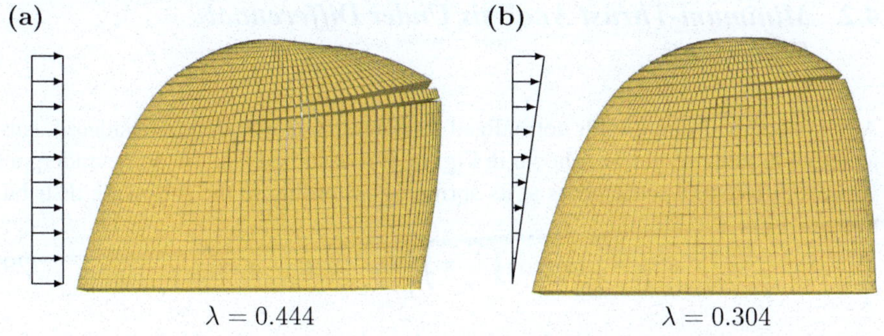

Fig. 7 Pseudo-static seismic analysis: incipient collapse mechanism of a catenary dome with rise-to-midspan ratio $f/l = 1.5$ and normalized thickness $h/l = 0.07$ under **a** uniformly or **b** linearly distributed pseudo-static seismic loads along the height (lateral view). Results obtained from finite volume discretization strategy are shown

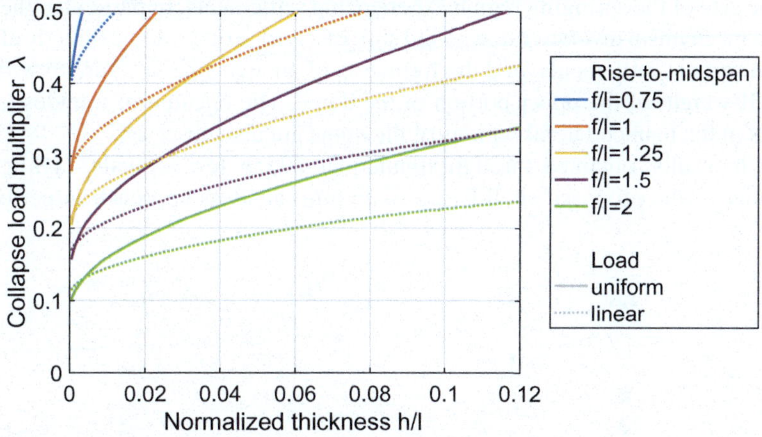

Fig. 8 Pseudo-static seismic analysis: collapse load multiplier λ of catenary domes versus normalized thickness h/l, for selected values of the rise-to-midspan ratio f/l and considering uniformly or linearly distributed seismic loads along the height of the dome. Results obtained from finite difference discretization strategy are shown

case of catenary domes under proportional horizontal loads. That is also reflected in the finite seismic resistance achieved for very small values of the normalized thickness h/l.

As a concluding remark, the obtained results confirm the capabilities of the proposed shell-based static limit analysis framework in addressing the pseudo-static limit analysis of masonry domes with arbitrary axially-symmetric geometry, also capturing the fine properties of their structural response and quantifying the influence of the seismic load distribution.

4.2 Minimum-Thrust Analysis Under Differential Settlements

A hemispherical dome with normalized thickness $h/R = 0.07$ is considered, subjected to its self-weight. As shown in Fig. 9a, it is assumed that the dome undergoes vertical differential settlements at its springing according to the following distribution:

$$\boldsymbol{\delta} = -\left[1 - \cos\left(\frac{\vartheta}{\alpha}\pi\right)\right]\delta\boldsymbol{k}, \quad \vartheta \in [-\alpha, \alpha], \tag{29}$$

in which δ and α respectively control the settlement magnitude and the size of the springing region where settlements occur. A minimum-thrust analysis is carried out considering the two values $\alpha = \{\pi/6, \pi/3\}$ of the size parameter. The settlement mechanisms obtained by the finite difference scheme are depicted in Fig. 9b, c, showing a shift from a localized to a semi-global mechanism on increasing the size of the springing region experiencing settlements. In particular, the semi-global mechanism involves a downward shift of a wedge of the dome directly affected by the vertical settlements, and the formation of an arch-like structure to withstand the self-weight of the upper portion of the dome. The opening of intrados parallel hinges in the haunch lateral regions of the dome are also observed.

Such results, obtained through simulations taking few seconds, highlight the potential of the proposed shell-based static limit analysis approach in performing

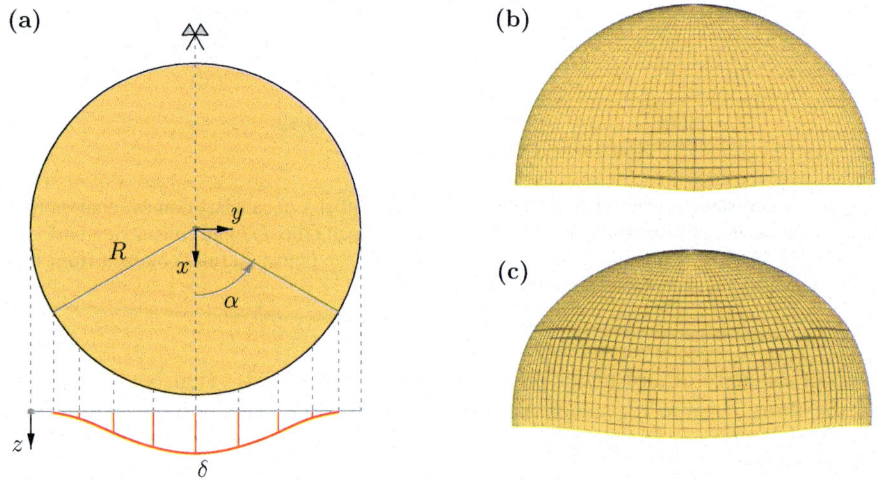

Fig. 9 Minimum thrust analysis under differential settlements: **a** distribution of vertical support settlements at the springing, and descending settlement mechanism assuming **b** $\alpha = \pi/6$ or **c** $\alpha = \pi/3$, for a hemispherical dome with normalized thickness $h/R = 0.07$ (frontal view). Results obtained from finite difference discretization strategy are shown

minimum-thrust analysis problems also characterized by non-standard settlement distributions in a completely automatic fashion and conciliating robustness, accuracy and effectiveness.

5 Conclusions

A shell-based static limit analysis approach has been proposed for the structural assessment of masonry domes subject to complex loading conditions, e.g. accounting for the presence of horizontal loads or related to differential settlements at the supports. In view of an application of the static theorem of limit analysis, a description of equilibrated and statically admissible stress states in the dome has been achieved by resorting to the classical statics of shells and to the shell stress resultants on the dome mid-surface, namely the normal-force and bending-moment tensors, and the shear-force vector. Collapse analysis and minimum-thrust analysis problems have been thus formulated exploiting differential or integral shell equilibrium conditions, alongside classical Heyman's assumptions. A finite difference and a finite volume discretization schemes have been presented to translate that formulation into a computational strategy, resulting in the efficient solution of a second-order programming problem by standard optimization software. The computational performances of the proposed framework have been explored in numerical simulations, including parametric analysis on the pseudo-static seismic collapse capacity of spherical and catenary domes under uniform or linear distributions of horizontal accelerations, and the minimum-thrust state of spherical domes under non-standard distributions of support differential settlements. Those results prove the potentialities of the proposed framework for an accurate and efficient structural assessment of masonry domes, including the prediction of collapse multiplier and incipient collapse mechanisms.

Acknowledgements The financial support of Ministero dell'Istruzione, dell'Università e della Ricerca, Italy, PRIN 2017 programme, project "3DP_Future" (grant 2017L7X3CS_004), and of University of Rome Tor Vergata, Italy, "Ricerca Scientifica di Ateneo 2021" program, project "OPTYMA" (CUP E83C22002290005), is gratefully acknowledged.

References

1. Aita, D., Barsotti, R., Bennati, S.: Studying the dome of Pisa cathedral via a modern reinterpretation of Durand-Claye's method. J. Mech. Mater. Struct. **14**(5), 603–619 (2019). https://doi.org/10.2140/jomms.2019.14.603
2. Angelillo, M., Babilio, E., Fortunato, A.: Singular stress fields for masonry-like vaults. Contin. Mech. Thermodyn. **15**(2–4), 423–441 (2013). https://doi.org/10.1007/s00161-012-0270-9
3. Babilio, E., Ceraldi, C., Lippiello, M., Portioli, F., Sacco, E.: Static analysis of a double-cap masonry dome. In: Carcaterra, A., Paolone, A., Graziani, G. (eds.) Proceedings of XXIV AIMETA Conference 2019. Lecture Notes in Mechanical Engineering, pp. 2082–2093. Springer, Cham (2020). https://doi.org/10.1007/978-3-030-41057-5_165

4. Baratta, A., Corbi, O.: On the statics of no-tension masonry-like vaults and shells: solution domains, operative treatment and numerical validation. Ann. Solid Struct. Mech. **2**, 107–122 (2011). https://doi.org/10.1007/s12356-011-0022-8
5. Beatini, V., Royer-Carfagni, G., Tasora, A.: The role of frictional contact of constituent blocks on the stability of masonry domes. Proc. R. Soc. A **474**, 20170740 (2018). https://doi.org/10.1098/rspa.2017.0740
6. Block, P., Lachauer, L.: Three-dimensional (3D) equilibrium analysis of gothic masonry vaults. Int. J. Archit. Herit. **8**(3), 312–335 (2014). https://doi.org/10.1080/15583058.2013.826301
7. Block, P., Lachauer, L.: Three-dimensional funicular analysis of masonry vaults. Mech. Res. Commun. **56**, 53–60 (2014). https://doi.org/10.1016/j.mechrescom.2013.11.010
8. Block, P., Ochsendorf, J.: Thrust network analysis: a new methodology for three-dimensional equilibrium. J. IASS **48**(3), 167–173 (2007)
9. Bruggi, M.: A constrained force density method for the funicular analysis and design of arches, domes and vaults. Int. J. Solids Struct. **193–194**, 251–269 (2020). https://doi.org/10.1016/j.ijsolstr.2020.02.030
10. Cascini, L., Gagliardo, R., Portioli, F.: LiABlock_3D: a software tool for collapse mechanism analysis of historic masonry structures. Int. J. Archit. Herit. **14**(1), 75–94 (2020). https://doi.org/10.1080/15583058.2018.1509155
11. Como, M.: Statics of historic masonry constructions, Springer Series in Solid and Structural Mechanics, 3 edn , vol. 9. Springer International Publishing, Cham. (2017). https://doi.org/10.1007/978-3-319-54738-1
12. Cusano, C., Cennamo, C., Angelillo, M.: Seismic vulnerability of domes: a case study. J. Mech. Mater. Struct. **13**(5), 679–689 (2018). https://doi.org/10.2140/jomms.2018.13.679
13. D'Altri, A.M., Sarhosis, V., Milani, G., Rots, J., Cattari, S., Lagomarsino, S., Sacco, E., Tralli, A., Castellazzi, G., de Miranda, S.: Modeling strategies for the computational analysis of unreinforced masonry structures: review and classification. Arch. Comput. Methods Eng. **27**, 1153–1185 (2020). https://doi.org/10.1007/s11831-019-09351-x
14. D'Ayala, D., Casapulla, C.: Limit state analysis of hemispherical domes with finite friction. In: Lourenço, P.B., Roca, P. (eds.) Historical Constructions 2001: possibilities of numerical and experimental techniques. In: Proceedings of the 3rd International Seminar, pp. 617–626. University of Minho, Guimarães, Portugal (2001)
15. Durand-Claye, A.: Vérification de la stabilité des voûtes et des arcs. Applications aux voûtes sphériques. Annales des Ponts et Chaussées **19**, 416–440 (1880)
16. Fraddosio, A., Lepore, N., Piccioni, M.D.: Thrust surface method: an innovative approach for the three-dimensional lower bound limit analysis of masonry vaults. Eng. Struct. **202**, 109846 (2020). https://doi.org/10.1016/j.engstruct.2019.109846
17. Fraternali, F.: A thrust network approach to the equilibrium problem of unreinforced masonry vaults via polyhedral stress functions. Mech. Res. Commun. **37**, 198–204 (2010). https://doi.org/10.1016/j.mechrescom.2009.12.010
18. Fraternali, F., Angelillo, M., Fortunato, A.: A lumped stress method for plane elastic problems and the discrete-continuum approximation. Int. J. Solids Struct. **39**(25), 6211–6240 (2002). https://doi.org/10.1016/S0020-7683(02)00472-9
19. Gould, P.L.: Analysis of Shells and Plates. Springer, New York (1988). https://doi.org/10.1007/978-1-4612-3764-8
20. Grillanda, N., Chiozzi, A., Milani, G., Tralli, A.: Collapse behavior of masonry domes under seismic loads: an adaptive NURBS kinematic limit analysis approach. Eng. Struct. **200**, 109517 (2019). https://doi.org/10.1016/j.engstruct.2019.109517
21. Heyman, J.: On shell solutions for masonry domes. Int. J. Solids Struct. **3**(2), 227–241 (1967). https://doi.org/10.1016/0020-7683(67)90072-8
22. Heyman, J.: The Stone Skeleton. Cambridge University Press, Cambridge (1995). https://doi.org/10.1017/CBO9781107050310
23. Kraus, H.: Thin Elastic Shells. Wiley, London (1967)
24. Lamé, M.G., Clapeyron, E.: Mémoire sur la stabilité des voûtes. Annales Des Mines **8**, 789–836 (1823)

25. Lau, W.: Equilibrium analysis of masonry domes (2006), MSc thesis, Massachusetts Institute of Technology
26. Lévy, M.: La statique graphique et ses applications aux constructions. Gauthier-Villars, Paris (1888)
27. Lucchesi, M., Padovani, C., Pasquinelli, G., Zani, N.: The maximum modulus eccentricities surface for masonry vaults and limit analysis. Math. Mech. Solids **4**(1), 71–87 (1999). https://doi.org/10.1177/108128659900400105
28. Lucchesi, M., Pintucchi, B., Zani, N.: Masonry-like material with bounded shear stress. Eur. J. Mech. A-Solids **72**, 329–340 (2018). https://doi.org/10.1016/j.euromechsol.2018.05.001
29. Marmo, F., Masi, D., Sessa, S., Toraldo, F., Rosati, L.: Thrust network analysis of masonry vaults subject to vertical and horizontal loads. In: Papadrakakis, M., Fragiadakis, M. (eds.) COMPDYN 2017 - Proceedings of the 6th International Conference on Computational Methods in Structural Dynamics and Earthquake Engineering. vol. 1, pp. 2227–2238 (2017). https://doi.org/10.7712/120117.5562.17018
30. Marmo, F., Rosati, L.: Reformulation and extension of the thrust network analysis. Comput. Struct. **182**, 104–118 (2017). https://doi.org/10.1016/j.compstruc.2016.11.016
31. MOSEK ApS: MOSEK Optimization Toolbox for MATLAB. Release 9.2.40 (2021). https://docs.mosek.com/9.2/toolbox.pdf
32. Naghdi, P.M.: The theory of shells and plates. In: Truesdell, C. (ed.) Linear Theories of Elasticity and Thermoelasticity, pp. 425–640. Springer, Berlin, Heidelberg (1973). https://doi.org/10.1007/978-3-662-39776-3_5
33. Navier, C.L.M.H.: Résumé des leçons données à l'École des Ponts et Chaussées sur l'application de la mécanique à l'établissement des constructions et des machines. Société Belge de Libraire, Brussels (1839)
34. Nodargi, N.A., Bisegna, P.: Thrust line analysis revisited and applied to optimization of masonry arches. Int. J. Mech. Sci. **179**, 105690 (2020). https://doi.org/10.1016/j.ijmecsci.2020.105690
35. Nodargi, N.A., Bisegna, P.: A unifying computational approach for the lower-bound limit analysis of systems of masonry arches and buttresses. Eng. Struct. **221**, 110999 (2020). https://doi.org/10.1016/j.engstruct.2020.110999
36. Nodargi, N.A., Bisegna, P.: Collapse capacity of masonry domes under horizontal loads: a static limit analysis approach. Int. J. Mech. Sci. **212**, 106827 (2021). https://doi.org/10.1016/j.ijmecsci.2021.106827
37. Nodargi, N.A., Bisegna, P.: Minimum thrust and minimum thickness of spherical masonry domes: a semi-analytical approach. Eur. J. Mech. A-Solids **87**, 104222 (2021). https://doi.org/10.1016/j.euromechsol.2021.104222
38. Nodargi, N.A., Bisegna, P.: A new computational framework for the minimum thrust analysis of axisymmetric masonry domes. Eng. Struct. **234**, 111962 (2021). https://doi.org/10.1016/j.engstruct.2021.111962
39. Nodargi, N.A., Bisegna, P.: A finite difference method for the static limit analysis of masonry domes under seismic loads. Meccanica **57**, 121–141 (2022). https://doi.org/10.1007/s11012-021-01414-3
40. Nodargi, N.A., Bisegna, P.: Generalized thrust network analysis for the safety assessment of vaulted masonry structures. Eng. Struct. E **270**, 114878 (2022). https://doi.org/10.1016/j.engstruct.2022.114878
41. Nodargi, N.A., Intrigila, C., Bisegna, P.: A variational-based fixed-point algorithm for the limit analysis of dry-masonry block structures with non-associative Coulomb friction. Int. J. Mech. Sci. **161–162**, 105078 (2019). https://doi.org/10.1016/j.ijmecsci.2019.105078
42. NTC2018: D.M. 17/01/2018 Aggiornamento delle "Norme tecniche per le costruzioni". S.O. alla G.U. n. 42 del 20/02/2018. Ministero delle Infrastrutture e dei Trasporti (2018)
43. O'Dwyer, D.W.: Funicular analysis of masonry vaults. Comput. Struct. **73**(1–5), 187–197 (1999). https://doi.org/10.1016/S0045-7949(98)00279-X
44. Oppenheim, I.J., Gunaratnam, D.J., Allen, R.H.: Limit state analysis of masonry domes. J. Struct. Eng. **115**(4), 868–882 (1989). https://doi.org/10.1061/(ASCE)0733-9445(1989)115:4(868)

45. Poleni, G.: Memorie Istoriche della Gran Cupola del Tempio Vaticano, e de' danni di essa, e de' Ristoramenti Loro. Stamperia del Seminario, Padua (1748)
46. Schwedler, J.W.: Die Konstruktion der Kuppeldächer. Zeitschrift für Bauwesen **16**, 7–34 (1866)
47. Simon, J., Bagi, K.: Discrete element analysis of the minimum thickness of oval masonry domes. Int. J. Archit. Herit. **10**(4), 457–475 (2016). https://doi.org/10.1080/15583058.2014.996921
48. Taroco, E.O., Blanco, P.J., Feijóo, R.A.: Introduction to the variational formulation in mechanics: Fundamentals and applications. Wiley, Chichester (2020). https://doi.org/10.1002/9781119600923
49. Zessin, J.: Collapse analysis of unreinforced masonry domes and curving walls. Ph.D. thesis, Massachusetts Institute of Technology (2012)

Robust Optimization Applied to Uncertain Limit Analysis

Jeremy Bleyer and Vincent Leclère

Abstract Limit analysis (LA) is an efficient tool for computing in a direct manner the ultimate load of a structure made of a perfectly plastic material. The lower bound static approach amounts to maximize the load factor such that one can find an optimal stress field in equilibrium with such loading and satisfying strength conditions at each point in the domain. In the deterministic case, the ultimate load is obtained via the resolution of a convex optimization problem. When loading or strength properties are random, the data of such an optimization problem become uncertain. Robust optimization theory is a branch of mathematical optimization which aim at finding an optimal solution of uncertain problems among all possible realizations of the uncertainty within a known uncertainty set. Applying the concepts of robust optimization to uncertain limit analysis, one may compute a worst-case ultimate load estimate associated with a given uncertainty set, for instance in the case of uncertain strength properties or uncertain load cases. This paper discusses how robust limit analysis problems can be reformulated, either exactly or approximately, into deterministic problems. In particular, the distinction between static and adjustable robust counterparts is introduced. In the former case, uncertain LA problems are replaced with a deterministic problem with reduced strength properties. In the latter case, additional optimization variables must be introduced in order to obtain an extended LA problem in much higher dimension.

Keywords Limit analysis · Ultimate load · Uncertainty · Robust optimization

---

J. Bleyer (✉)
Laboratoire Navier, Ecole des Ponts, Université Gustave Eiffel, CNRS, Marne-la-vallée, France
e-mail: jeremy.bleyer@enpc.fr

V. Leclère
CERMICS, Ecole des Ponts, Marne-la-vallée, France
e-mail: vincent.leclere@enpc.fr

G. Garcea and D. Weichert (eds.), *Direct Methods for Limit State of Materials and Structures*, Lecture Notes in Applied and Computational Mechanics 101,
https://doi.org/10.1007/978-3-031-29122-7_11

1 Introduction

Limit analysis [13, 22] is a powerful direct method used to estimate the collapse load of a structure consisting of a perfectly plastic material. The lower and upper bound approaches of limit analysis are naturally formulated as convex optimization problems for which given data consist of a known material yield criterion, a known reference loading and a known geometry [12]. However, in real-world applications, these parameters may be subject to uncertainty due to factors such as inaccurate load amplitude or direction, or variations in material strength. As a result, engineers often aim to design structures that are robust to such uncertainties, meaning that the collapse load must be safe for all possible combinations of uncertain parameters.

Traditionally, limit analysis has addressed this issue by either assuming a worst-case scenario for the uncertain parameters or by performing a stochastic analysis in which random realizations of the parameters are used. While the first approach can be overly conservative, it can also be challenging to determine the worst-case scenario in complex loading situations. The second approach, on the other hand, requires assuming a probability distribution for the parameters and solving a large number of problems to find the worst-case configuration, which may not be achievable in practice. General definitions of the probability of collapse have been given in [3, 21], later revisited by [2] using stochastic stress vectors. Various works have also considered the numerical computation of limit loads in a stochastic setting such as [23, 28] or [1, 10, 14, 17] for geotechnical applications. For instance, the reader can refer to [15] for a recent review of slope stability in spatially variable soils.

Alternative approaches have sought to evaluate the robustness or reliability of structures through non-probabilistic methods. In [19], the authors consider uncertain limit analysis of truss structures with very similar sources of uncertainties as those investigated in this work. For this purpose, they used the info-gap decision theory [4] which is however known to be difficult to apply in practice since robustness functions are very hard to compute in general. For the very specific case of truss structures investigated in [19], it can however be computed via the resolution of a linear programming problem. Similarly, mixed-integer programming approaches can also be used to compute a worst-case limit load [16] but solving such NP-hard problems is notoriously difficult and almost impossible for large-scale problems. Using a chance-constrained programming approach, [25, 26] considered limit analysis and shakedown theorems under normal or log-normal strength uncertainties for von Mises plasticity.

In this work, we propose an alternative approach that utilizes the principles of robust optimization theory [5, 7] to obtain a robust estimate of plastic limit loads in the presence of uncertainty. This approach allows us to design structures that are resistant to a wide range of uncertain parameters without relying on conservative assumptions or computationally intensive analyses. More precisely, uncertain limit analysis problems are formulated in the case of uncertain strength properties. A definition of the worst-case limit load is given using concepts of robust optimization theory. Then, in order to obtain computationally tractable formulations, different decision rules are introduced, in particular so-called *static* and *affinely adjustable* formulations. Static concepts are then applied to the definition of robust strength conditions and illustrated on the case of a Mohr-Coulomb criterion with uncertain cohesion and friction angle. Finally, the resolution of robust limit analysis at the structure scale is discussed for the case of strength uncertainties and loading uncertainties.

The manuscript is organized as follows: Sect. 2 introduces robust formulations of limit analysis theory in the case of strength uncertainty; Sect. 3 details the derivation of tractable robust counterparts of uncertain strength constraints arising in the previous formulations; Sect. 4 is devoted to the resolution of robust limit analysis problems with a specific emphasis on the case of loading uncertainties and the corresponding affinely adjustable robust formulations; finally, Sect. 5 draws some conclusions and perspectives for future research.

2 Robust Limit Analysis with Strength Uncertainties

2.1 Nominal and Uncertain Limit Analysis Problem

The *nominal* limit analysis problem amounts to computing the maximum load factor λ_N by solving the following convex maximization problem:

$$\begin{aligned} \lambda_N = \max_{\lambda,\boldsymbol{\sigma}} \; & \lambda \\ \text{s.t. } & \div \boldsymbol{\sigma} + \lambda \boldsymbol{f}^r + \boldsymbol{f}^f = 0 && \text{in } \Omega \\ & \boldsymbol{\sigma} \cdot \boldsymbol{n} = \lambda \boldsymbol{t}^r + \boldsymbol{t}^f && \text{on } \partial\Omega_T \\ & \boldsymbol{\sigma} \in G && \text{in } \Omega \end{aligned} \tag{N}$$

where λ is the load factor, $\boldsymbol{\sigma}$ the Cauchy stress field in Ω, $\boldsymbol{f}^r$ (resp. $\boldsymbol{f}^f$) is the reference (resp. fixed) body force, $\boldsymbol{t}^r$ (resp. $\boldsymbol{t}^f$) the reference (resp. fixed) contact force prescribed on some part $\partial\Omega_T$ of the boundary with unit normal $\boldsymbol{n}$ and G is the material yield/strength criterion which we assume to be a convex set (possibly unbounded) containing 0. In the above, the first two constraints correspond to the local balance equation and traction boundary conditions, whereas the last one corresponds to the strength condition which must be satisfied at all points $\boldsymbol{x} \in \Omega$. Note that formulation (N) corresponds to a static formulation which will result in a lower-bound estimate of the true collapse load when restricting to a finite-element subspace of statically admissible stress fields.

We now consider the case where the loading is certain but the material may possess uncertain properties such that the strength criterion is now written as $G(\boldsymbol{\zeta})$ where $\boldsymbol{\zeta} \in \mathbb{R}^m$ is a vector of uncertain parameters. Contrary to probabilistic approaches in which $\boldsymbol{\zeta}$ is a random variable with a given probability distribution, robust optimization approaches describe the uncertainty through the notion of an *uncertainty set* $\mathcal{U} \subseteq$

$\mathbb{R}^m$. It is assumed that any possible realization of the uncertainty belongs to the uncertainty set $\zeta \in \mathcal{U}$ without positing any probability distribution. The goal of robust optimization theory is to find an optimum solution to an uncertain optimization problem for any possible realization in this uncertainty set. Obviously, the choice of the uncertainty set is an important modeling step in such approaches and depends on our knowledge of the origins of the considered uncertainty. If probability distributions are known, uncertainty sets can be based on the size of the support or the shape of the probability distribution. For instance, its size can correspond to a certain confidence level of the probability distribution. It can also be built from available data.

This aspect is outside the scope of the present work, which presents a general methodology. One key assumption on the uncertainty used to obtain interesting results is that it is assumed to be convex. Although it can be more general, we assume, for simplicity, that $\mathcal{U}$ is a convex ball of unit radius for some norm i.e. $\mathcal{U} = \{\zeta \in \mathbb{R}^m \text{ s.t. } \|\zeta\| \leq 1\}$. In particular, we will note by $\mathcal{U}_p$ uncertainty sets corresponding to the L_p-ball (typically with $p = 1, 2$ or ∞).

The maximum load factor now becomes uncertain i.e. it depends on the value ζ of the uncertainty realization:

$$\begin{aligned} \lambda^+(\zeta) = \max_{\lambda,\boldsymbol{\sigma}}\ & \lambda \\ \text{s.t. } & \div\boldsymbol{\sigma} + \lambda \boldsymbol{f}^{\mathrm{r}} + \boldsymbol{f}^{\mathrm{f}} = 0 \\ & \boldsymbol{\sigma} \cdot \boldsymbol{n} = \lambda \boldsymbol{t}^{\mathrm{r}} + \boldsymbol{t}^{\mathrm{f}} \\ & \boldsymbol{\sigma} \in G(\zeta) \end{aligned} \tag{1}$$

The main purpose of robust optimization is to provide worst-case solutions to a given optimization problem. Our proposed theory of robust limit analysis therefore aims at evaluating the worst-case limit load among all possible realizations. In the remaining of this section, we discuss various robust formulations.

2.2 Adjustable Robust Optimization

For a given loading and two different given realizations of the uncertainty, one expects that the corresponding optimal stress fields will be different depending on the uncertainty realizations. The most natural approach therefore consists in considering the stress field and the corresponding load factor to be *recourse variables*, i.e. variables which depends on ζ. Thus, we are faced with an *adjustable robust counterpart* (ARC) to problem (1) defined as follows:

$$\begin{aligned} \lambda_{\mathrm{ARC}} = \min_{\zeta\in\mathcal{U}} \lambda^+(\zeta) = \min_{\zeta\in\mathcal{U}} \max_{\boldsymbol{\sigma}(\zeta),\lambda(\zeta)}\ & \lambda(\zeta) \\ \text{s.t. } & \div\boldsymbol{\sigma}(\zeta) + \lambda(\zeta) \boldsymbol{f}^{\mathrm{r}} + \boldsymbol{f}^{\mathrm{f}} = 0 \\ & \boldsymbol{\sigma}(\zeta) \cdot \boldsymbol{n} = \lambda(\zeta) \boldsymbol{t}^{\mathrm{r}} + \boldsymbol{t}^{\mathrm{f}} \\ & \boldsymbol{\sigma}(\zeta) \in G(\zeta) \end{aligned} \tag{ARC}$$

i.e. we find the largest load factor such that, for each uncertainty realization there exists an optimal stress field in equilibrium, with the corresponding collapse load factor, satisfying the strength criterion.

In the following, we also make use of the following equivalent formulation of the ARC problem [18, 24]:

$$\begin{aligned} \lambda_{\text{ARC}} = \max_{\bar{\lambda}} \ & \bar{\lambda} \\ \text{s.t. } & \forall \zeta \in \mathcal{U}, \exists \sigma, \lambda \text{ s.t. } \div \boldsymbol{\sigma} + \lambda \boldsymbol{f}^{\text{r}} + \boldsymbol{f}^{\text{f}} = 0 \\ & \qquad \boldsymbol{\sigma} \cdot \boldsymbol{n} = \lambda \boldsymbol{t}^{\text{r}} + \boldsymbol{t}^{\text{f}} \\ & \qquad \boldsymbol{\sigma} \in G(\zeta) \\ & \qquad \bar{\lambda} \leq \lambda \end{aligned} \tag{2}$$

where uncertainty of the objective function has been transferred to the constraints with the introduction of a static (non-adjustable) variable $\bar{\lambda}$.

2.3 *Static Robust Optimization*

Unfortunately, adjustable recourse problems are numerically challenging. Indeed, both formulations involve either a min/max problem (ARC) or an infinite number of constraints (2). To solve adjustable recourse problem, one typically makes a simplifying assumption on how recourse variables depend on the uncertainty, the so-called *decision rules*.

The most simple of such rules is to assume that recourse variables are in fact static, i.e. they do not depend on the uncertainty. This yields to a conservative static robust counterpart (RC) in which we look for a stress field $\boldsymbol{\sigma}$ and a load factor λ, independent of the exact realization of the uncertainty, which satisfy the strength condition $G(\zeta)$ for all $\zeta \in \mathcal{U}$. The corresponding problem can be formulated as follows:

$$\begin{aligned} \lambda_{\text{RC}} = \max_{\lambda, \sigma} \ & \lambda \\ \text{s.t. } & \div \boldsymbol{\sigma} + \lambda \boldsymbol{f}^{\text{r}} + \boldsymbol{f}^{\text{f}} = 0 \\ & \boldsymbol{\sigma} \cdot \boldsymbol{n} = \lambda \boldsymbol{t}^{\text{r}} + \boldsymbol{t}^{\text{f}} \\ & \boldsymbol{\sigma} \in G(\zeta) \qquad \forall \zeta \in \mathcal{U} \end{aligned} \tag{3}$$

What makes problem (3) a *robust optimization* problem is the condition $\forall \zeta \in \mathcal{U}$ in the last constraint. This implies that the constraint $\boldsymbol{\sigma} \in G(\zeta)$ must be fulfilled for any possible value of $\zeta \in \mathcal{U}$. It is therefore an infinite-dimensional constraint. One of the main goals of robust optimization theory is to make such a problem tractable using standard convex optimization algorithms.

For instance, the robust constraint can be reformulated as:

$$\boldsymbol{\sigma} \in G(\zeta) \quad \forall \zeta \in \mathcal{U} \quad \Leftrightarrow \quad \boldsymbol{\sigma} \in G_{\text{RC}} \tag{4}$$

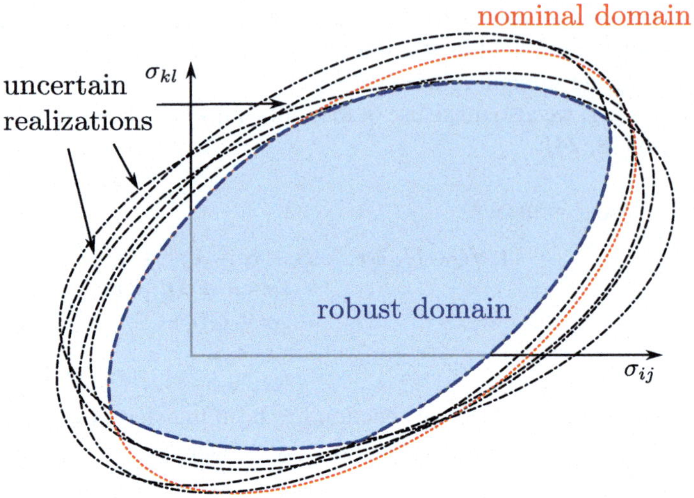

Fig. 1 Robust strength domain G_{RC} (in blue) obtained as the intersection of various uncertain realizations $G(\boldsymbol{\zeta})$ (in black) of a nominal domain (in red)

when introducing:

$$G_{\mathrm{RC}} = \bigcap_{\zeta \in \mathcal{U}} G(\boldsymbol{\zeta}) \tag{5}$$

the robust counterpart to the uncertain strength criterion. In order for a stress field to be admissible with respect to any possible realization of the uncertain strength criterion $G(\boldsymbol{\zeta})$, it has to belong to the intersection of all such domains (see Fig. 1).

Now, problem (3) writes as:

$$\begin{aligned} \lambda_{\mathrm{RC}} = \max_{\lambda,\sigma} \ & \lambda \\ \text{s.t. } & \div \boldsymbol{\sigma} + \lambda \boldsymbol{f}^{\mathrm{r}} + \boldsymbol{f}^{\mathrm{f}} = 0 \\ & \boldsymbol{\sigma} \cdot \boldsymbol{n} = \lambda \boldsymbol{t}^{\mathrm{r}} + \boldsymbol{t}^{\mathrm{f}} \\ & \boldsymbol{\sigma} \in G_{\mathrm{RC}} \end{aligned} \tag{RC}$$

which is now independent of the uncertainty realization. As a result, problem (RC) is a classical limit analysis problem with a different strength criterion given by (5). This makes problem (RC) very appealing provided that a simple expression for G_{RC} can be found. It is however very hard to determine a simple expression for the infinite-dimensional set intersection appearing in (5). Exact or approximate reformulation of strength criteria robust counterparts are discussed in Sect. 3.

2.4 Affinely Adjustable Robust Optimization

Unfortunately, if (RC) problems are numerically tractable, the obtained approximation might be unreasonably conservative [8]. A middle ground is the *affinely adjustable robust counterpart* (AARC), which consists in looking for adjustable variables $\boldsymbol{\sigma}(\boldsymbol{\zeta})$ and $\lambda(\boldsymbol{\zeta})$ that are affine functions of the uncertain variable, the so-called *affine decision rule* [6]:

$$\boldsymbol{\sigma}(\boldsymbol{\zeta}) = \boldsymbol{\sigma}_0 + \sum_{j=1}^{m} \boldsymbol{\sigma}_j \zeta_j \tag{6a}$$

$$\lambda(\boldsymbol{\zeta}) = \lambda_0 + \sum_{j=1}^{m} \lambda_j \zeta_j \tag{6b}$$

where the $\boldsymbol{\sigma}_i$ (resp. λ_i) represent $1 + m$ different stress fields (load factor variables) which are now static optimization variables. Inserting the affine decision rules (6a)–(6b) into (ARC), the corresponding AARC reads:

$$\begin{aligned}
\lambda_{\text{AARC}} = \max_{\boldsymbol{\sigma}_i, \lambda_i} \min_{\boldsymbol{\zeta} \in \mathcal{U}} \; & \lambda_0 + \sum_{j=1}^{m} \lambda_j \zeta_j \\
\text{s.t. } & \div \left(\boldsymbol{\sigma}_0 + \sum_{j=1}^{m} \boldsymbol{\sigma}_j \zeta_j \right) + \left(\lambda_0 + \sum_{j=1}^{m} \lambda_j \zeta_j \right) \boldsymbol{f}^{\text{r}} + \boldsymbol{f}^{\text{f}} = 0 \\
& \left(\boldsymbol{\sigma}_0 + \sum_{j=1}^{m} \boldsymbol{\sigma}_j \zeta_j \right) \cdot \boldsymbol{n} = \left(\lambda_0 + \sum_{j=1}^{m} \lambda_j \zeta_j \right) \boldsymbol{t}^{\text{r}} + \boldsymbol{t}^{\text{f}} \\
& \left(\boldsymbol{\sigma}_0 + \sum_{j=1}^{m} \boldsymbol{\sigma}_j \zeta_j \right) \in G(\boldsymbol{\zeta})
\end{aligned} \tag{7}$$

which can also be reformulated as follows:

$$\begin{aligned}
\lambda_{\text{AARC}} = \max_{\bar{\lambda}, \boldsymbol{\sigma}_i, \lambda_i} \; & \bar{\lambda} \\
\text{s.t. } & \div(\boldsymbol{\sigma}_j) + \lambda_j \boldsymbol{f}^{\text{r}} + \boldsymbol{f}^{\text{f}} = 0 \quad && \forall j = 0, \dots, m \\
& \boldsymbol{\sigma}_j \cdot \boldsymbol{n} = \lambda_j \boldsymbol{t}^{\text{r}} + \boldsymbol{t}^{\text{f}} && \forall j = 0, \dots, m \\
& \left(\boldsymbol{\sigma}_0 + \sum_{j=1}^{m} \boldsymbol{\sigma}_j \zeta_j \right) \in G(\boldsymbol{\zeta}) \; \forall \boldsymbol{\zeta} \in \mathcal{U} \\
& \bar{\lambda} \leq \lambda_0 + \sum_{j=1}^{m} \lambda_j \zeta_j && \forall \boldsymbol{\zeta} \in \mathcal{U}
\end{aligned} \tag{AARC}$$

in which we removed the uncertainty from the objective function and replaced the minimization over $\boldsymbol{\zeta}$ with robust constraints. Note that equality constraints depending on $\boldsymbol{\zeta}$ have been re-expressed by identifying the corresponding terms of the expansion in terms of ζ_i since $\mathcal{U}$ is full dimensional.

2.5 Comparison Between the Different Approaches

Summarizing, (RC) is the most conservative formulation yielding the smallest limit load. (AARC) is more flexible since it considers additional static variables $\boldsymbol{\sigma}_j, \lambda_j$ for $j = 1, \dots, m$ and reduces to (RC) if we fix all $\boldsymbol{\sigma}_j = 0$. As mentioned, (ARC) is less conservative than (AARC) since we allow for more general decision rules but is generally untractable. Finally, all of these formulations guard against all possible realizations of the uncertainty such that we have the following ordering:

$$\lambda_{\text{RC}} \leq \lambda_{\text{AARC}} \leq \lambda_{\text{ARC}} \leq \lambda^+(\boldsymbol{\zeta}) \quad \forall \boldsymbol{\zeta} \in \mathcal{U} \tag{8}$$

In the remainder of this work, the focus is put on the tractability of the different formulations. For (RC) to be tractable, the characterization of the safe domain G_{RC} must be tractable. Section 3 discusses conditions for which exact or approximate tractable formulations can be obtained. Tractable formulations of (AARC) are then discussed in Sect. 4.

3 Robust Strength Conditions

3.1 Uncertain Strength Conditions and a Tractable Approximation

Tractability of robust formulations such as (AARC) is essentially driven by how the uncertain strength criterion G depends on $\boldsymbol{\zeta}$. Unfortunately, we are not aware of any general results. However, in most applications, such uncertain constraints can be written in the following form:

$$g(\boldsymbol{\sigma} + \boldsymbol{\Sigma}\boldsymbol{\zeta}) \leq 1 - \boldsymbol{b}^{\text{T}}\boldsymbol{\zeta}, \quad \forall \boldsymbol{\zeta} \in \mathcal{U} \tag{9}$$

with $\boldsymbol{\sigma} \in \mathbb{R}^d$, $\boldsymbol{\Sigma} \in \mathbb{R}^{d\times m}$, d being the dimension of the stress space, $\boldsymbol{b} \in \mathbb{R}^m$ and g is a convex homogeneous function.

Exact reformulations of such a constraint are possible only if G or $\mathcal{U}$ is polyhedral. In the general case, one can benefit from the following safe approximation due to [9]: the robust constraint (9) can be safely approximated as follows:

$$g(\boldsymbol{\sigma}) + \|\boldsymbol{s}\|_* \leq 1 \tag{10}$$

where $\|\cdot\|_*$ is the dual norm of $\|\cdot\|$ defined as:

$$\|\boldsymbol{z}\|_* = \sup_{\|\boldsymbol{x}\|\leq 1} \boldsymbol{z}^{\mathrm{T}}\boldsymbol{x} \tag{11}$$

and where for $j = 1, \dots, m$:

$$s_j = \max\{g(\boldsymbol{\Sigma}_j) + b_j, g(-\boldsymbol{\Sigma}_j) - b_j\} \tag{12}$$

with $\boldsymbol{\Sigma}_j$ denoting the j-th column of $\boldsymbol{\Sigma}$.

3.2 Illustrative Application on a Robust Mohr-Coulomb Criterion

Let us consider the case of a Mohr-Coulomb strength criterion where the cohesion c and the friction angle ϕ are uncertain. A negative correlation is often encountered between both parameters, i.e. soils with low cohesion tend to exhibit higher friction angles than with higher cohesion. We denote by ρ the correlation coefficient between c and ϕ, with typical values ranging from -0.5 to -0.9 [27]. Let us therefore consider that $\boldsymbol{k} = (c, \phi)$ is given by:

$$\boldsymbol{k}(\boldsymbol{\zeta}) = \boldsymbol{k}_0 + \boldsymbol{K}\boldsymbol{\zeta}, \quad \text{for } \boldsymbol{\zeta} \in \mathcal{U} \tag{13}$$

where $\boldsymbol{k}_0$ corresponds to the nominal values and where the "correlation" matrix $\boldsymbol{K}$ is such that:

$$\boldsymbol{K}\boldsymbol{K}^{\mathrm{T}} = \begin{bmatrix} \Delta c^2 & \rho\Delta c\Delta\phi \\ \rho\Delta c\Delta\phi & \Delta\phi^2 \end{bmatrix} \quad \text{i.e. } \boldsymbol{K} = \begin{bmatrix} \Delta c & 0 \\ \rho\Delta\phi & \Delta\phi\sqrt{1-\rho^2} \end{bmatrix} \tag{14}$$

where Δc, $\Delta\phi$ are the parameters typical variations and are assumed to be positive. Note that if such variations were taken as the standard deviations of the corresponding parameters, $\boldsymbol{K}\boldsymbol{K}^{\mathrm{T}}$ would be the corresponding covariance matrix.

Figure 2a illustrates the corresponding uncertainty sets obtained in the case $c = 1$ MPa, $\phi_0 = 30°$, $\Delta c = 150$ kPa, $\Delta\phi = 5°$, $\rho = 0$ and for various choices for the norm involved in the definition of $\mathcal{U}$, resulting in a corresponding L_1 (diamond shape), L_2 (elliptic shape) or L_∞ (rectangular shape) ball in physical space. Figure 2b shows the same uncertainty sets in the case of a negative correlation $\rho = -0.5$ which results in similar polyhedral or elliptic sets skewed along the negative diagonal which encodes the negative correlation coefficient. Let us point out that the previous choices for the uncertainty set result in simple convex set but more complex sets could also be considered, based for instance on available data regarding cohesion and friction angle pairs.

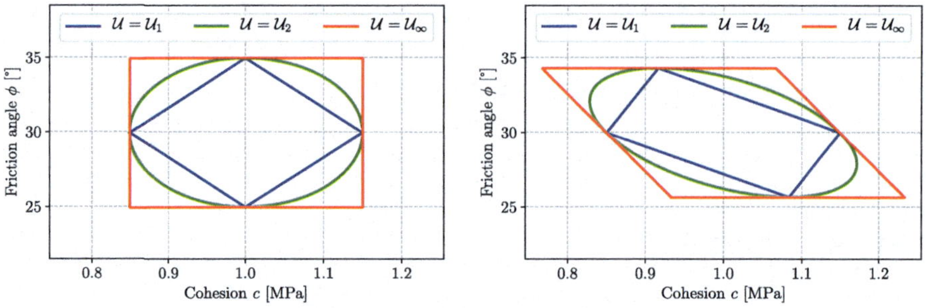

(a) $\rho = 0$ (no correlation)

(b) $\rho = -0.5$ (negative correlation)

Fig. 2 Uncertainty sets of cohesion and friction angles for $c = 1$ MPa, $\phi_0 = 30°$, $\Delta c = 150$ kPa, $\Delta\phi = 5°$ for various sets $\mathcal{U}_p$ corresponding to a L_p unit ball

The robust counterpart of the Coulomb criterion therefore reads:

$$\sigma_1 - \sigma_3 + (\sigma_1 + \sigma_3)\sin\phi(\boldsymbol{\zeta}) - 2c(\boldsymbol{\zeta})\cos\phi(\boldsymbol{\zeta}) \leq 0 \qquad \forall\boldsymbol{\zeta} \in \mathcal{U} \tag{15}$$

where σ_1 (resp. σ_3) is the maximum (resp. minimum) principal stress.

Assuming that the variations Δc, $\Delta\phi$ are small, linearization around $\boldsymbol{k}_0$ results in:

$$\begin{aligned}
&\sigma_1 - \sigma_3 + (\sigma_1 + \sigma_3)(\sin\phi_0 + \cos(\phi_0)(K_{21}\zeta_1 + K_{22}\zeta_2)) \\
&\quad - 2(c_0 + K_{11}\zeta_1)\cos\phi_0 \\
&\quad + 2c_0\sin\phi_0(K_{21}\zeta_1 + K_{22}\zeta_2) \leq 0 \qquad \forall\boldsymbol{\zeta} \in \mathcal{U}
\end{aligned} \tag{16}$$

with K_{ij} being the components of $\boldsymbol{K}$ defined in (14).

This yields the following robust counterpart:

$$\sigma_1 - \sigma_3 + (\sigma_1 + \sigma_3)\sin\phi_0 - 2c_0\cos\phi_0 + \|\boldsymbol{s}\|_* \leq 0 \tag{17}$$

where:

$$\boldsymbol{s} = \begin{pmatrix} |\left((\sigma_1 + \sigma_3)\cos\phi_0 + 2c_0\sin\phi_0\right)\rho\Delta\phi - 2\Delta c\cos\phi_0| \\ |(\sigma_1 + \sigma_3)\cos(\phi_0) + 2c_0\sin\phi_0|\sqrt{1-\rho^2}\Delta\phi \end{pmatrix} \tag{18}$$

Let us now investigate the simple case of no cross-correlation $\rho = 0$ with $\mathcal{U} = \{(\zeta_1, \zeta_2) \text{ s.t. } \|\boldsymbol{\zeta}\|_\infty \leq 1\}$. The previous expression reduces to:

$$\boldsymbol{s} = \begin{pmatrix} 2\Delta c\cos\phi_0 \\ ((\sigma_1 + \sigma_3)\cos(\phi_0) + 2c_0\sin\phi_0)\,\Delta\phi \end{pmatrix} \tag{19}$$

$$\|\boldsymbol{s}\|_* = \|\boldsymbol{s}\|_1 = 2\Delta c\cos\phi_0 + |(\sigma_1 + \sigma_3)\cos(\phi_0) + 2c_0\sin\phi_0|\,\Delta\phi \tag{20}$$

so that the robust Mohr-Coulomb criterion (17) reduces to:

$$\begin{aligned}&\sigma_1 - \sigma_3 + (\sigma_1 + \sigma_3)\sin\phi_0 \\ &+ |(\sigma_1 + \sigma_3)\cos(\phi_0) + 2c_0 \sin\phi_0|\,\Delta\phi \leq 2(c_0 - \Delta c)\cos\phi_0\end{aligned} \tag{21}$$

which can be further expressed as follows:

$$\begin{cases}\sigma_1 - \sigma_3 + (\sigma_1 + \sigma_3)(\sin\phi_0 + \cos(\phi_0)\Delta\phi) \leq 2c_{\text{min}}\cos\phi_0 - 2c_0\sin\phi_0\Delta\phi \\ \sigma_1 - \sigma_3 + (\sigma_1 + \sigma_3)(\sin\phi_0 - \cos(\phi_0)\Delta\phi) \leq 2c_{\text{min}}\cos\phi_0 + 2c_0\sin\phi_0\Delta\phi\end{cases} \tag{22}$$

where $c_{\text{min}} = c_0 - \Delta c$ is the worst-case cohesion. Introducing $\phi_{\text{min}} = \phi_0 - \Delta\phi$ the worst-case friction angle and $\phi_{\text{max}} = \phi_0 + \Delta\phi$ the best-case friction angle and using the fact that $\sin(\phi_{\text{max/min}}) \approx \sin\phi_0 \pm \cos(\phi_0)\Delta\phi$ and $\cos(\phi_{\text{max/min}}) \approx \cos\phi_0 \mp \sin(\phi_0)\Delta\phi$, the previous criterion is, in fact, a first-order approximation (in terms of Δc, $\Delta\phi$) to the following multi-surface criterion:

$$\begin{cases}\sigma_1 - \sigma_3 + (\sigma_1 + \sigma_3)\sin\phi_{\text{max}} \leq 2c_{\text{min}}\cos(\phi_{\text{max}}) \\ \sigma_1 - \sigma_3 + (\sigma_1 + \sigma_3)\sin\phi_{\text{min}} \leq 2c_{\text{min}}\cos(\phi_{\text{min}})\end{cases} \tag{23}$$

i.e. the obtained robust counterpart, for this specific case, (approximately) corresponds to the intersection of two Coulomb criteria with the worst-case cohesion and either the best or the worst-case friction angle. An illustration of such a result is given in Fig. 3. The yield surface corresponding to random realizations of $c(\zeta)$ and $\phi(\zeta)$ are also represented. One can indeed see that the obtained robust strength criterion forms a tight lower bound to the various realizations and is made of two sets of lines approximately characterized by the minimum and maximum friction angle ϕ_{min} and ϕ_{max}.

4 Solving Robust Limit Analysis Problems

4.1 Strength Uncertainty with Static Formulation

As discussed before, for a limit analysis problem with uncertain strength conditions, we can replace the original uncertain strength criterion by its robust counterpart when using a static decision rule for the stress field. This approximation is obviously conservative and can provide reasonable estimates of the robust limit load only when the uncertainty if of small amplitude so that the optimal stress field does not heavily depend on the uncertainty realization, making static decision rules relevant.

In this case, the resulting robust limit analysis problem is equivalent to a classical deterministic limit analysis problem in which the nominal strength criterion has been

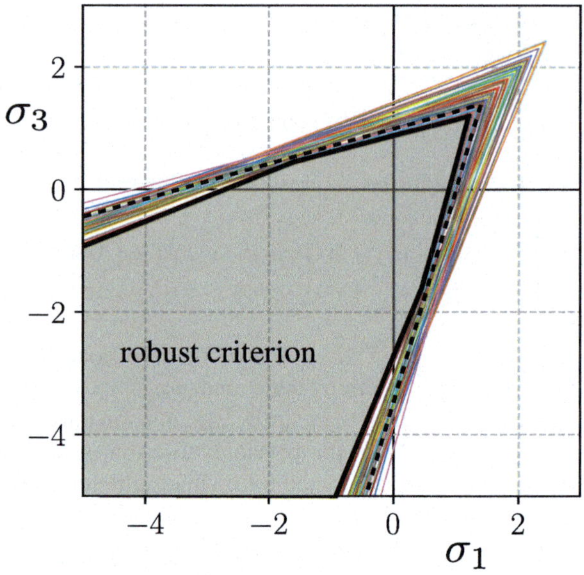

Fig. 3 Robust and uncertain Mohr-Coulomb criterion: $c_0 = 1$ MPa, $\phi_0 = 30°$, $\Delta c = 150$ kPa, $\Delta\phi = 5°$. Black dashed lines denote the nominal surface, thin coloured lines denote random realizations of the uncertain criterion. The robust domain is represented in gray and delimited by thick black lines

replaced by a smaller robust strength criterion. For a concrete implementation, the latter has to be formulated using tractable convex constraints.

As an illustration, we consider a slope stability problem for a cohesive-frictional soil with uncertain values for the cohesion and friction angle ($c = 1 \pm 0.1$ MPa and $\phi = (30 \pm 10)°$) for a pseudo-static earthquake loading $\boldsymbol{f} = (0.2g, -g)$. The corresponding load factor is interpreted here as the slope safety factor which should be larger than 1 to guarantee stability. The problem numerical resolution relies on a general-purpose domain-specific language (DSL), called `fenics_optim`, dedicated to automating the formulation and resolution of convex variational problems in a finite-element setting. The package is implemented as an add-on to the FEniCS Python interface and enables to easily formulate convex optimization problems using only a few lines of code and to discretize them in a very simple manner using various finite-element interpolation spaces. Their numerical resolution is performed efficiently using Mosek as the underlying conic programming solver [20]. More details regarding the package can be found in [11, 12] for its specific usage in the context of limit analysis.

Figure 4 represents the empirical distribution of the slope safety factor obtained for 200 random realizations of the material parameters. The nominal safety factor is slightly larger than 3 whereas the robust estimate is slightly less than 2 and indeed corresponds to the lower bound of the empirical distribution. This figure illustrates the advantage of using a robust formulation since, instead of running 200 LA computations, one is able to obtain an accurate estimate of the left part of the empirical distribution tail with a single computation. In this present case, only two uncertain parameters have been considered but the approach can be extended to a larger number of parameters. A typical example would be the modeling of soil spatial variability

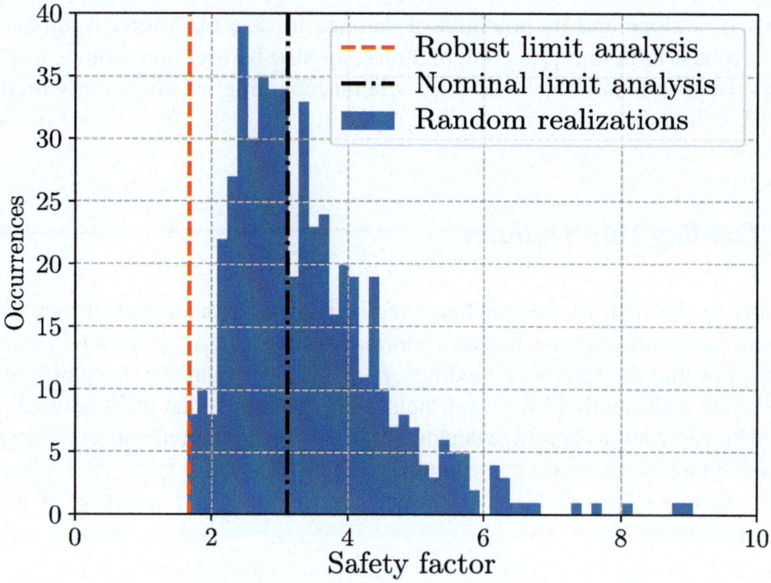

Fig. 4 Empirical distribution of the slope stability safety factor. The vertical black and red lines correspond to a single deterministic limit analysis with either nominal strength properties or using the corresponding robust strength condition

using random fields for instance. Besides, it can also be noted that the obtained estimate is not too conservative since a non-negligible number of uncertainty realizations are associated with a safety factor close to this robust estimate. Finally, it has to be pointed out that the variability on the friction angle induces a large variability on the obtained safety factor, explaining the difference between a nominal factor of 3 and a robust estimate around 2. This observation is further confirmed by the shape of the collapse mechanisms represented in Fig. 5. In the robust case, the collapse mechanism involves a much larger volume of soil than the nominal case since the most critical scenario corresponds to a smaller friction angle. Estimating the amount of soil mass mobilized during slope failure is an important point when assessing the

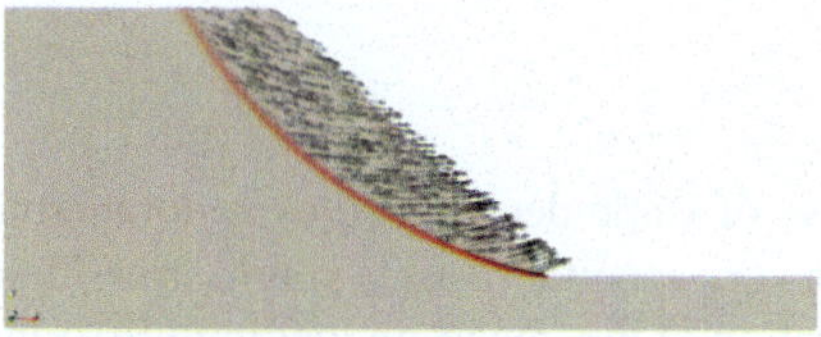

(a) Nominal collapse mechanism

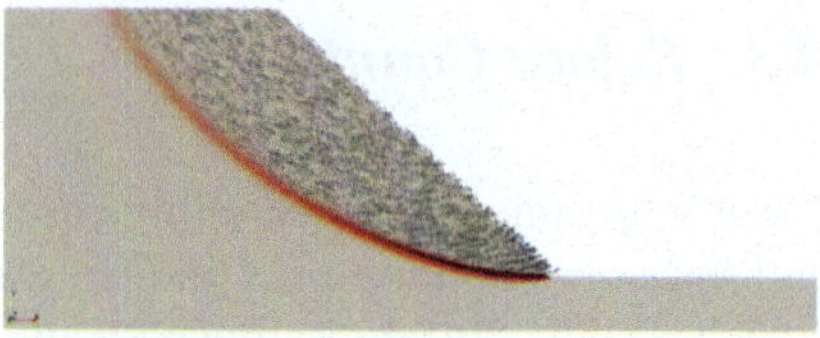

(b) Robust collapse mechanism

Fig. 5 Collapse mechanism and concentrated dissipation in slip lines for the nominal and robust case

stability of a slope and its potential of damage in case of failure. Again, one can see that robust limit analysis computations can also be used to obtain a worst-case estimate of such a mobilized soil mass when accounting for uncertainty on the soil material parameters.

4.2 Loading Uncertainties

Similarly to [16, 19], we assume here that the fixed distributed and surface loadings $\boldsymbol{f}^{\mathrm{f}}, \boldsymbol{t}^{\mathrm{f}}$ are uncertain and vary, around a nominal value, inside a convex set. In particular, we consider that the reference loadings $\boldsymbol{f}^{\mathrm{r}}, \boldsymbol{t}^{\mathrm{r}}$ are deterministic. Assuming them to be uncertain adds another layer of difficulty due to the fact that the loading direction along which one has to optimize depends on the uncertainty realization. This specific case will be left for a future contribution.

Without loss of generality, we characterize the uncertain variation of the fixed loadings as follows:

$$\boldsymbol{f}^{\mathrm{f}}(\zeta) = \boldsymbol{f}_0^{\mathrm{f}} + \sum_{j=1}^{m} \boldsymbol{f}_j^{\mathrm{f}} \zeta_j = \boldsymbol{f}_0^{\mathrm{f}} + \boldsymbol{F}^{\mathrm{f}} \zeta \tag{24a}$$

$$\boldsymbol{t}^{\mathrm{f}}(\zeta) = \boldsymbol{t}_0^{\mathrm{f}} + \sum_{j=1}^{m} \boldsymbol{t}_j^{\mathrm{f}} \zeta_j = \boldsymbol{t}_0^{\mathrm{f}} + \boldsymbol{T}^{\mathrm{f}} \zeta \tag{24b}$$

where we introduced the matrices $\boldsymbol{F}^{\mathrm{f}} = [(\boldsymbol{f}_j^{\mathrm{f}})_{j=1,\dots,m}]$ and $\boldsymbol{T}^{\mathrm{f}} = [(\boldsymbol{t}_j^{\mathrm{f}})_{j=1,\dots,m}]$ and where $\zeta \in \mathcal{U}$ with $\mathcal{U}$ a given convex uncertainty set. The corresponding uncertain limit analysis problem therefore reads:

$$\begin{aligned} \lambda^+(\zeta) = \max_{\lambda,\boldsymbol{\sigma}} \ & \lambda \\ \text{s.t. } & \div \boldsymbol{\sigma} + \lambda \boldsymbol{f}^{\mathrm{r}} + \boldsymbol{f}_0^{\mathrm{f}} + \boldsymbol{F}^{\mathrm{f}} \zeta = 0 \\ & \boldsymbol{\sigma} \cdot \boldsymbol{n} = \lambda \boldsymbol{t}^{\mathrm{r}} + \boldsymbol{t}_0^{\mathrm{f}} + \boldsymbol{T}^{\mathrm{f}} \zeta \\ & \boldsymbol{\sigma} \in G \end{aligned} \tag{25}$$

4.3 Robust Counterpart

Clearly, for this load uncertainty case, the use of static decision rules is doomed to fail since one cannot expect finding, except in very specific cases, a single stress field which is statically admissible with any realization of the uncertain loading (24). One must therefore resort to an adjustable robust optimization which, similarly to (2), reads:

$$\begin{aligned}\lambda_{\text{ARC}} = \max_{\bar{\lambda}}\ & \bar{\lambda} \\ \text{s.t. } & \forall \zeta \in \mathcal{U}, \exists \boldsymbol{\sigma}, \lambda \text{ s.t. } \div \boldsymbol{\sigma} + \lambda \boldsymbol{f}^{\mathrm{r}} + \boldsymbol{f}_0^{\mathrm{f}} + \boldsymbol{F}^{\mathrm{f}} \boldsymbol{\zeta} = 0 \\ & \qquad \boldsymbol{\sigma} \cdot \boldsymbol{n} = \lambda \boldsymbol{t}^{\mathrm{r}} + \boldsymbol{t}_0^{\mathrm{f}} + \boldsymbol{T}^{\mathrm{f}} \boldsymbol{\zeta} \\ & \qquad \boldsymbol{\sigma} \in G \\ & \qquad \bar{\lambda} \leq \lambda \end{aligned} \tag{26}$$

Again, in order to obtain a safe and tractable approximation to the above robust formulation, we resort to the use of the affine decision rules (6) and obtain the following AARC:

$$\begin{aligned}\lambda_{\text{AARC}} = \max_{\boldsymbol{\sigma}_i, \lambda_i} \min_{\zeta \in \mathcal{U}}\ & \lambda_0 + \sum_{j=1}^{m} \lambda_j \zeta_j \\ \text{s.t. } & \div \left(\boldsymbol{\sigma}_0 + \sum_{j=1}^{m} \boldsymbol{\sigma}_j \zeta_j \right) + \left(\lambda_0 + \sum_{j=1}^{m} \lambda_j \zeta_j \right) \boldsymbol{f}^{\mathrm{r}} + \boldsymbol{f}_0^{\mathrm{f}} + \boldsymbol{F}^{\mathrm{f}} \boldsymbol{\zeta} = 0 \\ & \left(\boldsymbol{\sigma}_0 + \sum_{j=1}^{m} \boldsymbol{\sigma}_j \zeta_j \right) \cdot \boldsymbol{n} = \left(\lambda_0 + \sum_{j=1}^{m} \lambda_j \zeta_j \right) \boldsymbol{t}^{\mathrm{r}} + \boldsymbol{t}_0^{\mathrm{f}} + \boldsymbol{T}^{\mathrm{f}} \boldsymbol{\zeta} \\ & \left(\boldsymbol{\sigma}_0 + \sum_{j=1}^{m} \boldsymbol{\sigma}_j \zeta_j \right) \in G \end{aligned} \tag{27}$$

which can be further formulated as follows:

$$\begin{aligned}\lambda_{\text{AARC}} = \max_{\bar{\lambda}, \boldsymbol{\sigma}_i, \lambda_i}\ & \bar{\lambda} \\ \text{s.t. } & \div(\boldsymbol{\sigma}_i) + \lambda_i \boldsymbol{f}^{\mathrm{r}} + \boldsymbol{f}_i^{\mathrm{f}} = 0 \ \forall i = 0, \ldots, m \\ & \boldsymbol{\sigma}_i \cdot \boldsymbol{n} = \lambda_i \boldsymbol{t}^{\mathrm{r}} + \boldsymbol{t}_i^{\mathrm{f}} \qquad \forall i = 0, \ldots, m \\ & \left(\boldsymbol{\sigma}_0 + \sum_{j=1}^{m} \boldsymbol{\sigma}_j \zeta_j \right) \in G \ \ \forall \zeta \in \mathcal{U} \\ & \bar{\lambda} \leq \lambda_0 + \sum_{j=1}^{m} \lambda_j \zeta_j \qquad \forall \zeta \in \mathcal{U} \end{aligned} \tag{28}$$

Clearly, (28) bears striking similarities with (AARC) in the sense that we look for $1+m$ stress fields statically admissible with a given loading (here we have an additional fixed loading for each $j = 1, \ldots, m$ compared to (AARC)). In particular, uncertainty has been removed from the equilibrium equations whereas only the last two constraints are robust ones which must be reformulated. In particular, the robust strength constraint can be reformulated, either exactly or approximately, using the results of Sect. 3. Finally, the last constraint can be reformulated as follows using the dual norm $\|\cdot\|_*$ to the norm involved in the definition of the uncertainty set $\mathcal{U}$. Indeed, introducing the vector $\boldsymbol{\Lambda} = (\lambda_j)_{j=1,\ldots,m}$, we can write:

$$\begin{aligned} &\bar{\lambda} \leq \lambda_0 + \boldsymbol{\Lambda}^{\mathrm{T}}\zeta \quad \forall \zeta \in \mathcal{U} \\ \Leftrightarrow \quad &\bar{\lambda} + \max_{\zeta \in \mathcal{U}} \{-\boldsymbol{\Lambda}^{\mathrm{T}}\zeta\} \leq \lambda_0 \\ \Leftrightarrow \quad &\bar{\lambda} + \|-\boldsymbol{\Lambda}\|_* \leq \lambda_0 \end{aligned} \tag{29}$$

which results in a tractable convex constraint for classical uncertainty sets.

In conclusion, we see that the robust reformulation of (28) is close to a classical limit analysis problem except that the number of stress fields and load factor is now $1 + m$ and that the strength criterion will couple all stress variables in a single constraint which would have been exactly or approximately reformulated to guarantee the robust constraint $\boldsymbol{\sigma}_0 + \sum_{j=1}^{m} \boldsymbol{\sigma}_j \zeta_j \in G, \ \forall \zeta \in \mathcal{U}$. As a result, the resulting robust problem will still be convex and representable using conic constraints. It will however be much larger in size than a deterministic problem.

5 Conclusions

In this work, we have proposed an extension of limit analysis theory to an uncertain setting using the robust optimization (RO) framework. Since limit analysis problems can be formulated as convex optimization programs, we can naturally apply robust optimization concepts when considering uncertain data. We covered two different sources of uncertainty, namely strength and loading uncertainty.

An important aspect of RO is related to the use of static or adjustable optimization variables. In the present LA case, it amounts to deciding whether we consider the stress field and load multiplier that we optimize for to be independent or dependent on the uncertain parameters. The main feature of RO is to propose tractable reformulations of uncertain constraints as standard deterministic constraints, possibly involving a much larger number of variables. Various results have been obtained for the two cases of static and adjustable formulations.

First, the use of static variables results in the *static robust counterpart* (RC) which deserves the following comments:

- (RC) is a standard deterministic LA problem where the uncertain strength criterion is replaced with a safe estimate called the *robust strength domain* G_{RC}.
- The robust strength domain is the smallest possible strength domain corresponding to all uncertainty realizations.
- Obtaining an explicit expression for the robust domain depends on how constraints depend on the uncertain parameters.
- Tractable approximations of the robust domain have been provided and illustrated on the case of a Mohr-Coulomb example.
- The resulting LA problem can be solved using standard tools and the resulting load estimate is a conservative safe approximation for all realization.

Clearly, this is a very conservative approach. In particular, finding such a stress field is not always possible. Our experience suggests that static formulations can be used only when considering strength uncertainty and in the case where this uncertainty is of small amplitude. Intuitively, this corresponds to the fact that the collapse stress field is only mildly perturbed by the realization of the uncertainty.

Second, in the general case where adjustable formulations are needed, simple decision rules must be chosen for the robust problem to be tractable. In particular, the case of loading uncertainty can only be tackled using adjustable formulations. More precisely:

- *Affine decision rules* assume an affine dependence of the load factor and stress field with respect to the uncertain parameters.
- Robust strength constraints take the form (9) which can be reformulated either exactly or approximately.
- The corresponding affinely adjustable problem can be reformulated to yield the deterministic optimization problem (AARC).
- The latter involves a much larger number of optimization variables compared to the nominal limit analysis problem. This number depends on the dimension of the uncertainty space.

Further research will focus on the numerical implementation of the proposed formulations in order to assess their efficiency on more involved examples. In this respect, specific strategies should probably be investigated in order to reduce the computational cost of the corresponding large-scale optimization problems, especially when considering AARC formulations. Analyzing such more advanced examples would therefore shed light on the necessity, or not, of considering more complex decision rules than affine rules such as piecewise-linear or nonlinear decision rules.

References

1. Ali, A.: Application of stochastic limit analysis to geotechnical stability problems. Ph.D. thesis, Centre of Excellence for Geotechnical Science and Engineering, University of Newcastle, Australia (2016)
2. Alibrandi, U., Ricciardi, G.: The use of stochastic stresses in the static approach of probabilistic limit analysis. Int. J. Numer. Meth. Eng. **73**(6), 747–782 (2008)
3. Augusti, G., Baratta, A., Casciati, F.: Probabilistic methods in structural engineering. CRC Press (1984)
4. Ben-Haim, Y.: Info-gap decision theory: decisions under severe uncertainty. Elsevier (2006)
5. Ben-Tal, A., El Ghaoui, L., Nemirovski, A.: Robust optimization. Princeton University Press (2009)
6. Ben-Tal, A., Goryashko, A., Guslitzer, E., Nemirovski, A.: Adjustable robust solutions of uncertain linear programs. Math. Program. **99**(2), 351–376 (2004)
7. Bertsimas, D., Brown, D.B., Caramanis, C.: Theory and applications of robust optimization. SIAM Rev. **53**(3), 464–501 (2011)
8. Bertsimas, D., Sim, M.: The price of robustness. Oper. Res. **52**(1), 35–53 (2004)
9. Bertsimas, D., Sim, M.: Tractable approximations to robust conic optimization problems. Math. Program. **107**(1), 5–36 (2006)
10. Bjerager, P., Ditlevsen, O.: Influence of uncertainty of local friction angle and cohesion on the stability of slope in coulomb soil. In: Reliability Theory and Its Application in Structural and Soil Mechanics, pp. 567–579. Springer, Berlin (1983)

11. Bleyer, J.: Automating the formulation and resolution of convex variational problems: applications from image processing to computational mechanics. ACM Trans. Math. Softw. (TOMS) **46**(3), 1–33 (2020)
12. Bleyer, J., Hassen, G.: Automated formulation and resolution of limit analysis problems. Comput. Struct. **243**, 106341 (2021)
13. Hill, R.: The Mathematical Theory of Plasticity. Clarendon Press, Oxford (1950)
14. Huang, J., Lyamin, A., Griffiths, D., Krabbenhoft, K., Sloan, S.: Quantitative risk assessment of landslide by limit analysis and random fields. Comput. Geotech. **53**, 60–67 (2013)
15. Jiang, S.H., Huang, J., Griffiths, D., Deng, Z.P.: Advances in reliability and risk analyses of slopes in spatially variable soils: a state-of-the-art review. Comput. Geotech. **141**, 104498 (2022)
16. Kanno, Y., Takewaki, I.: Worst case plastic limit analysis of trusses under uncertain loads via mixed 0–1 programming. J. Mech. Mater. Struct. **2**(2), 245–273 (2007)
17. Kasama, K., Whittle, A.J.: Effect of spatial variability on the slope stability using random field numerical limit analyses. Georisk: Assess. Manag. Risk Eng. Syst. Geohazards **10**(1), 42–54 (2016)
18. Marandi, A., Den Hertog, D.: When are static and adjustable robust optimization problems with constraint-wise uncertainty equivalent? Math. Program. **170**(2), 555–568 (2018)
19. Matsuda, Y., Kanno, Y.: Robustness analysis of structures based on plastic limit analysis with uncertain loads. J. Mech. Mater. Struct. **3**(2), 213–241 (2008)
20. MOSEK, A.: The MOSEK optimization API for Python 8.1.0 (2018). http://docs.mosek.com/8.1/pythonapi/index.htm
21. Salençon, J.: Calcul à la rupture et analyse limite. Presses de l'Ecole Nationale des Ponts et Chaussées (1983)
22. Salençon, J.: Yield Design. ISTE Ltd., Wiley, Inc., London, Hoboken (2013)
23. Staat, M.: Limit and shakedown analysis under uncertainty. Int. J. Comput. Methods **11**(03), 1343008 (2014)
24. Takeda, A., Taguchi, S., Tütüncü, R.: Adjustable robust optimization models for a nonlinear two-period system. J. Optim. Theory Appl. **136**(2), 275–295 (2008)
25. Tran, N.T., Staat, M.: Direct plastic structural design under lognormally distributed strength by chance constrained programming. Optim. Eng. **21**(1), 131–157 (2020)
26. Tran, N., Tran, T.N., Matthies, H., Stavroulakis, G., Staat, M.: Shakedown analysis under stochastic uncertainty by chance constrained programming. In: Advances in Direct Methods for Materials and Structures, pp. 85–103. Springer, Berlin (2018)
27. Wang, Y., Akeju, O.V.: Quantifying the cross-correlation between effective cohesion and friction angle of soil from limited site-specific data. Soils Found. **56**(6), 1055–1070 (2016)
28. Yang, L.F., Yu, B., Ju, J.W.: System reliability analysis of spatial variance frames based on random field and stochastic elastic modulus reduction method. Acta Mech. **223**(1), 109–124 (2012)

Advances of the RSDM-S: Robustness and Fast Convergence Issues

Ioannis A. Kapogiannis and Konstantinos V. Spiliopoulos

Abstract The Residual Stress Decomposition Method (RSDM) is an iterative numerical procedure which has been developed to estimate, in a direct way, the kind of asymptotic stress states under cyclic loading of inelastic structures. The method was the basis to formulate another numerical procedure, which was named RSDM-S, to establish safety margins for elastic shakedown under mechanical and/or thermal loads. The method exploits the expected cyclic nature of the residual stresses of the asymptotic cycle state. Starting from a load factor high above shakedown an iterative procedure shrinks the loading domain until the conditions of the limit cycle, which marks the shakedown state, are met. The procedure consists of two loops an external incremental that reduces the load factor and an internal iterative loop that establishes a cyclic state for the current load factor. The current work refers to advancements of the method in terms of robustness and fast convergence. It discusses the efficiency of the numerical scheme used which is proved to have a continuous descent towards the shakedown factor with superlinear convergence. Examples of application of structures undergoing various kinds of cyclic actions like, mechanical or thermomechanical loads or cyclic imposed displacements are presented, and shakedown domains are constructed.

1 Introduction

The last decades, structures and structural components are designed to operate beyond the elastic limit in favor of material savings. Especially in case of cyclic thermomechanical loadings the allowable stresses may be greater than the yield limit. The amplitude of the cyclic load will determine the magnitude of the inelastic strains, which will be responsible either for the failure due to alternating plastic straining (low cycle fatigue) and/or incremental plastic straining (ratcheting), or for safety, through

---

I. A. Kapogiannis · K. V. Spiliopoulos (✉)
Department of Civil Engineering, Institute of Structural Analysis and Antiseismic Research, National Technical University of Athens, Zografou Campus, 157-80, Athens, Greece
e-mail: kvspilio@central.ntua.gr

G. Garcea and D. Weichert (eds.), *Direct Methods for Limit State of Materials and Structures*, Lecture Notes in Applied and Computational Mechanics 101,
https://doi.org/10.1007/978-3-031-29122-7_12

elastic shakedown. Thus, the post-elastic response of structures due to cyclic thermal and mechanical loads is always a major concern for the designer engineer.

Apart from the thermomechanical loads, support excitations due to repeated accidental loads, such as the earthquakes, may establish a pattern of cyclic imposed displacements to the structures.

The induced stresses, due to seismic actions, undergo many complete reversals in a small period like the duration of an earthquake. Designing such structures to behave elastically during earthquakes, without damage, may render the project economically unviable. Consequently, it may be necessary for the structure to suffer some damage and therefore dissipate energy input, during the earthquake. Thus, the same question arises whether after a sequence of imposed cyclic displacements, the post-elastic response will lead to a long-term stabilization of the damage, with an effect to extend the life cycle of a structure.

Previous years, in order to study the post-elastic response of a structure, one should perform step-by-step inelastic analysis based on a specific time-history. In this way one could be sure that, the structure would end up to a safe or unsafe asymptotic state. Besides the fact that this approach is time consuming and may have convergence problems, no general answer of safety will be given except for the specific load history. However, a class of numerical methods, called Direct Methods exists, (a most recent compilation of these methods may be found in [13], which may provide safety margins for any load combinations. These methods have a much lower computational cost as they bypass the transient deformation stages and search the asymptotic states, in a direct way, right from the start of the calculations.

Most direct methods deal with the shakedown problem as being a constrained optimization problem, described by the theorems of [11] and [6]. The structure is discretized with many finite elements (FE) and large-scale nonlinear mathematical programming (MP) problems must be solved. Towards this direction, general-purpose efficient optimization algorithms, like the interior point method (IPM) or conic programming, are often employed, as part of the method. These may be combined with other algorithms to assess the behavior of materials that require a high degree of intensive computational burden (e.g. [3]).

A strain driven algorithm that converts the MP problem of the lower bound shakedown theorem to an equivalent incremental-iterative problem of fictitious elastoplastic steps has been proposed and recently applied to the analysis of fiber-based 3D framed structures [10]

The linear matching method (LMM) is an iterative method that produces a sequence of linear elastic solutions by modifying the elastic moduli of the various parts of the structure so that the stress equals the yield stress. Thus, in this respect each time it matches a linear problem to a plasticity problem. The method was developed in the context of shakedown [14]. The method was expanded and employed, since then, to many applications in problems of structural mechanics, e.g. to find ratcheting limits [2, 8], and more recently in shakedown with hardening and thermal effects [9]

A simplified Direct Method originally conceived for creep problems [15, 16] was further developed and applied to account for plastic behavior [18]. The metod was called Residual Stress Decomposition method (RSDM) [19]. It is an iterative stress driven direct method which can predict the final asymptotic state of a cyclically loaded structure (either shakedown, or alternating plasticity or incremental collapse). In this asymptotic stress state, the residual stresses have cyclic behavior, thus they can be decomposed in Fourier series whose coefficients may be calculated in an iterative manner by satisfying compatibility and equilibrium at time points inside the cycle. The method was formulated to calculate the shakedown limit (RSDM-S) of structures subjected to cyclic thermomechanical loads and multidimensional loading domains [20–22].

An upgrade of the RSDM-S appeared quite recently [17]. The upgrade was both in the robustness and the numerical efficiency. The robustness is guaranteed as the sequence of the iterative steps is theoretically proved to be monotonically decreasing towards the final solution. A numerical scheme that possesses a superlinear convergence makes it a very fast procedure. The method was also formulated to cater for cyclic imposed displacements.

In the present work, these issues of robustness and convergence are further discussed and analyzed. The efficiency of the approach is further demonstrated through new applications to structures under mechanical, thermomechanical, or imposed displacements, simulating earthquake loading.

2 Theoretical Background

Let a body of volume V with surface S be subjected to mechanical load P, applied on a part of the surface S_f, prescribed displacements $\overline{\boldsymbol{u}}$, applied on another part S_{pr} and fixed displacements on another part S_u (Fig. 1).

The mechanical load and prescribed displacements are applied periodically with period T. One may assume that the minimum values of the cyclic load or prescribed

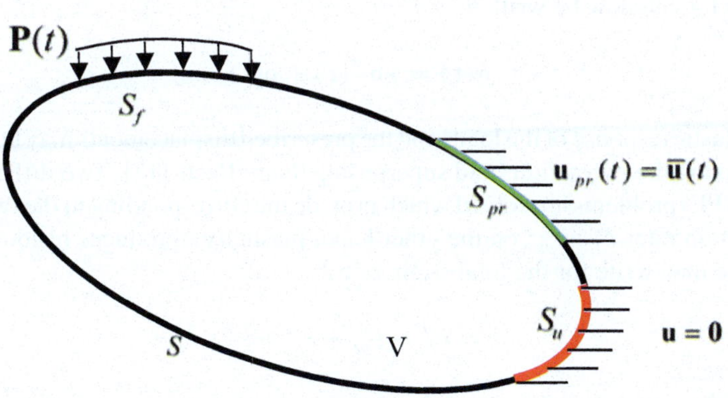

Fig. 1 Body subjected to forces and imposed displacements. Reproduced from [17]. Copyright © Elsevier Masson SAS. All rights reserved

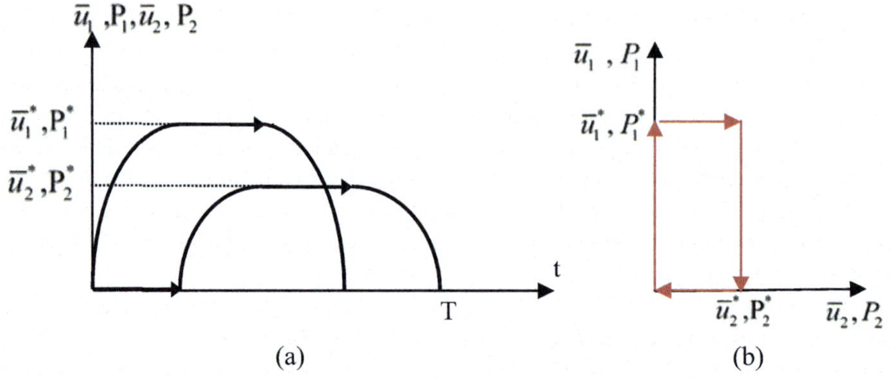

Fig. 2 Independent cyclic loading (mechanical (and/or) imposed displacement) variation over one time period **a** in time domain, **b** in loading domain. Reproduced from [17].

displacements are zero and the starred quantities represent the maximum values (Fig. 2). It has been proved [7] that if a structure shakes down under a cyclic loading program containing the vertices of the loading domain, then it will shake down for any loading path contained in this domain. Such a cyclic program may be seen in Fig. 2, in either the time domain (a), or the loading domain (b).

This domain may be isotropically varied if multiplied with a load factor γ. Thus, the idea behind RSDM-S is to find the largest loading domain for which shakedown occurs, by moving from a large value of γ to smaller ones.

In response to the cyclic loading program, the stresses in the structure at a cycle point $\tau = t/T$ (where this point is either a point in the time domain or a vertex in the loading domain) are decomposed into an elastic part $\boldsymbol{\sigma}^{el}$, in response to the applied external cyclic actions, and a residual stress part $\boldsymbol{\rho}$. In the search for the shakedown factor γ, the elastic stresses are themselves multiplied by this factor. Thus, the total stress vector can now be written:

$$\boldsymbol{\sigma}(\tau) = \gamma\boldsymbol{\sigma}^{el}(\tau) + \boldsymbol{\rho}(\tau) \tag{1}$$

The elastic response of the loads and the prescribed displacements may be obtained by separating the two actions and superposing their effects [17]. Two different finite element (FE) problems are solved which provide the corresponding to the two actions elastic strain rates $\dot{\varepsilon}_L^{el}$, $\dot{\varepsilon}_{pr}^{el}$; on the other hand, plasticity introduces residual strains. Thus, one may write for the total strain rate $\dot{\varepsilon}$:

$$\dot{\varepsilon} = \dot{\varepsilon}_L^{el} + \dot{\varepsilon}_{pr}^{el} + \dot{\varepsilon}_r^{el} + \dot{\varepsilon}^{pl} \tag{2}$$

Where $\dot{\varepsilon}^{pl}$ are the plastic strains and $\dot{\varepsilon}_r^{el}$ is the residual elastic straining. Since both terms $\dot{\varepsilon}$ and $\dot{\varepsilon}_L^{el} + \dot{\varepsilon}_{pr}^{el}$ in (2) are kinematically admissible, the sum:

$$\dot{\varepsilon}_r = \dot{\varepsilon}_r^{el} + \dot{\varepsilon}^{pl} \tag{3}$$

is also kinematically admissible. In a FE environment this may be expressed as $\dot{\varepsilon}_r = \mathbf{B}\dot{\mathbf{r}}_r$, where $\mathbf{B}$ is the well-known FE compatibility matrix between strains and FE nodal displacements.

The elastic term $\dot{\varepsilon}_r^{el}$ is related to the residual stress via the elastic material matrix $\mathbf{D}$. Thus, one may write:

$$\dot{\varepsilon}_r = \mathbf{D}^{-1}\dot{\boldsymbol{\rho}} + \dot{\varepsilon}^{pl} \rightarrow \dot{\boldsymbol{\rho}} = \mathbf{D}\dot{\varepsilon}_r - \mathbf{D}\dot{\varepsilon}^{pl} \tag{4}$$

Expressing residual strain compatibility and equilibrium of residual stresses with zero loads, one may write, from the principle of virtual work (PVW):

$$\int_V \mathbf{B}^{\mathrm{T}} \cdot \dot{\boldsymbol{\rho}} dV = 0$$

$$\rightarrow \int_V (\mathbf{B}^{\mathrm{T}}\mathbf{D}\mathbf{B}) dV \dot{\mathbf{r}}_r = \int_V \mathbf{B}^T \mathbf{D}\dot{\varepsilon}^{pl} dV \rightarrow \mathbf{K}\mathbf{P}_r = \int_V \mathbf{B}^T \mathbf{D}\dot{\varepsilon}^{pl} dV \tag{5}$$

with $\mathbf{K}$ being the standard stiffness matrix.

The cyclic nature of the residual stresses at the asymptotic cycle (e.g., [4]) allows their decomposition in Fourier series.

$$\boldsymbol{\rho}(\tau) = \frac{1}{2}\mathbf{a}_0 + \sum_{k=1}^{n} \{\cos(2k\pi\tau) \cdot \mathbf{a}_k + \sin(2k\pi\tau) \cdot \mathbf{b}_k\} \tag{6}$$

with the values of the Fourier coefficients being given, [18–20].

$$\mathbf{a}_k = -\frac{1}{k\pi} \int_0^1 \{[\dot{\rho}(\tau)](\sin 2k\pi\tau)\} d\tau \tag{7}$$

$$\mathbf{b}_k = \frac{1}{k\pi} \int_0^1 \{[\dot{\rho}(\tau)](\cos 2k\pi\tau)\} d\tau \tag{8}$$

The basis of the RSDM-S are the Eqs. (6)–(8). Very good accuracy was attained by keeping just three terms of the series, i.e. n = 3.

An upgraded numerical scheme of the RSDM-S has been very recently presented [17]. It consists of an inner and an outer loop. The outer incremental type loop updates the shakedown factor, which is then used in the inner loop to iteratively update the Fourier coefficients found by performing time integration over the values of $\dot{\boldsymbol{\rho}}$ evaluated (Eqs. (4) and (5)) at the vertices of the loading domain. The iterations of the internal loop stop when a cyclic solution has been established. This is manifested when two successive values of φ, defined by (9), coincide within a certain accuracy:

$$\varphi = \sum_{k=1}^{n} \|\mathbf{a}_k\| + \sum_{k=1}^{n} \|\mathbf{b}_k\| \tag{9}$$

If we denote by $\gamma^{(\mu)}$, the value of the current shakedown factor inside an outer iteration μ, the following formula is used to update it for the first two iterations:

$$\gamma^{(\mu+1)} = \gamma^{(\mu)} - \varphi(\gamma^{(\mu)}) \tag{10}$$

whereas for the next outer iterations the following formula is used:

$$\gamma^{(\mu+1)} = \gamma^{(\mu)} - \frac{\gamma^{(\mu)} - \gamma^{(\mu-1)}}{\varphi\big(\gamma^{(\mu)}\big) - \varphi\big(\gamma^{(\mu-1)}\big)} \cdot \varphi\big(\gamma^{(\mu)}\big) \tag{11}$$

The proposed relationship is a regular falsi procedure for finding the zero of the function $\varphi(\gamma)$, defined at the points of the convergence of the inner loops. Thus, the convergence of the outer loops is superlinear (e.g., [5]).

Given that $\gamma^{(\mu)} > \gamma^{(\mu+1)}$, for the corresponding values of φ, it will hold that $\varphi(\gamma^{(\mu)}) > \varphi(\gamma^{(\mu+1)})$. This is an important assumption to prove that φ is a monotonously descending function, as assumed in Fig. 3.

The proof of the monotonicity may be found in [17] and is illustrated in the present work (Fig. 4). It is related to the fact that because of (7) and (8) the norms of the vectors of the coefficients of the Fourier series are directly related to the norms of the residual stress rate vectors which in turn are directly related to the plastic strain vector (Eq. (4)). The value of φ, on the other hand is proportional to the length of this vector (Eq. 9), which, since it is measured through the radial return rule, is the distance from the yield surface when the total stress exceeds it. Thus, the proof is obvious, from Fig. 4, where one may see the total stress vectors OA and OB at the

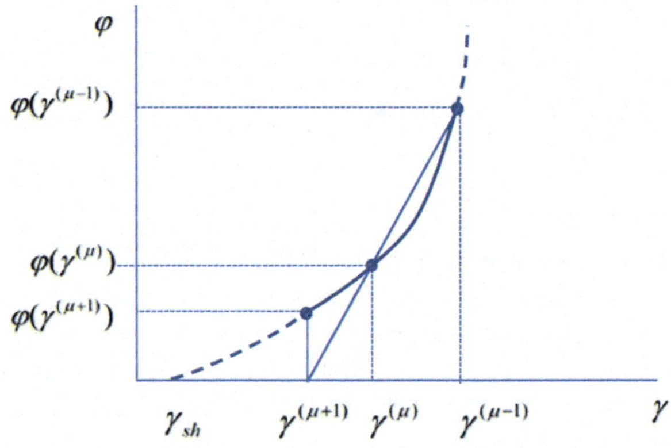

Fig. 3 Convergent sequence of solutions. Reproduced from [17].

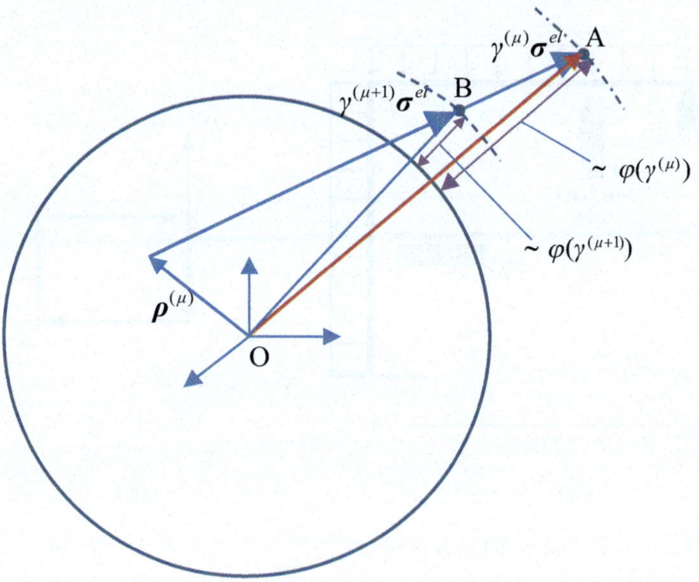

Fig. 4 Proof of the descending sequence of $\varphi(\gamma)$

end of iteration μ and at the start of iteration $\mu + 1$, respectively, with both points A and B located on the elastic stress vector $\boldsymbol{\sigma}^{el}$.

Initial starting point of the descending algorithm may be considered as three or four times the maximum elastic limit which is located at one of the vertices of the loading domain.

3 Examples of Application

In the present work, the updated RSDM-S is used to evaluate load and displacement shakedown limits in new examples. The results are validated either by performing step-by-step analyses or by comparing with the corresponding results of the bibliography. All examples highlight the speed and the accuracy of the RSDM-S.

3.1 The Simple Frame

The first example is the simple sway frame of Fig. 5a, as introduced in [12].

Two distributed loads (P_1 and P_2) act independently, varying from the value "0" to the maximum values P_1^* and P_2^*, as shown in Fig. 5b. The mechanical properties were E = 20,000 kN/cm^2, ν =0.3, σ_y = 10 kN/cm^2. The RSDM-S was run considering

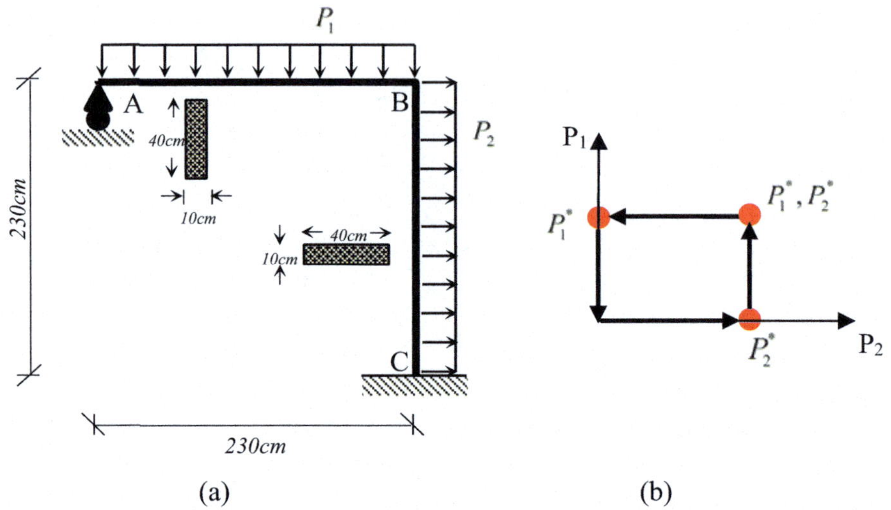

Fig. 5 **a** Geometry and loads, **b** Loading cycle

five time points of the loading cycle (the vertices of the loading domain). 350 brick elements were used for the discretization (Fig. 6).

Different shakedown limits were calculated, considering different ratios of P_1/P_2. The shakedown domain is presented in Fig. 7.

In all the cases the convergence appeared quite smooth. In Fig. 8 one may see such a convergence at point A, which was accomplished in twelve iterations.

It is pointed out that, the results are in good agreement with those presented [12].

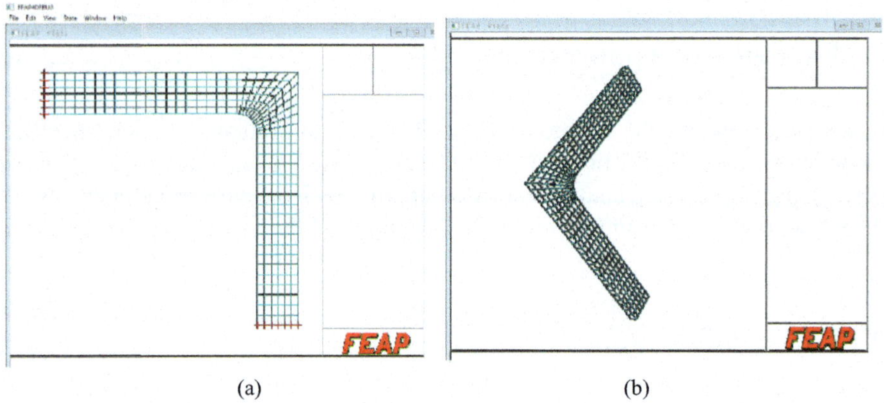

Fig. 6. 2D view and 3D view of the frame using 350 brick elements. Reproduced from [17].

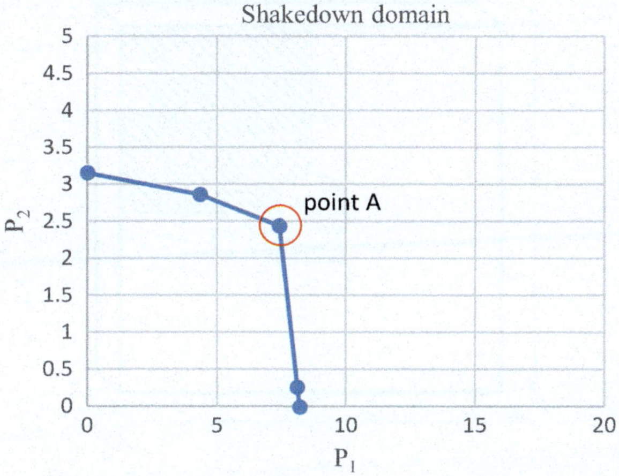

Fig. 7 Shakedown domain

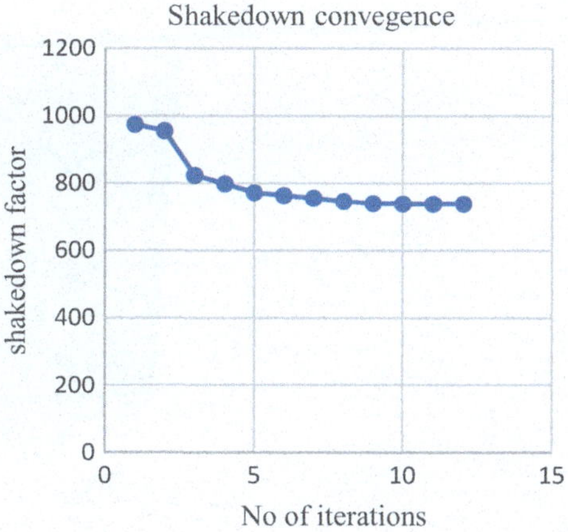

Fig. 8 Convergence diagram for point A

3.2 *The Slab with the Hole*

The numerical efficiency of the upgraded RSDM-S is further demonstrated also in the case of thermomechanical loading. The benchmark problem of the square plate with a circular hole in its center is selected. The structure is subjected to both thermal and distributed mechanical loads (Fig. 9a). The mechanical load is applied at the edge of the slab and is uniformly distributed. The temperature ranges from the inner to the outer edge according to the formulae:

$$\theta(r, t) = \theta_0 + \frac{\Delta\theta * \ln\left(\frac{2.5D}{r}\right)}{\ln(5)} \tag{12}$$

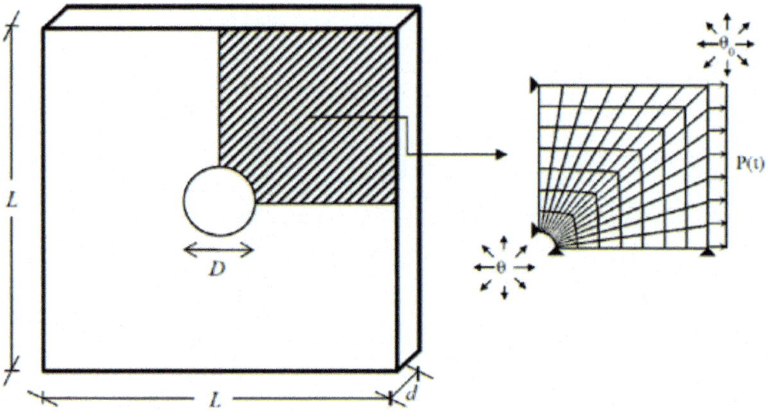

(a)

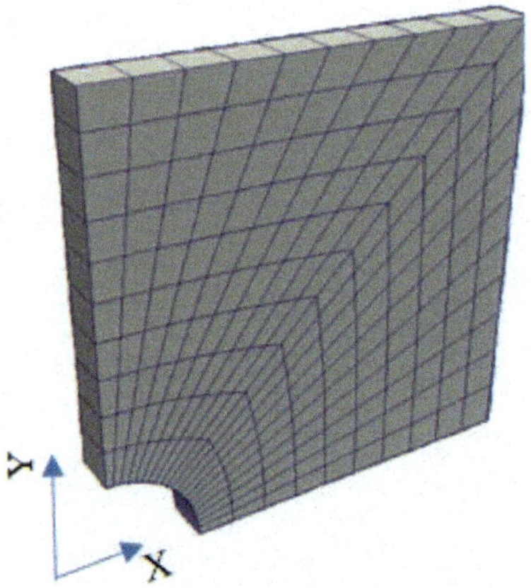

(b)

Fig. 9 **a** Geometry, **b** loading Discretization of the slab

Due to the symmetry of the geometry and the loading, only one quarter of the plate is discretized (Fig. 9b). Let D be the diameter of the circle, L the length of the slab and d the thickness, then $D/L = 0.2, d/L = 0.05$. In the present work, L is equal to 20 cm. The boundary conditions along the X-axis and the Y-axis are considered rolled. Results for the cyclic thermal load θ and the cyclic load P varying proportionally from *0* to θ^* and P^* will be investigated. The material properties are E = 180 GPa, v = 0.3 and $\sigma_y = 200$ MPa. The model consists of 220 brick elements.

The RSDM-S converged in 15 external iterations. The corresponding shakedown domain for different ratios of θ^*/P^* is presented in Fig. 10.

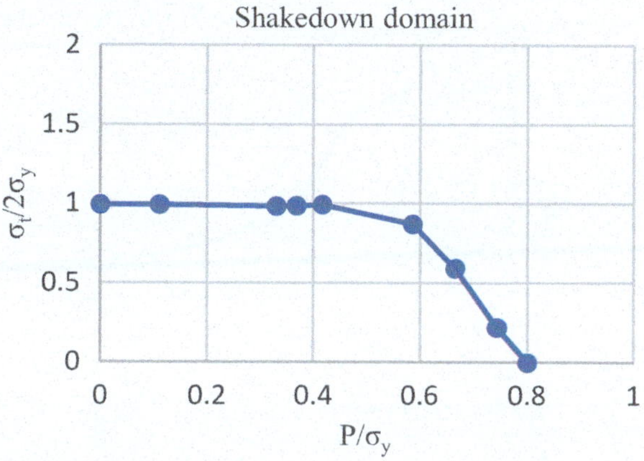

Fig. 10 Shakedown domain for the holed slab. The value σ_t corresponds to the maximum thermal stress developed in the structure

3.3 90° Pipe Elbow

Pipe elbows are met in almost all types of piping systems. Being parts of machinery configurations, pipelines, and industrial facilities, they are often subjected to earthquake loads. Thus, these components usually undergo cyclic loads and/or imposed cyclic displacements.

In the present example, the shakedown domain for a typical 90° steel pipe bend subjected to cyclic out of plane-imposed displacement is investigated. The steel elbow, of a yield stress of 360 MPa, consists of two straight pipes with an 8-inch outer diameter, 8.18 mm depth and of an equal length of 1.10 m. The left end of the pipe is considered fixed and the imposed displacement is applied at the right support with direction along the horizontal axis Z, as shown in Fig. 11.

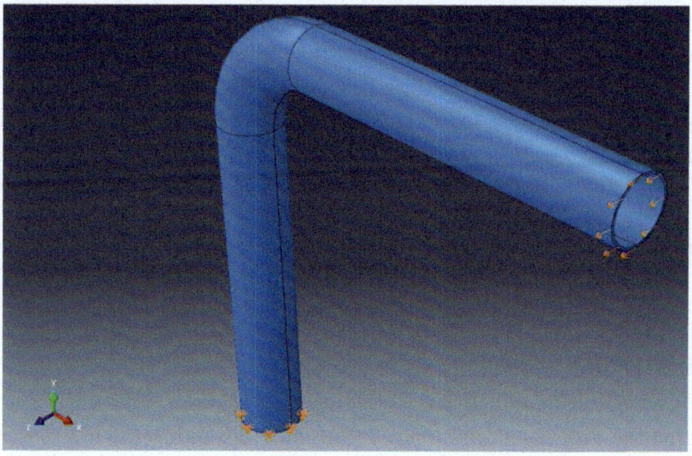

Fig. 11 The configuration of the pipe elbow

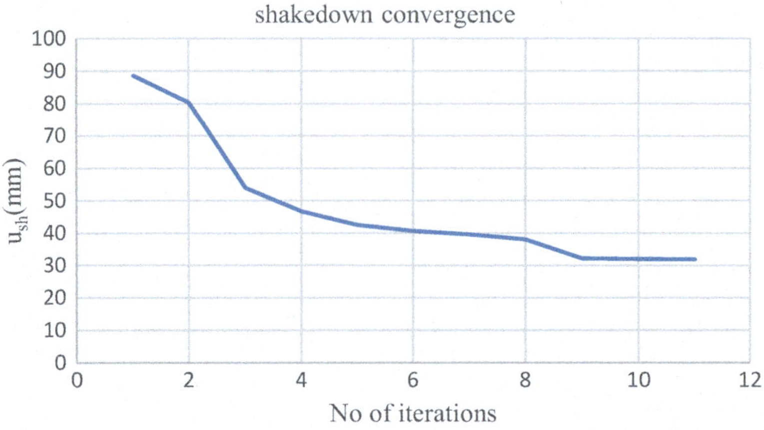

Fig. 12 Convergence diagram of the RSDM-S in the case of imposed displacement

The cross section was divided into three layers along the thickness with the structure being discretized with 10,528 hexagonal brick elements. The elbow is subjected to horizontal cyclic imposed displacement varying from 0 to u*. The yield displacement turned out to be $u_y = 18$ mm.

According to the RSDM-S the shakedown displacement is equal to $u_{sh} = 32$ mm. The convergence is quite good and the procedure is completed in 11 iterations, as shown in Fig. 12.

The shakedown limit produced by the RSDM-S was validated with results obtained by step-by-step analyses using the Abaqus software [1]. Two analyses were run, considering the amplitude of the imposed displacement higher and lower than the shakedown limit.

In the first case, the imposed displacement was equal to 60 mm and the plastic strain was found to be always increasing (ratcheting). The most stressed point was point A located near the fixed support. For this point, the accumulated plastic strain at the end of the 16th cycle of loading is shown in Fig. 13.

The evolution of the equivalent plastic strain from cycle to cycle for the point A may be seen in Fig. 14.

In the case where the imposed displacement was set equal to 20 mm, the plastic deformation appeared and locked around 1%. 50 cycles were used and the shakedown condition met right from the first cycle as presented in Fig. 15.

4 Convergence Issues

To underline the numerical efficiency of the method, the following remarks can be made concerning its convergence characteristics:

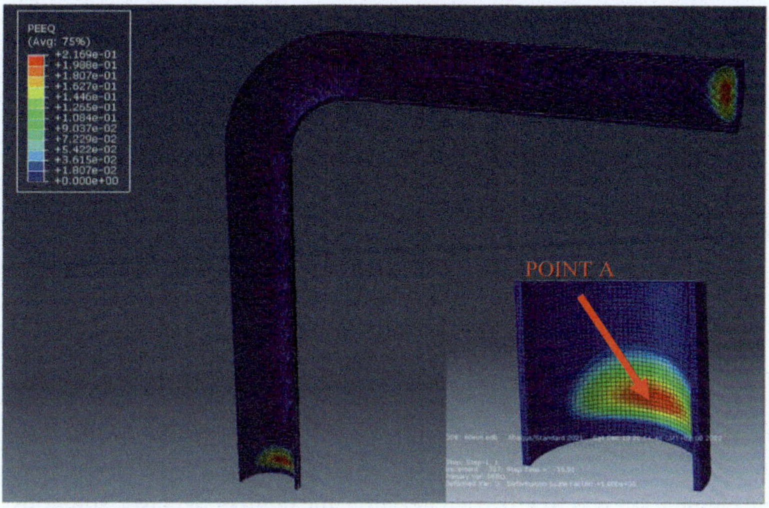

Fig. 13 Contour of equivalent plastic strain at the last loading cycle

Fig. 14 Plastic strain diagram in case of 60 mm

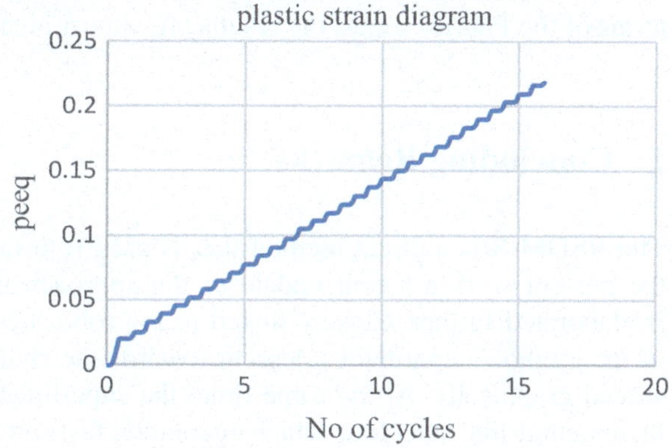

Both the internal and the external loops are controlled through the iterative values of φ (Eq. (9)) [17]. Internal loop iterations stop when the relative difference between two successive values of φ is of a tolerance of 10^{-3}, whereas the external loop iterations stop when the value of φ reaches the tolerance of 10^{-4}.

Convergence evolution towards the shakedown factor is plotted against the external loop iterations, where the load factor changes value. There is no standard number of internal loop iterations (internal loops have linear convergence), but due to the external superlinear convergence, the number of external incremental-like loops limits the total number of iterations. For all the different loading cases of the examples considered herein, to reach the final shakedown factor, this total number never exceeded 120 iterations (for example, for the elbow problem with the larger number

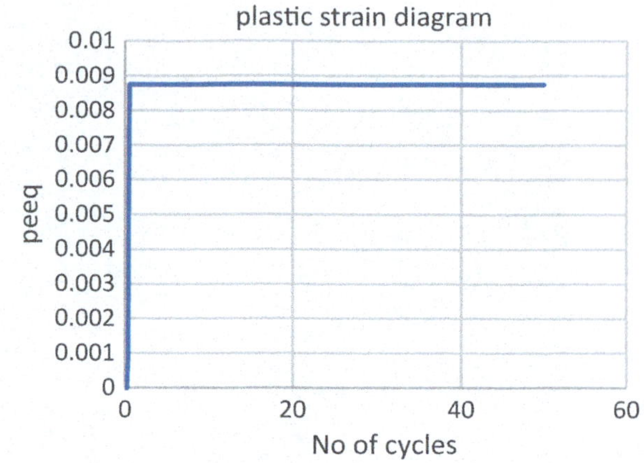

Fig. 15 Plastic strain diagram in the case of 20 mm

of finite elements, it was 87). In each iteration, the number of elastic solutions (Eq. 5) equals the number of vertices of the loading domain; the stiffness matrix used to find these solutions is formed and decomposed only once (at the very beginning of the procedure). Additionally, it should be noted that even with $n = 1$, as the number of terms of the Fourier series, the results are almost identical.

5 Concluding Remarks

The RSDM-S is a direct method that is used to establish shakedown domains. In the present work a recent update of the approach that appeared in the literature is elaborated further. Closely linked to the robustness of the procedure, the proof of the method's continuous descent towards the shakedown load factor is demonstrated graphically. At the same time, the superlinear convergence of the external incremental-like outer loop which guarantees fast convergence is underlined. Further structural examples to the already published ones, undergoing diverse cyclic loading actions from thermomechanical loading to imposed displacements are discussed and construction of corresponding shakedown domains are presented.

References

1. ABAQUS: Analysis user's manual. Dassault Systems Simulia Inc, Version **6**, 14 (2014)
2. Chen, H.F., Ponter, A.R.S.: A method for the evaluation of a ratchet limit and the amplitude of plastic strain for bodies subjected to cyclic loading. Eur. J. Mech. Solid. **20**, 555–571 (2001)
3. Chen, G., Wang, H., Bezold, A., Broeckmann, C., Weichert, D., Zhang, L.: Strengths prediction of particulate reinforced metal matrix composites (PRMMCs) using direct method and artificial neural network. Compos. Struct. **223**, 110951 (2019)

4. Gokhfeld, D.A., Cherniavsky, O.F.: Limit Analysis of Structures at Thermal Cycling, Sijthoff & Noordhoff (1980)
5. Isaakson, E., Keller, H.B.: Analysis of Numerical Methods. Wiley, New York (1966)
6. Koiter, W.: General problems for elastic–plastic solids. In: Sneddon, H. (ed.) Progress in Solid Mechanics, vol. 4, pp. 165–221. North-Holland, Amsterdam (1960)
7. König, J.: Shakedown of Elastic-Plastic Structures. Elsevier (1987)
8. Lytwin, M., Chen, H.F., Ponter, A.R.S.: A generalized method for ratchet analysis of structures undergoing arbitrary thermo-mechanical load histories. Int. J. Numer. Methods Eng. **104**(2), 104–124 (2015)
9. Ma, Z., Chen, H., Liu, Y., Xuan, F.Z.: A direct approach to the evaluation of structural shakedown limit considering limited kinematic hardening and nonisothermal effect. Eur. J. Mech. Solid. **79**, 103877 (2020)
10. Magisano, D., Garcea, G.: Fiber-based shakedown analysis of three-dimensional frames under multiple load combinations: mixed finite elements and incremental iterative solution. Int. J. Numer. Methods Eng. **121**, 3743–3767 (2020)
11. Melan, E.: Zur Plastizität des räumlichen Kontinuums. Ing. Arch. **9**, 116–126 (1938)
12. Pham, T.P.: Upper bound limit and shakedown Analysis of elastic-plastic bounded linearly kinematic hardening structures, Ph.D. thesis, RWTH Aachen University (2011)
13. Pisano, A.A., Spiliopoulos, K.V., Weichert, D. (eds.): Direct Methods: Methodological Progress and Engineering Applications. Springer Nature Switzerland (2021)
14. Ponter, A.R.S., Carter, K.F.: Shakedown state simulation techniques based on linear elastic solutions. Comput. Methods Appl. Mech. Eng. **140**, 259–279 (1997)
15. Spiliopoulos, K.V.: Simplified methods for the steady state inelastic analysis of cyclically loaded structures. In: Weichert, D., Maier, G. (eds.) Inelastic Analysis of Structures under Variable Loads, pp. 213–232. Kluwer Academic Publishers (2000)
16. Spiliopoulos, K.V.: A simplified method to predict the steady cyclic stress state of creeping structures. ASME J. App. Mech. **69**, 148–153 (2002)
17. Spiliopoulos, K.V., Kapogiannis, I.A.: Fast and robust RSDM shakedown solutions of structures under cyclic variation and imposed displacements. Eur. J. Mech. Solids **95**, 104657 (2022)
18. Spiliopoulos, K.V., Panagiotou, K.D.: A direct method to predict cyclic steady states of elastoplastic structures. Comput. Methods Appl. Mech. Eng. **223–224**, 186–198 (2012)
19. Spiliopoulos, K.V., Panagiotou K.D.: The residual stress decomposition Method (RSDM): A novel direct method to predict cyclic elastoplastic states. In: Spiliopoulos, K., Weichert, D. (eds.), Direct Methods for Limit States in Structures and Materials. Springer Science + Business Media, Dordrecht, pp. 139–155 (2014)
20. Spiliopoulos, K.V., Panagiotou, K.D.: A residual stress decomposition-based method for the shakedown analysis of structures. Comput. Methods Appl. Mech. Eng. **276**, 410–430 (2014)
21. Spiliopoulos, K.V., Panagiotou, K.D.: A numerical procedure for the shakedown analysis of structures under cyclic thermomechanical loading. Arch. Appl. Mech. **85**(9), 1499–2151 (2015)
22. Spiliopoulos, K.V., Panagiotou, K.D.: An enhanced numerical procedure for the shakedown analysis in multidimensional loading domains. Comput. Struct. **193**, 155–171 (2017)

Mixed Fiber Elements and Incremental-Iterative Algorithm for Shakedown and Limit Fire Analysis of 3D Frames

Domenico Magisano and Giovanni Garcea

Abstract This work presents an efficient fiber analysis for evaluating the shakedown safety factor of 3D frames under multiple load combinations. Mixed beam elements are employed for an accurate discretization of the structures. A continuation method, similar to an incremental elasto-plastic analysis, is used at structural level. It evaluates a fictitious equilibrium path made of a sequence of safe states with a converging non-decreasing load factor. Each point of the path is obtained by finding kinematic variables corresponding to self-equilibrated stresses satisfying Melan's condition for the current load factor to be safe. The stress admissible domain is defined at fiber level as a function of the load factor using the maximum and minimum effect due to all loads. The overall analysis allows a direct application of the Newton method easily implemented in commercial codes. The same numerical strategy can solve also a different problem, named limit fire analysis. It consists in evaluating for an assigned external load the maximum time of fire exposure that leads to the structural collapse as a consequence of the strength reduction of the materials.

Keywords Frame structures · Fire analysis · Limit analysis · Mixed finite elements

1 Introduction

Despite its technical implications, shakedown analysis seems still currently confined to the research community or particular applications instead of being used as a common tool in structural design. One of the main open problems is the difficulty in managing the large number of load conditions [17, 18, 21] and the complex

---

D. Magisano (✉) · G. Garcea
Dipartimento di Ingegneria Informatica, Modellistica, Elettronica e Sistemistica, Università della Calabria, 87030 Rende (Cosenza), Italy
e-mail: domenico.magisano@unical.it

G. Garcea
e-mail: giovanni.garcea@unical.it

G. Garcea and D. Weichert (eds.), *Direct Methods for Limit State of Materials and Structures*, Lecture Notes in Applied and Computational Mechanics 101,
https://doi.org/10.1007/978-3-031-29122-7_13

combination rules required for a realistic definition of the load domain, which can lead to a very time-consuming analysis. For example, we can consider the simple case of linear combination rules where the load domain is defined as a combination of basic actions varying between a minimum and a maximum value. According to the König [5] (convexity) theorem, we can consider the set of vertexes of the load domain, in this case a convex polytope with 2^p vertexes, where p is the number of basic loads. In fact, the plastic admissibility of the elastic stresses due to all possible loads within the load domain, known as *stress envelope*, requires the admissibility of the stresses corresponding to the load vertexes only. However, for values of p in the order of tens, as typically occurs in structural design, the number of vertexes becomes so large to prevent the solution process.

This work presents an accurate, efficient and very simple fiber-based analysis for evaluating the shakedown safety factor of 3D frames subjected to multiple load combinations. Mixed FEs with equilibrated stress interpolation are used for an accurate discretization of the structure. The first novelty point is the stress admissible domain defined at fiber level as a function of the load factor accounting for the elastic stress envelope due to all possible loads. Compared to stress resultants approaches, the stress envelope is now defined by two scalar values, namely the maximum and minimum effect on the fiber, regardless of dimension and complexity of the load domain. This key feature guarantees that, unlike in existing proposals [6, 17, 18, 21], the number of load combinations has no influence on the cost of the nonlinear part of the analysis. A decomposition method inspired by the one proposed in [2, 3] is applied to solve the global lower bound optimization problem through an incremental-iterative method, similar to a standard elasto-plastic analysis. Following a strain-driven approach, each point of a pseudo-equilibrium path is obtained by finding kinematic discrete variables corresponding to a self-equilibrated stress field able to satisfy the admissibility condition required by Melan's theorem for the current load factor to be safe. The structural residual requires, at FE level, the definition of plastically admissible stresses, corresponding to assigned kinematic FE variables and load factor. To this aim, a novel decomposition method is proposed for mixed FEs, that avoids the solution of the FE optimization problem [4, 6] replaced by section state determinations at each integration point allowing a simple imposition of the fiber plastic admissibility for the assigned load factor. An iterative element state determination is then applied to preserve the equilibrated stress interpolation and the FE compatibility. The proposed shakedown analysis is based on the direct application of the full Newton method for both global and local equations without any need for optimization tools and, then, easy to implement in existing FE codes. Moreover, the incremental-iterative solution is competitive with other general optimization methods due to the large number of constraints, which are not involved in the global iterations of our approach.

In addition, this work introduces the limit fire analysis for assessing the ultimate fire resistance of 3D frames under given dead loads. It is assumed that the fire model is characterized by a non-decreasing temperature over time as suggested by technical standards, the strength of the materials does not increase with heating as typical of most structural materials and the structure undergoes small deformations. Under these hypotheses, we derive a lower bound theorem that makes it possible to define the limit fire duration as a synthetic safety limit based only on the concepts of equilibrium and strength. The problem is very similar to the shakedown one.

2 Modeling of 3D Frame Structures

2.1 *The 3D Beam Model*

Let us consider a cylinder in a reference configuration $\mathcal{B}$, characterized by length ℓ and confined by a lateral boundary, $\partial\mathcal{B}$, and two terminal bases, Ω_0 and Ω_ℓ. The cylinder is referred to a Cartesian frame $(\mathcal{O}, x_1 \equiv s, x_2, x_3)$ with unit vectors $\{\mathbf{i}_1, \mathbf{i}_2, \mathbf{i}_3\}$ and $\mathbf{i}_1$ aligned with the cylinder axis. In this system (see Fig. 1), we denote with $\mathbf{X} = s\mathbf{i}_1 + \mathbf{x}$ the position of a point P, where s is an abscissa which identifies the generic cross-section $\Omega(s)$ of the beam, while $\mathbf{x} = x_2\mathbf{i}_2 + x_3\mathbf{i}_3$ is the position of P inside $\Omega(s)$.

The displacement field $\mathbf{v}[\mathbf{X}]$ of the model is expressed, as usual, as a rigid motion of the section

$$\mathbf{v}(\mathbf{X}) = \mathbf{u}(s) + \boldsymbol{\varphi}(s) \times \mathbf{x} \tag{1}$$

where $\mathbf{u}(s)$ and $\boldsymbol{\varphi}(s)$ are the mean translation and rotation of the section and the operator $\times$ denotes the cross product. The kinematics assumed in Eq. (1) allows us to evaluate, using a linear Cauchy continuum, the stress-strain work $\mathcal{W}$ in terms of the generalized strains and stresses of the section as

$$\mathcal{W} := \int_\ell \left(\mathbf{n}(s)^T \boldsymbol{\varepsilon}(s) + \boldsymbol{\tau}(s)^T \boldsymbol{\gamma}(s)\right) ds \tag{2}$$

with the generalized strains $\boldsymbol{\varepsilon}$, collecting axial strain and bending curvatures, and $\boldsymbol{\gamma}$, collecting shear strains and torsional curvature, defined as

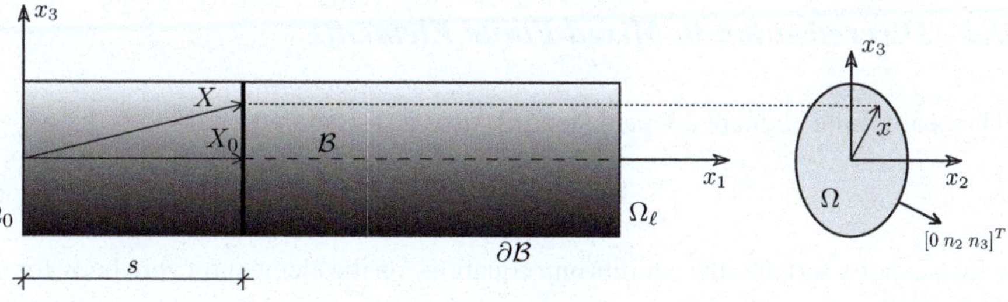

Fig. 1 The cylindrical solid

$$\boldsymbol{\varepsilon}(s) \equiv \begin{bmatrix} e \\ \chi_2 \\ \chi_3 \end{bmatrix} = \begin{bmatrix} u_{1,s} \\ \varphi_{2,s} \\ \varphi_{3,s} \end{bmatrix} \qquad \boldsymbol{\gamma}(s) \equiv \begin{bmatrix} \gamma_2 \\ \gamma_3 \\ \chi_1 \end{bmatrix} = \begin{bmatrix} u_{2,s} - \varphi_3 \\ u_{3,s} + \varphi_2 \\ \varphi_{1,s} \end{bmatrix} \tag{3}$$

where a comma stands for derivative. The stress resultants on the section $\mathbf{n}$ and $\boldsymbol{\tau}$ due to normal and tangential stresses respectively are defined as

$$\mathbf{n}(s) \equiv \begin{bmatrix} N_1(s) \\ M_2(s) \\ M_3(s) \end{bmatrix} = \int_{\Omega} \sigma_{11}(s, \mathbf{x})\mathbf{a}(\mathbf{x}) d\Omega, \qquad \boldsymbol{\tau}(s) \equiv \begin{bmatrix} N_2(s) \\ N_3(s) \\ M_1(s) \end{bmatrix} = \int_{\Omega} \begin{bmatrix} \sigma_{12} \\ \sigma_{13} \\ \sigma_{13}x_2 - \sigma_{12}x_3 \end{bmatrix} d\Omega \tag{4}$$

with $\mathbf{a} = [1, x_3, -x_2]^T$. In particular, $\mathbf{n}$ collects axial force and bending moments while $\boldsymbol{\tau}$ collects shear forces and torque. Introducing the vectors

$$\mathbf{t}(s) = \begin{bmatrix} \mathbf{n}(s) \\ \boldsymbol{\tau}(s) \end{bmatrix} \qquad \boldsymbol{\rho}(s) = \begin{bmatrix} \boldsymbol{\varepsilon}(s) \\ \boldsymbol{\gamma}(s) \end{bmatrix} \tag{5}$$

the elastic constitutive law can be written as

$$\boldsymbol{\rho}(s) = \mathbf{F}(s)\mathbf{t}(s) \quad \text{with} \quad \mathbf{F}(s)^{-1} = \mathbf{C}(s) = \begin{bmatrix} \mathbf{E} & \mathbf{0} \\ \mathbf{0} & \boldsymbol{G} \end{bmatrix} \tag{6}$$

where the coefficients of the cross-section compliance matrix $\mathbf{F}$ can be obtained as in [13]. For a future use, we introduce also the extraction operator $\mathbf{T}_n$ and $\mathbf{T}_\tau$ such that

$$\begin{aligned} \mathbf{n}(s) &= \mathbf{T}_n\mathbf{t}(s), \quad \boldsymbol{\tau}(s) = \mathbf{T}_\tau\mathbf{t}(s), \quad \mathbf{t}(s) = \mathbf{T}_n^T\mathbf{n}(s) + \mathbf{T}_\tau^T\boldsymbol{\tau}(s) \\ \boldsymbol{\varepsilon}(s) &= \mathbf{T}_n\boldsymbol{\rho}(s), \quad \boldsymbol{\gamma}(s) = \mathbf{T}_\tau\boldsymbol{\rho}(s), \quad \boldsymbol{\rho}(s) = \mathbf{T}_n^T\boldsymbol{\varepsilon}(s) + \mathbf{T}_\tau^T\boldsymbol{\gamma}(s). \end{aligned} \tag{7}$$

Finally, using the kinematics in Eq. (1), the normal strain over the section, work-conjugated to $\sigma \equiv \sigma_{11}$, is

$$\epsilon(s, \mathbf{x}) \equiv v_{1,s} = \mathbf{a}(\mathbf{x})^T\boldsymbol{\varepsilon}(s, \mathbf{x}). \tag{8}$$

2.2 Discretization in Mixed Finite Elements

The beam finite element adopted (see [6]) uses a stress interpolation

$$\mathbf{t}(s) = \mathbf{L}(s)\boldsymbol{\beta}_e \tag{9}$$

which exactly satisfies the equilibrium equations on the element for zero body forces, that is

$$\boldsymbol{\mathcal{N}}_{,s} = \mathbf{0}, \qquad \boldsymbol{\mathcal{M}}_{,s} + \mathbf{i}_1 \wedge \boldsymbol{\mathcal{N}} = \mathbf{0}, \tag{10}$$

while body load effects can be exactly included as a particular solution. Equation (10) states that $\mathcal{N} \equiv [N_1, N_2, N_3]^T$ and the torsional moment component M_1 are constant, while the two flexural components $M_2(s)$ and $M_3(s)$ of $\mathcal{M}(s) \equiv [M_1, M_2, M_3]^T$ are linear with s and linked to the shear resultants so that $N_2\ell = -(M_3(\ell) - M_3(0))$ and $N_3\ell = (M_2(\ell) - M_2(0))$. The internal work in Eq. (2) becomes

$$\mathcal{W} \equiv \mathcal{N}^T (\mathbf{u}(\ell) - \mathbf{u}(0)) + \mathcal{M}(\ell)^T \varphi(\ell) - \mathcal{M}(0)^T \varphi(0) = \mathbf{d}_e^T \mathbf{Q}_e^T \boldsymbol{\beta}_e \tag{11}$$

allowing us to directly obtain the discrete form of $\mathcal{W}$ without any FEM interpolation for the kinematic variables $\mathbf{u}(s)$ and $\varphi(s)$. In Eq. (11) the vectors that collect the kinematics $\mathbf{d}_e$ and static $\boldsymbol{\beta}_e$ FE generalized parameters and the compatibility operator $\mathbf{Q}_e$ are defined as

$$\boldsymbol{\beta}_e = \begin{bmatrix} N_1 \\ M_2(0) \\ M_3(0) \\ M_2(\ell) \\ M_3(\ell) \\ M_1 \end{bmatrix}, \quad \mathbf{d}_e = \begin{bmatrix} \mathbf{u}(0) \\ \varphi(0) \\ \mathbf{u}(\ell) \\ \varphi(\ell) \end{bmatrix}, \quad \mathbf{Q}_e = \frac{1}{\ell} \begin{bmatrix} -\ell\, \mathbf{i}_1^T & \mathbf{0} & \ell\, \mathbf{i}_1^T & \mathbf{0} \\ \mathbf{i}_3^T & -\ell\, \mathbf{i}_2^T & -\mathbf{i}_3^T & \mathbf{0} \\ -\mathbf{i}_2^T & -\ell\, \mathbf{i}_3 & \mathbf{i}_2^T & \mathbf{0} \\ -\mathbf{i}_3^T & \mathbf{0} & \mathbf{i}_3^T & \ell\, \mathbf{i}_2^T \\ \mathbf{i}_2^T & \mathbf{0} & -\mathbf{i}_2^T & \ell\, \mathbf{i}_3^T \\ \mathbf{0} & -\ell\, \mathbf{i}_1^T & \mathbf{0} & \ell\, \mathbf{i}_1^T \end{bmatrix}. \tag{12}$$

2.3 Fiber Analysis

For frames with medium to large span to depth ratios of the members, it is a reasonable approximation [1, 6, 19, 20] to define the elasto-plastic behavior only in terms of normal stress component $\sigma \equiv \sigma_{11}$, while plastic deformations due to shear and torsion are assumed to be negligible. The plastic admissibility condition can be written as

$$\sigma_-(s, \mathbf{x}) \le \sigma(s, \mathbf{x}) \le \sigma_+(s, \mathbf{x}), \qquad \forall (s, \mathbf{x}) \tag{13}$$

where $\sigma_+(s, \mathbf{x})$ and $\sigma_-(s, \mathbf{x})$ are the stress limits of the material in traction (positive) or compression (negative) respectively. Omitting the dependence on the point coordinates s and $\mathbf{x}$ to simplify the notation, the admissibility condition (13) can be rewritten by introducing the yield function $f(\sigma)$ as

$$f(\sigma) \equiv |\sigma - c| - r \le 0 \tag{14}$$

where

$$c = \frac{\sigma_+ + \sigma_-}{2} \qquad r = \frac{\sigma_+ - \sigma_-}{2} \tag{15}$$

The constitutive law is imposed at a discrete number of points belonging to the FE domain. To this end, a certain number of integration points (IPs) are used along the beam axis. Gauss-Lobatto rules are preferred in order to include the end sections.

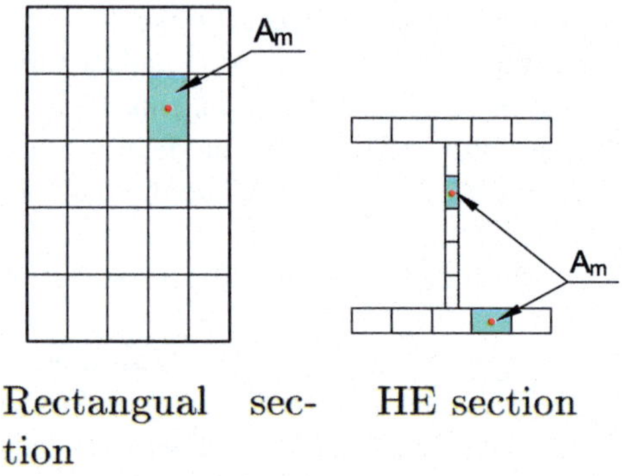

Fig. 2 Typical fiber discretization of the sections

For each IP, the cross section is discretized using a finite number n_f of fibers with a normal stress σ constant over each fiber domain of area A and equal to the mid point one (cf. Fig. 2).

2.4 Linear Elastic Solution

The linear elastic problem corresponds to the stationarity of the Hellinger-Reissner functional Π_{HR} that, at the element level, can be expressed as

$$\Pi_{HR} = \mathbf{d}_e^T \mathbf{Q}_e^T \boldsymbol{\beta}_e - \frac{1}{2} \boldsymbol{\beta}_e^T \mathbf{H}_e \boldsymbol{\beta}_e - \mathbf{d}_e^T \mathbf{p}_e$$

where $\mathbf{p}_e$ is the elemental contribution of the external loads and, introducing the FE interpolation for the generalized stress, the elastic compliance matrix of the element is obtained from the equivalence

$$\int_\ell \mathbf{t}(s)^T \mathbf{F} \mathbf{t}(s) ds = \boldsymbol{\beta}_e^T \mathbf{H}_e \boldsymbol{\beta}_e \quad \Rightarrow \quad \mathbf{H}_e = \int_\ell \mathbf{L}(s)^T \mathbf{F} \mathbf{L}(s)\, ds. \tag{16}$$

Since the stress interpolation is discontinuous across element boundaries, the stationarity of Π_{HR} with respect to $\boldsymbol{\beta}_e$ furnishes the FE elastic constitutive law

$$\boldsymbol{\beta}_e(\mathbf{d}_e) = \mathbf{H}_e^{-1} \mathbf{Q}_e \mathbf{d}_e \tag{17}$$

which allows us to express the elastic problem in terms of displacement variables only. The stationarity condition with respect to $\mathbf{d}_e$ furnishes the equilibrium equations on the element as

$$\mathbf{Q}_e^T \boldsymbol{\beta}_e(\mathbf{d}_e) - \mathbf{p}_e = \mathbf{0} \tag{18}$$

which, in the elastic case, become

$$\mathbf{K}_e\mathbf{d}_e - \mathbf{p}_e = \mathbf{0} \quad \text{with} \quad \mathbf{K}_e = \mathbf{Q}_e^T\mathbf{H}_e^{-1}\mathbf{Q}_e.$$

The linear elastic solution can be then achieved by solving the global linear equations obtained by assembling the global elastic stiffness matrix **K** and load vector **p**:

$$\mathbf{K}\mathbf{d} - \mathbf{p} = \mathbf{0}.$$

At least 3 Gauss-Lobatto IPs are needed to exactly integrate the elastic complementary energy along the element

$$\frac{1}{2}\boldsymbol{\beta}_e^T\mathbf{H}_e\boldsymbol{\beta}_e = \frac{1}{2}\sum_g \mathbf{n}_g^T\mathbf{E}_g^{-1}\mathbf{n}_g w_g + \frac{1}{2}\sum_g \boldsymbol{\tau}_g^T\boldsymbol{G}_g^{-1}\boldsymbol{\tau}_g w_g \tag{19}$$

A generic integration point quantity $(.)_g$ belongs to a FE e but the subscript e is omitted to simplify the notation.

Using the fiber discretization, the complementary energy density of the section, on the basis of Eq. (8), can then be evaluated as

$$\frac{1}{2}\mathbf{n}_g^T\mathbf{E}_g^{-1}\mathbf{n}_g = \frac{1}{2}\sum_m \frac{1}{E_m}\sigma_m^2 A_m \tag{20}$$

where E is the Young modulus and the subscript m denotes a generic fiber. Also in this case, a generic fiber quantity $(.)_m$ belongs to an IP g of a FE e, but subscripts g and e are omitted for a simpler notation.

Remembering that the fiber strain is linked to the generalized ones as $\epsilon_m = \mathbf{a}_m^T\boldsymbol{\varepsilon}$ and the elastic law $\sigma = E\epsilon$, the previous equation allows us to evaluate $\mathbf{E}_g$ as

$$\sum_m E_m\epsilon_m^2 A_m = \frac{1}{2}\boldsymbol{\varepsilon}_g^T\left(\sum_m E_m A_m \mathbf{a}_m\mathbf{a}_m^T\right)\boldsymbol{\varepsilon}_g \quad \Rightarrow \quad \mathbf{E}_g = \left(\sum_m E_m A_m \mathbf{a}_m\mathbf{a}_m^T\right)$$

For the fiber model, the normal stress resultants in Eq. (4) on the section can be computed as a function of the normal stresses of the fibers at IP, collected in vector $\boldsymbol{\sigma}_g = [\sigma_1, \cdots, \sigma_{nf}]$, as

$$\mathbf{n}_g(\boldsymbol{\sigma}_g) = \sum_m \sigma_m A_m \mathbf{a}_m. \tag{21}$$

We have that the vector collecting all stress resultants can be written, exploiting Eq. (7), as

$$\mathbf{t}_g(\boldsymbol{\sigma}_g, \boldsymbol{\tau}_g) = \begin{bmatrix}\mathbf{n}_g\\ \boldsymbol{\tau}_g\end{bmatrix} \equiv \mathbf{T}_n^T\sum_m(\sigma_m A_m \mathbf{a}_m) + \mathbf{T}_\tau^T\boldsymbol{\tau}_g. \tag{22}$$

2.5 Elasto-plastic Solution

Starting from an initial stress state $\boldsymbol{\sigma}_g^{(0)}$ and $\boldsymbol{\tau}_g^{(0)}$ stored at each IP g, the elasto-plastic solution can be obtained, as demonstrated in [8, 12, 22], by solving the following 3 groups of equations:

$$\text{Global equations} \quad \mathbf{Q}^T\boldsymbol{\beta} - \mathbf{p} = \mathbf{0} \tag{23a}$$

$$\text{Element equations} \quad \begin{cases} \mathbf{Q}_e\Delta\mathbf{d}_e - \sum_g \mathbf{L}_g^T \Delta\boldsymbol{\rho}_g w_g = \mathbf{0} \\ \mathbf{t}_g(\boldsymbol{\sigma}_g, \boldsymbol{\tau}_g) - \mathbf{L}_g\boldsymbol{\beta}_e - \bar{\mathbf{t}}_g = \mathbf{0} \quad \forall g \end{cases} \tag{23b}$$

$$\begin{cases} \text{Fiber normal stress} \quad \begin{cases} \Delta\sigma_m - E_m\left(\Delta\epsilon_m - \mu_m \dfrac{\partial f_m(\sigma_m)}{\partial\sigma_m}\right) = 0 \\ \mu_m \geq 0 \quad \mu_m f_m(\sigma_m) = 0 \end{cases} \\ \text{Tangential stress resultants} \quad \Delta\boldsymbol{\tau}_g = \mathbf{G}_g\mathbf{T}_\tau\Delta\boldsymbol{\rho}_g \end{cases} \tag{23c}$$

where $\Delta\epsilon_m = \mathbf{a}_m^T\Delta\varepsilon$, with $\Delta\varepsilon = \mathbf{T}_n\Delta\boldsymbol{\rho}_g = [\Delta e, \Delta\chi_2, \Delta\chi_3]^T$, denotes the increment of strain work-conjugate to σ_m. Global equations, obtained assembling element counterpart Eq. (18), impose the equilibrium of internal and external forces. Element-wise equations link the stress DOFs $\boldsymbol{\beta}_e$ with the displacement increment $\Delta\mathbf{d}_e$, preserving the kinematic compatibility of the generalized strain increment $\Delta\boldsymbol{\rho}_g$ at each IP g with the element displacement increment $\Delta\mathbf{d}_e$ and the FE equilibrated stress interpolation in Eq. (9). Vector $\bar{\mathbf{t}}_g$ collects the generalized stresses at the gth IP corresponding to the particular solution of the equilibrium equations due to distributed loads. Finally, constitutive laws at each IP provide the section stress resultants $\mathbf{t}_g$ corresponding to the generalized strain increment $\Delta\boldsymbol{\rho}_g$. The first of Eq. (23c) is the elasto-plastic constitutive law of the mth fiber integrated by a backward Euler scheme for an elastic-perfectly plastic Drucker material. This can be seen as the first order condition of the following closest point projection problem:

$$\begin{cases} \text{minimize} & \dfrac{1}{2}E_m(\sigma_m - \sigma_m^*)^2 \\ \text{subject to} & f_m(\sigma_m) \leq 0 \end{cases} \tag{24}$$

with the fiber elastic predictor $\sigma_m^* = \sigma_m^{(0)} + E_m\Delta\varepsilon_m$. The solution of this one-dimensional optimization problem is provided explicitly by a simple min-max operation:

$$\sigma_m = \max\left(\sigma_-, \min\left(\sigma_m^*, \sigma_+\right)\right). \tag{25}$$

3 Shakedown Analysis

We say that a structure shakes down to an elastic state or, simply, shakedown occurs if, after an initial phase during which the occurrence and the accumulation of plastic strain increments are possible, the structural response, for every load path, tends to be purely elastic with a finite total plastic work.

3.1 Load Domain

The FE external load vector $\mathbf{p}(t)$, varying with the time t, can be expressed as a combination of p load patterns $\mathbf{p}_i$ belonging to an admissible *load domain* $\mathbb{P}$. We assume

$$\mathbb{P} := \bigcup_{k=1}^{q} \mathbb{P}^{(k)}\ , \quad \mathbb{P}^{(k)} := \left\{ \mathbf{p} = \sum_{i=1}^{p} \alpha_{ki}(t)\mathbf{p}_i \ :\ \alpha_{ki}^{min} \leq \alpha_{ki}(t) \leq \alpha_{ki}^{max} \right\} \tag{26}$$

where α_{ki} defines the range of variability in time of $\mathbf{p}_i$ for the kth load combination accounting for the probability of simultaneous actions. Each combination $\mathbb{P}^{(k)}$ is a convex polytope with 2^p vertexes.

3.2 Melan's Shakedown Theorem and Admissibility Condition

A sufficient condition for shakedown is given by the classic Bleich-Melan static theorem: shakedown occurs if there exists an additional time-independent self equilibrated stress field such that the yield condition is satisfied at every point in the body for any possible loading.

For frames with medium to large span to depth ratios of the members, it is a reasonable approximation [1, 6, 19, 20] to define the elastic domain only in terms of normal stress components $\sigma \equiv \sigma_{11}$. In this case, it is useful to define the envelope of elastic stress $\mathbb{S}(s, \mathbf{x})$ associated to a generic point of the body $(s, \mathbf{x})$ as the set of elastic stresses $\hat{\sigma}(s, \mathbf{x})$ produced by all loads $\mathbf{p} \in \mathbb{P}$.

With these assumptions, the static shakedown theorem can be formulated as follows: shakedown occurs if there exists an additional time-independent self-equilibrated stress $\sigma(s, \mathbf{x})$ such that

$$f_c(s, \mathbf{x}) \leq \lambda\hat{\sigma}(s, \mathbf{x}) + \sigma(s, \mathbf{x}) \leq f_t(s, \mathbf{x}), \qquad \forall\hat{\sigma}(s, \mathbf{x}) \in \mathbb{S}(s, \mathbf{x}) \tag{27}$$

where $f_t(s, \mathbf{x})$ and $f_c(s, \mathbf{x})$ are the stress limits of the material in traction (positive) or compression (negative) respectively. An amplification factor λ of the reference load domain is introduced. This makes it possible to define the shakedown safety factor of the reference load domain as the maximum value of λ for which shakedown occurs.

3.3 Elastic Stress Envelope

For each point $\mathbf{x}$ of the cross section s the elastic stress envelope $\mathbb{S}(s, \mathbf{x})$ becomes the closed line segment

$$\mathbb{S}(s, \mathbf{x}) = \{\hat{\sigma} : \hat{\sigma}^{\min} \leq \hat{\sigma}(t) \leq \hat{\sigma}^{\max}\} \tag{28}$$

where $\hat{\sigma}^{\max}$ and $\hat{\sigma}^{\min}$ are the maximum and minimum values respectively of the normal elastic stress $\hat{\sigma}$ of the point for each possible load in $\mathbb{P}$. It is worth noting that $\mathbb{S}(s, \mathbf{x})$ is always defined by two vertexes independently from the number of load combinations and from the complexity of $\mathbb{P}$.

From now on, to simplify the notation, we avoid to report the dependence of the stress on $(s, \mathbf{x})$. The endpoints $\hat{\sigma}^{\min}$ and $\hat{\sigma}^{\max}$ can be easily evaluated according to the load definition in Eq. (26) which implies that

$$\mathbb{S}(s, \mathbf{x}) = \bigcup_{k=1}^{q} \mathbb{S}^{(k)}(s, \mathbf{x}) \,, \quad \mathbb{S}^{(k)} := \left\{ \hat{\sigma}^{(k)} = \sum_{i=1}^{p} \alpha_{ki}(t)\hat{\sigma}_i \, : \, \alpha_{ki}^{min} \leq \alpha_{ki}(t) \leq \alpha_{ki}^{max} \right\} \tag{29}$$

We can first evaluate the stress envelope $\mathbb{S}^{(k)}$ for $\mathbf{p} \in \mathbb{P}^{(k)}$ as

$$\mathbb{S}^{(k)}(s, \mathbf{x}) = \{\hat{\sigma}^{(k)} : \hat{\sigma}^{k,\min} \leq \hat{\sigma}^{(k)} \leq \hat{\sigma}^{k,\max}\}$$

where, letting $\hat{\sigma}_i$ the elastic normal stress of the fiber due to $\mathbf{p}_i$,

$$\hat{\sigma}^{k,I} = \begin{cases} \sum_i \alpha_{ki}^{\max} \hat{\sigma}_i & \text{if } n^I \hat{\sigma}_i \geq 0 \\ \sum_i \alpha_{ki}^{\min} \hat{\sigma}_i & \text{if } n^I \hat{\sigma}_i < 0 \end{cases} \qquad I = \min, \max \tag{30}$$

with $n^{\min} = -1$ and $n^{\max} = 1$. The endpoints of the overall envelope can be then obtained as

$$\hat{\sigma}^{\max} = \max_{(k)} \hat{\sigma}^{k\,\max} \quad \text{and} \quad \hat{\sigma}^{\min} = \min_{(k)} \hat{\sigma}^{k\,\min}. \tag{31}$$

Equation (30) represents a particularization to the one-dimensional stress space of the formulation reported in [6] for the selection of the meaningful stress vertexes.

3.4 Shakedown Yield Function

The shakedown admissibility condition (27) for a load domain amplified by λ, omitting the dependence from s and $\mathbf{x}$, can be written in terms of stress envelope endpoints as

$$\begin{cases} \sigma + \lambda\hat{\sigma}^{\max} \leq f_t \\ \sigma + \lambda\hat{\sigma}^{\min} \geq f_c. \end{cases} \tag{32}$$

This rewriting allows us to introduce the *shakedown yield function* $f(\sigma, \lambda)$

$$f(\sigma, \lambda) \equiv |\sigma - c(\lambda)| - r(\lambda) \leq 0 \tag{33}$$

where

$$\begin{aligned} c &= c_0 - \lambda\hat{c} \quad \text{with} \quad c_0 = \frac{f_t + f_c}{2} \quad \hat{c} = \frac{\hat{\sigma}^{\max} + \hat{\sigma}^{\min}}{2} \\ r &= r_0 - \lambda\hat{r} \quad \text{with} \quad r_0 = \frac{f_t - f_c}{2} \quad \hat{r} = \frac{\hat{\sigma}^{\max} - \hat{\sigma}^{\min}}{2} \end{aligned} \tag{34}$$

Condition in Eq. (33) defines the admissible domain of the additional self-equilibrated stress σ that depends on the load amplifier λ. We can note how this domain is unrelated to the complexity of the load variation, but depends only on the maximum $\hat{\sigma}^{\max}$ and minimum $\hat{\sigma}^{\min}$ elastic effect on the point, apart from the material strength.

It is important to remark the first important feature of the proposed approach: only the two vertexes of the elastic envelope of each fiber affect the admissible domain, independently from the number of basic actions or the complexity of the load domain.

3.5 Shakedown Safety Factor According to the Lower Bound Theorem

The shakedown safety factor λ_S of the structure is the maximum value of the load factor which satisfies the sufficient condition of the Bleich–Melan static theorem [5, 6, 24]. This can be written in discrete form as the following constrained optimisation problem

$$\begin{aligned} \text{maximize} \quad & \lambda \\ \text{subject to} \quad & \mathbf{Q}^T\boldsymbol{\beta} = \mathbf{0} \\ & f(\sigma_m, \lambda) \leq 0 \qquad \forall e, g, m \\ & \mathbf{t}_g(\boldsymbol{\sigma}_g, \boldsymbol{\tau}_g) - \mathbf{L}_g\boldsymbol{\beta}_e = \mathbf{0} \quad \forall g, e \end{aligned} \tag{35}$$

where the last equality constraint is necessary to preserve the equilibrated stress interpolation (9) over the element.

We can note that (35) is very similar to the static limit analysis optimization problem. In particular, the only relevant difference is that the plastic admissibility is imposed for the two vertexes of $\mathbb{S}(s, \mathbf{x})$. Indeed, shakedown analysis reduces to limit analysis when the stress envelope degenerates into a single point [4].

3.6 Elastic Limit and Non-empty Limit

It is useful to define two load factors named elastic limit λ_E and non-empty limit $\lambda_\emptyset$. λ_E is the maximum load factor for which the elastic stress produced by all loads satisfies the admissibility condition without additional self-equilibrated stress fields, i.e. $f_m(0, \lambda) \leq 0$ for any fiber m in the structure.

$\lambda_\emptyset$ is the maximum load factor for which the shakedown admissible domain is a non-empty set at any fiber in the structure. This means that for $\lambda = \lambda_\emptyset$ there exists a fiber m in the structure whose shakedown admissible domain $\{\sigma_m : f_m(\sigma_m, \lambda) \leq 0\}$ reduces to a single point, i.e. its domain radius $r(\lambda)$ vanishes. According to the expression of $r(\lambda)$ in Eq. (34), we have

$$\lambda_\emptyset = \min_{e,g,m} \left(\frac{r_{0m}}{\hat{r}_m} \right).$$

The shakedown safety factor λ_S must satisfy the condition

$$\lambda_E \leq \lambda_S \leq \lambda_\emptyset. \tag{36}$$

3.7 Incremental Strategy for Fiber-Based Shakedown Analysis

Let us denote with $\mathbf{z} = \{\lambda, \boldsymbol{\beta}, \mathbf{d}, \boldsymbol{\tau}_g, \boldsymbol{\sigma}_g, \boldsymbol{\rho}_g\}$ the set of all the problem variables. It is possible to define a sequence of states $\mathbf{z}^{(n)}$ corresponding to a non-decreasing sequence of load factors $\lambda^{(n)}$. A superscript $(n-1)$ is used to denote the consolidated quantities found at the previous step while (n) denotes the searched solution at the new step of the sequence. $\Delta(.) = (.)^{(n)} - (.)^{(n-1)}$ represents the difference between quantities of step (n) and $(n-1)$. We introduce the following non-negative energy term

$$\Psi_g(\Delta\boldsymbol{\sigma}_g, \Delta\boldsymbol{\tau}_g) = \frac{1}{2}\sum_m \frac{1}{E_m}\Delta\sigma_m^2 A_m w_g + \frac{1}{2}\Delta\boldsymbol{\tau}_g^T \mathbf{G}_g^{-1} \Delta\boldsymbol{\tau}_g w_g \tag{37}$$

which represents a norm of the stress increment. The problem variables at the new step $\mathbf{z}^{(n)}$, omitting the superscript (n) to simplify the notation, that correspond to an assigned $\lambda = \bar{\lambda} \geq \lambda^{(n-1)}$, are obtained as the solution of the following optimization problem

$$\begin{aligned}
\text{minimize} \quad & \sum_{e,g} \Psi_g(\Delta\boldsymbol{\sigma}_g, \Delta\boldsymbol{\tau}_g) \\
\text{subject to} \quad & \mathbf{Q}^T \boldsymbol{\beta} = \mathbf{0} \\
& f_m(\bar{\lambda}, \sigma_m) \leq 0 \qquad \forall m, g, e \\
& \mathbf{t}_g(\boldsymbol{\sigma}_g, \boldsymbol{\tau}_g) - \mathbf{L}_g \boldsymbol{\beta}_e = \mathbf{0} \quad \forall g, e
\end{aligned} \tag{38}$$

i.e. by finding the increment of self-equilibrated stress with minimum norm (37) which makes it possible to fulfill the sufficient condition of the lower bound shakedown theorem for $\bar{\lambda} = \lambda^{(n)}$ to be safe. Clearly, the sequence $\lambda^{(n)}$ can start from $\lambda^{(0)} = \lambda_E$ since no additional stresses are needed for $\lambda < \lambda_E$. Let us introduce the Lagrangian function

$$\begin{aligned}
\mathcal{L} = & \sum_{e,g} \Psi_g(\Delta\boldsymbol{\sigma}_g, \Delta\boldsymbol{\tau}_g) - \Delta\mathbf{d}^T \mathbf{Q}^T \boldsymbol{\beta} + \sum_{e,g,m} \mu_m f_m(\bar{\lambda}, \sigma_m) A_m w_g \\
& - \sum_{e,g} \Delta\boldsymbol{\rho}_g^T \left(\mathbf{t}_g(\boldsymbol{\sigma}_g, \boldsymbol{\tau}_g) - \mathbf{L}_g \boldsymbol{\beta}_e\right) w_g
\end{aligned}$$

where the Lagrangian multipliers are the increment of kinematic variables $\Delta\mathbf{d}$, the increment of generalized strains $\Delta\boldsymbol{\rho}_g$ at each IP g and the plastic multiplier μ_m of each fiber m. The optimization problem in Eq. (38) is then equivalent to

$$\begin{cases} \mathcal{L} = \text{stationary} \\ \mu_m \geq 0 \quad \forall m, g, e \end{cases} \tag{39}$$

that is the stationarity of the Lagrangian with non-negative multipliers μ_m of the inequality constraints. It is necessary to note that, system in Eq. (39) admits solution only if $\lambda^{(n)} \leq \lambda_S$ because, by definition, no additional self-equilibrated stress able to satisfy the admissibility condition can be found beyond λ_S. This suggests that it is not convenient to directly assign the values of $\lambda^{(n)}$. Instead, we can use a continuation strategy replacing $\lambda = \bar{\lambda}$ with an arc-length equation for defining $\mathbf{z}^{(n)}$. Indeed, the sequence points $\mathbf{z}^{(n)}$ belongs to a curve in the hyperspace of $\mathbf{z}$. A particular $\mathbf{z}^{(n)}$ along this curve can be identified by a constraint equation

$$g(\Delta\mathbf{z}) - \Delta\xi = 0 \tag{40}$$

by assigning a value to the positive parameter $\Delta\xi$, with $g(\Delta\mathbf{z})$ defining a suitable distance of $\mathbf{z}^{(n)}$ from $\mathbf{z}^{(n-1)}$, known as arc-length in continuation methods [9, 15]. A proper arc-length definition must assure the existence of a solution for (39) for all values of $\Delta\xi$ and $\Delta\lambda \geq 0$.

The stationarity of $\mathcal{L}$ with respect to $\mathbf{d}$ imposes the self-equilibrium of the additional stress and can be collected in the group of global equations together with Eq. (40):

$$\text{Global equations} \quad \begin{cases} \mathbf{Q}^T\beta = \mathbf{0} \\ g(\Delta\mathbf{z}) - \Delta\xi = 0. \end{cases} \tag{41a}$$

The stationarity of $\mathcal{L}$ with respect to $\boldsymbol{\beta}_e$ and $\boldsymbol{\rho}_g$ furnishes element-wise equations preserving the kinematic compatibility of $\Delta\boldsymbol{\rho}_g$ with $\Delta\mathbf{d}_e$ and the FE equilibrated stress interpolation in Eq. (9):

$$\text{Element equations} \quad \begin{cases} \mathbf{Q}_e\Delta\mathbf{d}_e - \sum_g \mathbf{L}_g^T\Delta\boldsymbol{\rho}_g w_g = \mathbf{0} \\ \mathbf{t}_g(\boldsymbol{\sigma}_g, \boldsymbol{\tau}_g) - \mathbf{L}_g\boldsymbol{\beta}_e = \mathbf{0} \quad \forall g \end{cases} \quad \forall e. \tag{41b}$$

Finally, the stationarity of $\mathcal{L}$ with respect to σ_m and $\boldsymbol{\tau}_g$ provides constitutive laws at each IP linking the section stress resultants $\mathbf{t}_g$ to the section generalized strain increment $\Delta\boldsymbol{\rho}_g$:

$$\begin{cases} \text{Fiber stress } \sigma \quad \begin{cases} \Delta\sigma_m - E_m\left(\Delta\epsilon_m - \mu_m\dfrac{\partial f_m(\lambda, \sigma_m)}{\partial\sigma_m}\right) = 0 \\ \mu_m \geq 0 \quad \mu_m f_m(\lambda, \sigma_m) = 0 \end{cases} \\ \text{Tangential stress resultants} \quad \Delta\boldsymbol{\tau}_g = \boldsymbol{G}_g\mathbf{T}_\tau\Delta\boldsymbol{\rho}_g \end{cases} \tag{41c}$$

where $\Delta\epsilon_m = \mathbf{a}_m^T\Delta\varepsilon$, with $\Delta\varepsilon = \mathbf{T}_n\Delta\boldsymbol{\rho}_g = [\Delta e, \Delta\chi_2, \Delta\chi_3]^T$, denotes the increment of strain work-conjugate to σ_m

$$\Delta\epsilon_m = \Delta e + \Delta\chi_2 x_{3m} - \Delta\chi_3 x_{2m}.$$

The first of Eq. (41c), for an assigned $\Delta\boldsymbol{\rho}_g$ and then $\Delta\epsilon_m$, is the elasto-plastic constitutive law of the fiber m integrated by a backward Euler scheme and can be seen as the first order condition of the following CPP problem:

$$\begin{cases} \text{minimize} \quad \dfrac{1}{2}E_m(\sigma_m - \sigma_m^*)^2 \\ \text{subject to} \quad f_m(\lambda, \sigma_m) \leq 0 \end{cases} \quad \forall m \tag{42}$$

with the fiber elastic predictor $\sigma_m^* = \sigma_m^{(n-1)} + E_m\Delta\varepsilon_m$. The solution of this one-dimensional optimization problem is straightforward and just requires a min-max scalar operation:

$$\sigma_m = \max\left(f_c - \lambda\hat{\sigma}_{min},\ \min\left(\sigma_m^*,\ f_t - \lambda\hat{\sigma}_{max}\right)\right). \tag{43}$$

Equations (41) define the *elasto-plastic* shakedown step since are very similar to the step equations of a standard incremental elasto-plastic analysis [12, 20, 22]. The main difference is that the admissible domain for the self-equilibrated stress σ is a function of the load factor λ.

4 Limit Fire Analysis

4.1 Model of the RC Frames in Fire

Temperature Distribution For the particular simple case of rectangular concrete sections exposed to fire, we adopt the handy formulas to estimate the 2D temperature distribution proposed by Wickstrom [23]. Details on their application are reported in [10, 11]. The temperature distribution is assumed to be constant with respect to the longitudinal axis s of the heated structural elements, i.e. it is equal for all sections of the considered element. It is worth highlighting that the fire model is characterized by a non-decreasing temperature $T(\mathbf{x}, t)$ at any point of the structure with respect to the fire duration t.

Strength Reduction for Concrete and Steel Rebars The concrete tensile strength is considered negligible at any temperature. The concrete compressive strength experiences significant degradation at elevated temperatures. The reduced compressive strength for concretes f_{cT} can be estimated from its ambient value f_c as

$$f_{cT} = k_c(T) f_c \tag{44}$$

where the dimensionless reduction factor $k_c(T)$ is given by Eurocode 2. Similarly, for the reduced strength of reinforcing bars f_{yT} is computed from its ambient value f_y using $k_s(T)$ taken from Lie et al.'s model [7]

$$f_{yT} = k_s(T) f_y. \tag{45}$$

Both $k_c(\mathbf{x}, t)$ and $k_s(\mathbf{x}, t)$ are non-increasing function of t, i.e. the strength of the materials at t_2 is less than or equal to the strength at t_1 for any $t_2 > t_1$.

4.2 Time-Dependent Admissibility Condition

The admissibility condition at a given fire duration t can be written as

$$\sigma_-(s, \mathbf{x}, t) \leq \sigma(s, \mathbf{x}) \leq \sigma_+(s, \mathbf{x}, t), \qquad \forall(s, \mathbf{x}) \tag{46}$$

where $\sigma_+(s, \mathbf{x}, t)$ and $\sigma_-(s, \mathbf{x}, t)$ are now the time-dependent stress limits of the material in traction (positive) or compression (negative). According to the material model just presented, for concrete

$$\sigma_+(s, \mathbf{x}, t) = 0 \quad \text{and} \quad \sigma_-(s, \mathbf{x}, t) = -k_c(\mathbf{x}, t) f_c,$$

while for steel reinforcing bars

$$\sigma_+(s, \mathbf{x}, t) = k_s(\mathbf{x}, t) f_y \quad \text{and} \quad \sigma_-(s, \mathbf{x}, t) = -k_s(\mathbf{x}, t) f_y.$$

Omitting the dependence on the position $(s, \mathbf{x})$ to simplify the notation, it is possible to define the admissible domain of the stress σ introducing the time-dependent yield function $f(\sigma, t)$ as

$$f(\sigma, t) \equiv |\sigma - c(t)| - r(t) \leq 0 \tag{47}$$

where

$$c(t) = \frac{\sigma_+(t) + \sigma_-(t)}{2} \qquad r(t) = \frac{\sigma_+(t) - \sigma_-(t)}{2}. \tag{48}$$

4.3 Safety Condition in Case of Fire

A sufficient condition for a structure to be safe at ambient temperature is given by the classic static theorem of limit analysis [14, 16].

Theorem 1 (Static theorem of limit analysis) *The structure is safe if there exists a stress field in equilibrium with the external loads such that the yield condition is satisfied at every point in the body.*

In case of a generic fire event which changes the strength of the materials, we can say, as simple corollary, that a structure is safe at a given fire duration t if the hypotheses of Theorem 1 are satisfied at any time less than or equal to t.

However, the fire model adopted in this work has a particular feature: the strength of the materials is a non-increasing function of the fire duration. In this context, the following sufficient condition holds.

Theorem 2 (Static theorem of limit fire analysis) *The structure is safe at a given fire duration t if there exists a stress field in equilibrium with the external loads such that the yield condition at t is satisfied at every point in the body and the strength of the materials is a non-increasing function of the fire duration.*

The proof of Theorem 2 exploits the hypothesis that the strength of the materials is a non-increasing function of the fire duration. This means that if a stress field is admissible at t then it is admissible also at any time less than t. It follows, according to Theorem 1, that if the structure is safe at t then it is safe also at any time less than t.

4.4 Lower Bound Theorem of Limit Fire Analysis

Under the hypothesis that the strength of the materials is a non-increasing function of the fire duration, the limit fire duration t_L, i.e. the time of fire exposure which leads to the collapse of the structure, can be defined as the maximum value of t that satisfies the sufficient condition of the static theorem of limit fire analysis 2. In discrete form, this can be formalized as follows.

Theorem 3 (Lower bound theorem of limit fire analysis in discrete form) *The limit fire duration t_L of the structure discretized in mixed fiber finite elements is the solution of the following constrained optimization problem*

$$\begin{array}{lll} \textit{maximize} & t & \\ \textit{subject to} & \boldsymbol{Q}^T\boldsymbol{\beta} - \boldsymbol{p} = \boldsymbol{0} & \\ & f(\sigma_m, t) \leq 0 & \forall e, g, m \\ & \boldsymbol{t}_g(\boldsymbol{\sigma}_g, \boldsymbol{\tau}_g) - \boldsymbol{L}_g\boldsymbol{\beta}_e - \bar{\boldsymbol{t}}_g = \boldsymbol{0} & \forall g, e \end{array} \tag{49}$$

where the last equality constraint preserves the equilibrated stress interpolation (9) *along each element.*

We can note that (49) is a nonlinear optimization problem, convex in stress variables, very similar to the static limit and shakedown analyses. In particular, instead of a load amplifier, we have now a time parameter that reduces the admissible domain.

4.5 Incremental Strategy for Limit Fire Analysis

Let us denote with $\mathbf{z} = \{t, \boldsymbol{\beta}, \mathbf{d}, \boldsymbol{\tau}_g, \boldsymbol{\sigma}_g, \boldsymbol{\rho}_g\}$ the set of all the problem variables. It is possible to define a sequence of states $\mathbf{z}^{(n)}$ corresponding to a non-decreasing time of fire exposure $t^{(n)}$. In the following, a superscript $(n-1)$ is used to denote quantities at the known state found at the previous step of the incremental process, while (n) denotes the searched solution at the new step of the sequence. $\Delta(.) = (.)^{(n)} - (.)^{(n-1)}$ represents the difference between quantities at (n) and $(n-1)$. The sequence of points $\mathbf{z}^{(n)}$ belongs to the curve in the hyperspace of $\mathbf{z}$ of equilibrated and admissible solutions. Starting from a known $\mathbf{z}^{(n-1)}$, a particular $\mathbf{z}^{(n)}$ along this curve can be identified by a constraint equation

$$g(\Delta\mathbf{z}) - \Delta\xi = 0 \tag{50}$$

which defines a suitable distance from $\mathbf{z}^{(n-1)}$ to $\mathbf{z}^{(n)}$. At each step of the incremental analysis, $\mathbf{z}^{(n)}$ is found by solving, for an assigned $\Delta\xi$, a generalization of Eqs. (23) which takes account of the time-dependent yield function of the fibers:

$$\text{Global equations} \quad \begin{cases} \mathbf{Q}^T\beta - \mathbf{p} = \mathbf{0} \\ g(\Delta\mathbf{z}) - \Delta\xi = 0. \end{cases} \tag{51a}$$

$$\text{Element equations} \quad \begin{cases} \mathbf{Q}_e\Delta\mathbf{d}_e - \sum_g \mathbf{L}_g^T\Delta\boldsymbol{\rho}_g w_g = \mathbf{0} \\ \mathbf{t}_g(\boldsymbol{\sigma}_g, \boldsymbol{\tau}_g) - \mathbf{L}_g\boldsymbol{\beta}_e - \bar{\mathbf{t}}_g = \mathbf{0} \qquad \forall g \end{cases} \tag{51b}$$

$$\begin{cases} \text{Fiber normal stress} \quad \begin{cases} \Delta\sigma_m - E_m\left(\Delta\epsilon_m - \mu_m \dfrac{\partial f_m(\sigma_m, t)}{\partial \sigma_m}\right) = 0 \\ \mu_m \geq 0 \quad \mu_m f_m(\sigma_m, t) = 0 \end{cases} \\ \text{Tangential stress resultants} \quad \Delta\boldsymbol{\tau}_g = \mathbf{G}_g\mathbf{T}_\tau\Delta\boldsymbol{\rho}_g \end{cases} \tag{51c}$$

with $\Delta\epsilon_m = \mathbf{a}_m^T\mathbf{T}_n\Delta\boldsymbol{\rho}_g$. The first of Eq. (51c) is the elasto-plastic time-dependent constitutive law of the fiber m integrated by a backward Euler scheme, which is equivalent to

$$\sigma_m = \max\left(\sigma_-(t),\ \min\left(\sigma_m^*,\ \sigma_+(t)\right)\right). \tag{52}$$

with the fiber elastic predictor $\sigma_m^* = \sigma_m^{(n-1)} + E_m\Delta\varepsilon_m$. Note that thermal strains and variations of elastic properties over time, according to the lower bound theorem Eq. (49), can be neglected because they do not affect the limit fire duration.

Equations (51) define the *elasto-plastic* fire step since are very similar to the Eq. (23) of a standard incremental elasto-plastic analysis. The main difference is that the admissible domain for the stress σ is a function of the fire duration t, while the loads are kept constant. From this point of view, the algorithm is similar to the shakedown analysis seen in the previous Sect. [8].

5 Incremental Algorithm: Strain-Driven Implementation

An incremental strain-drive algorithm is presented for the limit fire duration problem, but it can be applied also to the shakedown problem. The only difference is that a constant load vector is involved in the structural equilibrium for the limit fire analysis, while it is not involved in the equilibrium equations of the shakedown problem since this is expressed in terms of additional stress with respect to the elastic envelope. Also, the admissible domain is a function of the load factor λ in the shakedown analysis, while it is a function of the fire duration t in the limit fire analysis. Instead of solving Eqs. (41) or (51) all together, it is possible to apply a strain-driven decomposed strategy based on the solution of the following three nested sub-steps.

- A *section state determination* provides the stress resultants $\mathbf{t}_g$ at each IP along the element axis as a function of an assigned section generalized strain increment $\Delta\boldsymbol{\rho}_g$ and the current fire duration t using the constitutive law (51c).

- An *element state determination* finds the element stress interpolation variables $\boldsymbol{\beta}_e$ corresponding to an assigned increment of element nodal kinematic variables $\Delta\mathbf{d}_e$ and the current fire duration estimate t by means of the element Eq. (51b). This require an iterative solution at FE level involving, at each iteration, the section state determination at the IPs.
- A *global incremental-iterative process* solves the global equilibrium Eq. (51a) step-by-step with the stress interpolation variables $\boldsymbol{\beta}$ expressed as functions of the unknown kinematic degrees of freedom $\mathbf{d}$ and the current fire duration t (or load factor λ) through the element state determination.

5.1 *Section State Determination*

The section state determination expresses the stress resultants $\mathbf{t}_g$ of a section g as a function $\mathbf{t}_g(\Delta\boldsymbol{\rho}_g, t)$ of the generalized strain increment $\Delta\boldsymbol{\rho}_g$ and the time of fire exposure t by means of the constitutive Eq. (51c). To this end, the following simple procedure is used.

- $\Delta\boldsymbol{\varepsilon}_g$ and $\Delta\boldsymbol{\gamma}_g$ are extracted from $\Delta\boldsymbol{\rho}_g$ (see Eq. (7)).
- $\Delta\boldsymbol{\varepsilon}_g = [\Delta e_g, \Delta\chi_{2g}, \Delta\chi_{3g}]^T$ is used to evaluate the normal strain increment in each fiber

$$\Delta\epsilon_m = \Delta e_g + \Delta\chi_{2g} x_{3m} - \Delta\chi_{3g} x_{2m}.$$

- Using the elastic predictor $\sigma_m^* = \sigma_m^{(n-1)} + E_m \Delta\epsilon_m$, the new value of the fiber normal stress σ_m is computed as

$$\sigma_m = \max\left(\sigma_-(t),\ \min\left(\sigma_m^*,\ \sigma_+(t)\right)\right).$$

- The normal stress resultants are then evaluated as

$$\mathbf{n}_g = \sum_m \sigma_m A_m \mathbf{a}_m.$$

- An elastic update is used for the tangential stress resultants:

$$\boldsymbol{\tau}_g = \boldsymbol{\tau}_g^{(n-1)} + \Delta\boldsymbol{\tau}_g \quad \text{with} \quad \Delta\boldsymbol{\tau}_g = \mathbf{G}_g \Delta\boldsymbol{\gamma}_g$$

5.2 *Element State Determination*

The FE stress interpolation variables $\boldsymbol{\beta}_e$ corresponding to an assigned increment of element kinematic variables $\Delta\mathbf{d}_e$ and a fire duration t can be obtained by solving the element-wise system

$$\begin{cases} \mathbf{r}_g \equiv \mathbf{t}_g(\Delta\boldsymbol{\rho}_g, t) - \mathbf{L}_g\boldsymbol{\beta}_e - \bar{\mathbf{t}}_g = \mathbf{0}, & \forall g \\ \mathbf{r}_e \equiv \mathbf{Q}\Delta\mathbf{d}_e - \sum_g \mathbf{L}_g^T \Delta\boldsymbol{\rho}_g w_g = \mathbf{0} \end{cases} \tag{53}$$

The vanishing of $\mathbf{r}_g$ for each IP (section) g along the element axis assures that the stress resultants $\mathbf{t}_g(\Delta\boldsymbol{\rho}_g, t)$ given by the section state determination match the equilibrated stress interpolation in Eq. (9). The condition $\mathbf{r}_e = \mathbf{0}$ is necessary to preserve kinematic compatibility between section strain increments $\Delta\boldsymbol{\rho}_g$ and $\Delta\mathbf{d}_e$. The solution of Eq. (53) is achieved using Newton's iterations, where $\Delta\mathbf{d}_e$ and t are assigned while $\Delta\boldsymbol{\rho}_g$ and $\boldsymbol{\beta}_e$ are the problem unknowns. The element internal forces can be the computed as

$$\mathbf{s}_e(\mathbf{d}_e, t) \equiv \mathbf{Q}_e^T \boldsymbol{\beta}_e(\mathbf{d}_e, t). \tag{54}$$

5.3 Incremental-Iterative Global Solution

The global Eq. (51a) can written in the same form as in a standard strain-driven elasto-plastic analysis with an arc-length incremental method providing a path of equilibrium points $(t^{(n)}, \mathbf{d}^{(n)})$. The difference is that the load factor is now replaced by a non-decreasing t converging to the limit fire duration. For each increment $\Delta\xi$, starting from the previous known equilibrium configuration $\{t^{(n-1)}, \mathbf{d}^{(n-1)}\}$, the new point $\{t, \mathbf{d}\}$ at (n) is obtained by solving

$$\begin{cases} \mathbf{r}(\mathbf{d}, t) \equiv \mathbf{s}(\mathbf{d}, t) - \mathbf{p} = \mathbf{0}. \\ q(\mathbf{d}, t) \equiv g(\Delta\mathbf{d}, \Delta t) - \Delta\xi = 0 \end{cases} \tag{55}$$

with $\Delta\mathbf{d} = \mathbf{d} - \mathbf{d}^{(n-1)}$ and $\Delta t = t - t^{(n-1)}$, while $\mathbf{s}(\mathbf{d}, t) \equiv \mathbf{Q}^T\boldsymbol{\beta}(\mathbf{d}, t)$ is the internal force vector that can be directly obtained by assembling the elemental contribution $\mathbf{s}_e(\mathbf{d}_e, t)$ provided by the element state determination, instead of using the equivalent global operation $\mathbf{Q}^T\boldsymbol{\beta}(\mathbf{d}, t)$. The first equations in (55) imposes the global equilibrium of internal forces $\mathbf{s}(\mathbf{d}, t)$ and external load vector. The last of Eq. (55) assigns the increment size $\Delta\xi$, with $g(\Delta\mathbf{d}, \Delta t)$ the arc-length measure. Both $\mathbf{s}$ and q are functions of $\mathbf{d}$ and t. A Newton iterative scheme can be used to solve (55) [8, 10]. The incremental analysis is stopped when $t^{(n)} \approx t^{(n-1)}$ with $\Delta\mathbf{d} \neq 0$, i.e. when the current $\Delta\mathbf{d}$ is a mechanism and no more stress redistribution is possible. The last value $t^{(n)}$ of the sequence is the limit fire duration t_L. An adaptive increment size $\Delta\xi$, based on the number of iterations required to achieve convergence at the previous step, is useful to minimize the overall computational cost. In the numerical tests, this is achieved using the simple algorithm reported in [9].

6 Numerical Tests on Shakedown Analysis

In this section, some numerical applications are illustrated. We start with a very simple test, for which the shakedown safety factor is easy to obtain analytically, to validate the proposed numerical procedure. A simple 3D frame and a large building are then analyzed to show the effectiveness of the proposal and the accuracy of the fiber model. Three material types are used for the cross sections:

- structural steel with $E = 210000MPa$, $f_c = -300MPa$ and $f_t = 300MPa$;
- concrete with $E = 25000MPa$, $f_c = -30MPa$ and $f_t = 0$;
- reinforcing bars with $E = 210000MPa$, $f_c = -400MPa$ and $f_t = 400MPa$.

IPE and HE sections are used for structural steel, while rectangular section are employed for reinforced concrete (RC) buildings. In this last case, the section is supposed to be confined in such a way that concrete has a sufficient ductility in compression. Typical fiber discretizations of the sections is shown in Fig. 2. Flanges and web of HE and IPE sections are discretized using $n_l \times n_t$ fibers, n_t through the thickness and n_l along the orthogonal direction. A uniform $n_f \times n_f$ discretization is used for the concrete part of RC sections, while reinforcing bars are modeled using a single fiber per bar with area concentrated at the bar center. All the results reported are obtained using 3 Gauss-Lobatto IPs per FE.

6.1 A Continuous Beam Supported on Four Points

The first test regards the continuous beam supported on 4 points (Fig. 3), with $L = 2500$ mm and M_y the plastic moment of the IPE360 steel section. The beam is subjected to two load pattern whose variability in time is given by the coefficients α_1 and α_2. Four loading cases are considered: case 1) $\alpha_1 = 2$ and $\alpha_2 = 0$; case 2) $\alpha_1 = 0$ and $\alpha_2 = 1$; case 3) $\alpha_1 = 2$ and $\alpha_2 = 1$; case 4) $\alpha_1 \in [0; 1]$ and $\alpha_2 \in [0; 2]$. In the first 3 cases α_1 and α_2 are constant and this means that the shakedown analysis coincides with a limit analysis. In the forth case, loads vary within known ranges and, in the present context, it is the most interesting situation. This example is chosen to validate the proposed numerical analysis since the exact solution can be easily derived analytically. The proposed FE shakedown analysis is now tested. The structure is discretized using 6 FEs to deal with the concentrated forces. Figure 4 illustrates

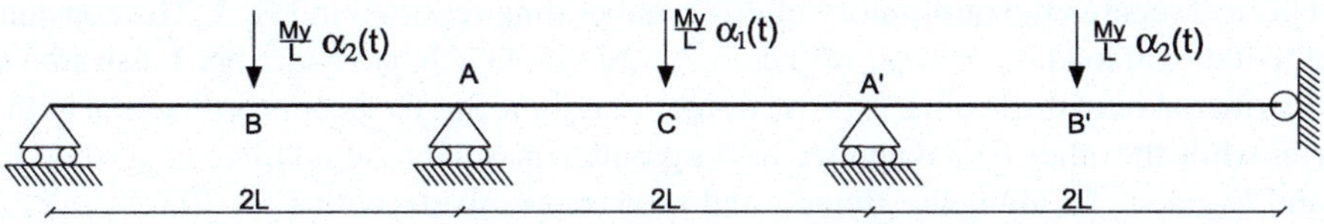

Fig. 3 Continuous beam: geometry, loads and boundary condition

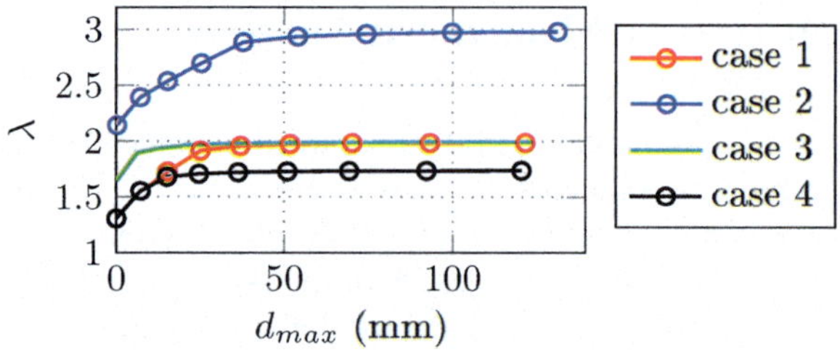

Fig. 4 Continuous beam: pseudo-equilibrium path for all load cases and 10x1 fibers per wall of the section

Table 1 Continuous beam: shakedown safety factor for all load cases

Fibers	λ_{S1}	λ_{S2}	λ_{S3}	λ_{S4}
5×1	1.98	2.96	1.98	1.72
10×1	2.00	3.00	2.00	1.74
Exact	2.00	3.00	2.00	1.74

the pseudo-equilibrium path provided by the proposed incremental-iterative analysis using 10×1 fibers for each wall of the section. All paths, for constant (case 1, 2 and 3) and variable (case 4) loads, exhibit an initial behaviour without additional displacements as no stress redistribution occurs up to the elastic limit. Also, the curve has a plateau in its final part. This means that the sequence of non-decreasing load factors, for all load cases, converges to a value representing the safety factor, for which the system becomes a mechanism. The safety factor λ_{Sk} associated to each load case k is reported in Table 1. It is possible to note that the proposed fiber-based numerical analysis furnishes the exact safety factor using 10×1 fibers per wall, while the solution for 5×1 fibers is affected by a small error. Indeed, every even number of fibers provides an exact plastic moment in simple bending problems, while an odd number cannot capture exactly the stress discontinuity at the neutral axis. Interestingly, the shakedown safety factor for variable loads λ_{S4} is lower than all the limit load factors λ_{S1}, λ_{S2} and λ_{S3} corresponding to the vertexes of the load domain.

6.2 A Large Building

This test regards the multi-story multi-span building reported in Fig. 5. The constant gravitational load $g_i = \alpha_i g_0$, with $g_0 = 4$ KN/mq of each one way ribbed slab area is distributed onto the beams: 90% of weight is assigned to the beams orthogonal to the ribs while the other 10% is applied on the parallel ones. Horizontal forces $F_{xj} = \alpha_x F_j$ and $F_{yj} = \alpha_y F_j$, along the global x and y axis respectively, with $F = 30 KN/m \cdot z_j$, act at each story located at z_j from the ground. The story forces are distributed as

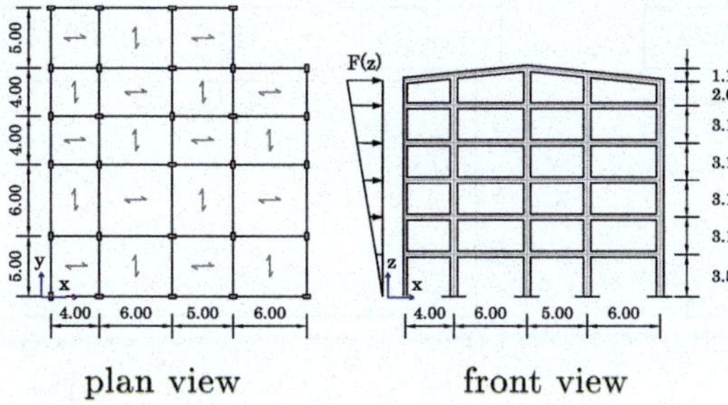

RC building
Columns section: RS1
Beams section: RS2
Steel building
Columns section: HEB400
Beams section: IPE360

Fig. 5 Large building: geometry and cross sections

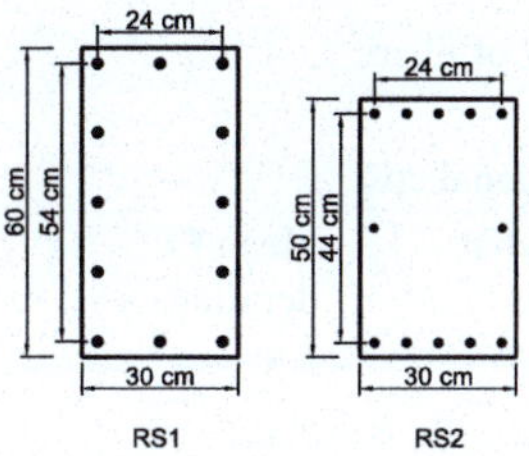

Reinforcing bars
RS1: $12\phi 20\,mm$
RS2 top and bottom: $10\phi 16\,mm$
RS2 lateral: $2\phi 14\,mm$

Fig. 6 RC sections: geometry and materials

concentrated forces on the top of each column along the global x and y axis. Three load combinations are considered:

- combination 1, $\alpha_i \in [0.5; 1.4]\,\forall i$, $\alpha_x = 0$, $\alpha_y = 0$;
- combination 2, $\alpha_i \in [0.5; 1]\,\forall i$, $\alpha_x \in [-1; 1]$, $\alpha_y \in [-0.3; 0.3]$;
- combination 3, $\alpha_i \in [0.5; 1]\,\forall i$, $\alpha_x \in [-0.3; 0.3]$, $\alpha_y \in [-1; 1]$.

More complex combination rules could be also adopted as in [6]. Two material configurations are considered. In the first one, columns and beams have the RC sections denoted RS1 and RS2 in Fig. 6 respectively. In the second one, steel IPE360 and HEB400 sections are used for beams and columns respectively. A single mixed FE for all beams and columns provides a converged solution.

Figure 7 shows the pseudo-equilibrium path for the 2 material configurations varying the number of fibers. It is possible to observe how the incremental-iterative process quickly converges to the shakedown safety factor also for a large building. However, two different behaviors characterize the two material configurations in this test. In the RC case, the curve has a plateau in its final part ($\lambda^n \approx \lambda^{n-1}$), i.e. no more stress redistribution is possible because an additional displacement increment would be a mechanism. Conversely, for the steel building under consideration, the incremental analysis stops without a plateau because λ^n reaches the value $\lambda_\emptyset$ for which there exists one fiber where the admissible domain becomes an empty set. Table 2 shows that a small number of fibers provides an accurate safety factor in the RC case: the solution accuracy is satisfactory also with 5×5 fibers. In the same table, we can see also the low global and element iterative effort. Similar considerations

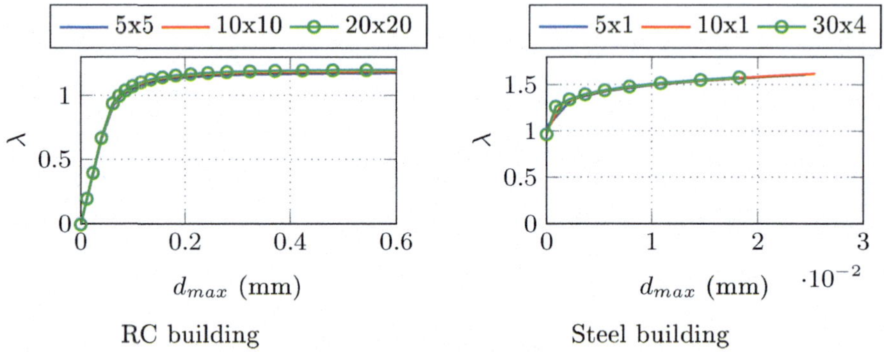

Fig. 7 Large building: pseudo-equilibrium path varying the number of fibers

Table 2 Large RC building: shakedown safety factor and computational cost

Fibers	λ_S	Steps	Total global iterations	Mean FE iterations
5 × 5	1.18	20	73	2.32
10 × 10	1.19	21	77	2.59
20 × 20	1.20	20	72	2.60
100 × 100	1.20	21	77	2.77

Table 3 Large steel building: shakedown safety factor and computational cost

Fibers	λ_E	λ_S	Steps	Total global iterations	Mean FE iterations
5 × 1	1.05	1.61	6	18	1.87
10 × 1	1.01	1.62	8	22	1.91
30 × 1	0.98	1.59	8	21	1.94
30 × 4	0.97	1.58	8	19	2.01

hold for the steel case, as reported in Table 3. However, in this case the number of steps of the incremental process is lower because the analysis stops when $\lambda = \lambda_{\emptyset}$. As a final remark, in both the RC and steel case, the proposed continuation method for fiber-based shakedown analysis requires an overall computational cost comparable to that of a single elasto-plastic analysis.

7 Numerical Tests on Limit Fire Analysis

The real scale building reported in Fig. 8 is analyzed. The vertical story load p of the one way ribbed slab as depicted in Fig. 8 is distributed on the beams. For each floor

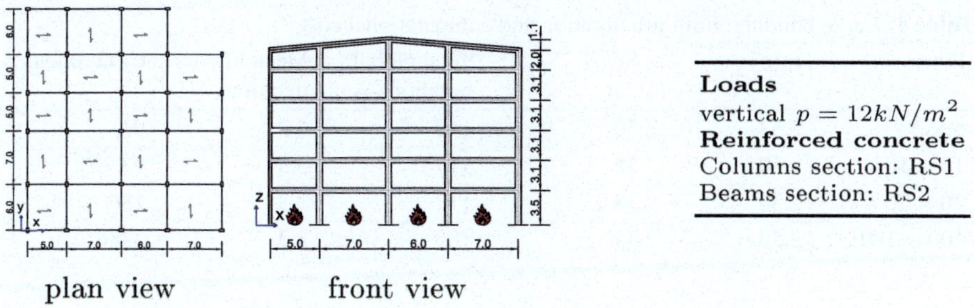

Fig. 8 Large building: geometry (lengths in meters), loads and cross-sections

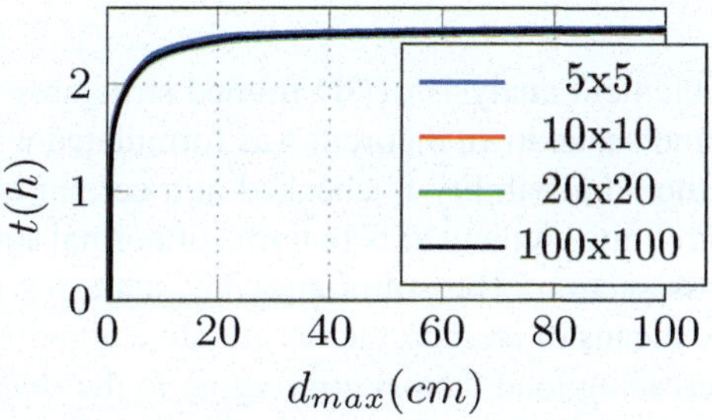

Fig. 9 Large building: equilibrium path varying the number of fibers

area, 90% of weight is assigned to the beams orthogonal to the ribs, while the other 10% is applied on the parallel ones. A concrete compressive strength $f_c = 20\,MPa$ and reinforcing bars with $f_y = 391\,MPa$ are considered. The fire event involves the ground floor only, with columns exposed to fire around the entire perimeter and beams on three edges excluding the top one. A single FE with 3 Gauss-Lobatto IPs is employed for each beam and column. A mesh refinement does not affect the limit fire duration. The fire duration-displacement incremental process is reported in Fig. 9. Table 4 shows that a small number of fibers provides an accurate limit fire duration: the solution accuracy is satisfactory also with 5×5 fibers, while 10×10 give practically the exact result. As a final remark, the proposed numerical procedure is very efficient. Just to get an image of it, the nonlinear analysis of this real scale example takes about 1.5 min using 10×10 fibers and our non-optimized in-house MATLAB code.

Table 4 Large building: limit fire duration and computational cost

Fibers	$t_L(h)$	Steps	Total global iterations	Mean FE iterations	CPU time (s)
5 × 5	2.53	50	140	3.2	91
10 × 10	2.50	35	113	3.3	93
20 × 20	2.50	38	117	3.3	186
100 × 100	2.50	34	106	3.3	3486

8 Conclusions

A shakedown finite element analysis of 3D framed structures was proposed in this work. The lower bound shakedown theorem was formulated with a distributed plasticity model where the admissibility is checked at a certain number of integration points along the beam finite element axis in terms of normal stress of the fibers used to discretize the cross section. The solution of the shakedown optimization problem was achieved by means of an incremental-iterative algorithm, which provides a sequence of non-decreasing load factors converging to the shakedown safety factor. At each step, the method searches for an admissible self-equilibrated stress field required by Melan's theorem for the current load factor to be safe. The incremental analysis can be performed in a strain-driven incremental-iterative manner very similar to a standard elasto-plastic analysis. The main differences regard the admissible domain, which is a function of the load factor, and the structural equilibrium which is imposed for self-equilibrated stresses. Importantly, loads affect only the admissible domain, defined once and for all in terms of a single minimum and maximum effect on each fiber due to all possible loads. This work introduced also the limit fire analysis concept and its numerical formulation for assessing the fire resistance of 3D frame structures. The starting assumption is that the fire event produces a monotonously non-increasing strength over time. This makes it possible to derive a lower bound theorem which defines the limit fire duration, i.e. the time of fire exposure which leads to the structural collapse, based only on equilibrium and time-dependent strength. The same incremental-iterative algorithm proposed for the shakedown analysis can be used for evaluating the limit fire duration. Additional details are available in [8] for the shakedown problem and in [10] for the fire analysis.

References

1. Bleyer, J., De Buhan, P.: Yield surface approximation for lower and upper bound yield design of 3d composite frame structures. Comput. Struct. **129**, 86–98 (2013)
2. Casciaro, R., Garcea, G.: An iterative method for shakedown analysis. Comput. Methods Appl. Mech. Eng. **191**(49–50), 5761–5792 (2002)

3. Garcea, G., Armentano, G., Petrolo, S., Casciaro, R.: Finite element shakedown analysis of two-dimensional structures. Int. J. Numer. Methods Eng. **63**(8), 1174–1202 (2005). https://doi.org/10.1002/nme.1316
4. Garcea, G., Leonetti, L.: A unified mathematical programming formulation of strain driven and interior point algorithms for shakedown and limit analysis. Int. J. Numer. Methods Eng. **88**(11), 1085–1111 (2011). https://doi.org/10.1002/nme.3188
5. König, J.: Shakedown of Elastic–Plastic Structures. Elsevier (1987)
6. Leonetti, L., Casciaro, R., Garcea, G.: Effective treatment of complex statical and dynamical load combinations within shakedown analysis of 3D frames. Comput. & Struct. **158**, 124–139 (2015). https://doi.org/10.1016/j.compstruc.2015.06.002
7. Lie, T.: Structural Fire Protection: ASCE Manuals and Reports on Engineering Practice, vol. 78, p. 241 (1992)
8. Magisano, D., Garcea, G.: Fiber-based shakedown analysis of three-dimensional frames under multiple load combinations: mixed finite elements and incremental-iterative solution. Int. J. Numer. Methods Eng. **121**(17), 3743–3767 (2020). https://doi.org/10.1002/nme.6380
9. Magisano, D., Leonetti, L., Garcea, G.: How to improve efficiency and robustness of the newton method in geometrically non-linear structural problem discretized via displacement-based finite elements. Comput. Methods Appl. Mech. Eng. **313**, 986–1005 (2017). https://doi.org/10.1016/j.cma.2016.10.023
10. Magisano, D., Garcea, G.: Limit fire analysis of 3d frame structures. Eng. Struct. **233**, 111762 (2021). https://doi.org/10.1016/j.engstruct.2020.111762
11. Magisano, D., Liguori, F., Leonetti, L., de Gregorio, D., Zuccaro, G., Garcea, G.: A quasi-static nonlinear analysis for assessing the fire resistance of reinforced concrete 3d frames exploiting time-dependent yield surfaces. Comput. & Struct. **212**, 327–342 (2019). https://doi.org/10.1016/j.compstruc.2018.11.005
12. Nodargi, N.A., Bisegna, P.: A mixed finite element for the nonlinear analysis of in-plane loaded masonry walls. Int. J. Numer. Methods Eng. **120**(11), 1227–1248 (2019). https://doi.org/10.1002/nme.6179
13. Petrolo, A., Casciaro, R.: 3D beam element based on saint venant's rod theory. Comput. Struct. **82**(29–30), 2471–2481 (2004)
14. Ponter, A.R., Fuschi, P., Engelhardt, M.: Limit analysis for a general class of yield conditions. Eur. J. Mech. A/Solids. **19**(3), 401–421 (2000). https://doi.org/10.1016/S0997-7538(00)00170-4
15. Riks, E.: An incremental approach to the solution of snapping and buckling problems. Int. J. Solids Struct. **15**(7), 529–551 (1979). https://doi.org/10.1016/0020-7683(79)90081-7
16. Save, M., Massonnet, C., de Saxce, G.: Plastic Limit Analysis of Plates, Shells, and Disks. North Holland (1997)
17. Simon, J.W., Weichert, D.: Shakedown analysis with multidimensional loading spaces. Comput. Mech. **49**(4), 477–485 (2012)
18. Simon, J.W.: Limit states of structures in n-dimensional loading spaces with limited kinematical hardening. Comput. & Struct. **147**, 4–13 (2015). https://doi.org/10.1016/j.compstruc.2014.09.019, CIVIL-COMP
19. Skordeli, M.A., Bisbos, C.: Limit and shakedown analysis of 3D steel frames via approximate ellipsoidal yield surfaces. Eng. Struct. **32**(6), 1556–1567 (2010)
20. Spacone, E., Filippou, F.C., Taucer, F.F.: Fibre beam-column model for non-linear analysis of R/C frames: part I. Formulation. Earthq. Eng. & Struct. Dyn. **25**(7), 711–725 (1996). https://doi.org/10.1002/(SICI)1096-9845(199607)25:7<711::AID-EQE576>3.0.CO;2-9
21. Spiliopoulos, K., Panagiotou, K.: An enhanced numerical procedure for the shakedown analysis in multidimensional loading domains. Comput. & Struct. **193**, 155–171 (2017). https://doi.org/10.1016/j.compstruc.2017.08.008
22. Taylor, R.L., Filippou, F.C., Saritas, A., Auricchio, F.: A mixed finite element method for beam and frame problems. Comput. Mech. **31**(1), 192–203 (2003). https://doi.org/10.1007/s00466-003-0410-y. May

23. Wickstrom, U.: A very simple method for estimating temperature in fire exposed concrete structures. Fire Technol. Tech. Rep. SP-RAPP. **46**, 186–194 (1986)
24. Zouain, N.: Encyclopedia of Computational Mechanics. chap. Shakedown and safety assessment. pp. 291–334. John Wiley & Sons (2004)

The manufacturer's authorised representative in the EU is Springer Nature Customer Service Centre GmbH, Europaplatz 3, 69115 Heidelberg, Germany. If you have any concerns regarding our products, please contact ProductSafety@springernature.com

Printed and bound by CPI Group (UK) Ltd, Croydon, CR0 4YY
01/10/2026
02225902-0008